Integrated Remediation Processes toward Heavy Metal-Contaminated Environment

Integrated Remediation Processes toward Heavy Metal-Contaminated Environment

Guest Editors

Luhua Jiang
Liang Hu
Zhigang Yu

MDPI

Basel • Beijing • Wuhan • Barcelona • Belgrade • Novi Sad • Cluj • Manchester

Guest Editors
Luhua Jiang
School of Minerals Processing and Bioengineering
Central South University
Changsha
China

Liang Hu
School of Minerals Processing and Bioengineering
Central South University
Changsha
China

Zhigang Yu
Australian Centre for Water and Environmental Biotechnology
The University of Queensland
Brisbane
Australia

Editorial Office
MDPI AG
Grosspeteranlage 5
4052 Basel, Switzerland

This is a reprint of the Special Issue, published open access by the journal *Toxics* (ISSN 2305-6304), freely accessible at: https://www.mdpi.com/journal/toxics/special_issues/X6MN8374EP.

For citation purposes, cite each article independently as indicated on the article page online and as indicated below:

Lastname, A.A.; Lastname, B.B. Article Title. *Journal Name* **Year**, *Volume Number*, Page Range.

ISBN 978-3-7258-5745-6 (Hbk)
ISBN 978-3-7258-5746-3 (PDF)
https://doi.org/10.3390/books978-3-7258-5746-3

Contents

About the Editors

Luhua Jiang

Luhua Jiang is an Associate Professor at the School of Minerals Processing and Bioengineering at Central South University. He has extensive experience in directing research that examines the fate and behavior of metal(loid)s in nonferrous metal mining and soils and their interactions with the soil microbial community. He has also been active in research involving remediation of toxic metal(loid)-contaminated soils by bioleaching, biomineralization, and environmental–functional materials. His research has resulted in more than 50 journal articles. He has published over 70 papers in internationally renowned journals, such as *Environmental Science & Technology*, *Chemical Engineering Journal*, and *Journal of Hazardous Materials*.

Liang Hu

Liang Hu holds a Doctor of Engineering degree and is an associate professor and doctoral supervisor. The main research directions include environmental microbial technology, heavy metal pollution remediation, microbial–mineral interface reactions, etc. More than 80 SCI papers have been published, and over 20 national invention patents have been authorized. Among them, the total citations of the papers exceed 4,800, and the H-index is 35. In recent years, I have presided over one National Natural Science Foundation of China general project, one National Natural Science Foundation of China Youth project, one Central University Research Start-up Fund project, and one Central South University Large-scale Instrument Open Sharing Fund Project. I have participated in one National Natural Science Foundation of China Innovation Group Project and one National Key Research and Development Program project.

Zhigang Yu

Zhigang Yu is a researcher at the Australian Centre for Water and Environmental Biotechnology (ACWEB) at The University of Queensland, where his work focuses on critical areas of public health microbiology, including antimicrobial resistance, bacterial evolution, and the role of plasmids in horizontal gene transfer. His research investigates microbial interactions, the fate of pathogens in engineered systems, and the efficacy of disinfection processes, utilizing environmental surveillance to understand and mitigate the spread of resistance in the environment.

Preface

The rapid advancement of urban industrialization and mineral resource exploitation has triggered the widespread release of toxic metallic pollutants, particularly heavy metals, into global ecosystems. These contaminants persistently accumulate in aquatic systems, agricultural lands, and the atmospheric environment, posing substantial risks to both ecological stability and public health. This reprint responds to this challenge by presenting a collection of pioneering research focused on the theoretical foundations and practical applications of heavy metal remediation. It explores emerging methodologies spanning microbial remediation, advanced functional materials, and integrated multi-technology approaches. The contributions provide significant insights into the biogeochemical cycling of heavy metals, elucidate fundamental mechanisms of contaminant transformation and stabilization, and establish robust frameworks for ecological risk assessment. By synthesizing these advancements, this reprint aims to accelerate the development of sustainable remediation technologies that align with broader environmental goals, ultimately contributing to ecological restoration efforts.

Luhua Jiang, Liang Hu, and Zhigang Yu
Guest Editors

toxics

MDPI

Editorial

Integrated Remediation Processes Toward Heavy Metal-Contaminated Environment

Luhua Jiang [1,2,*], Liang Hu [1,2] and Zhigang Yu [3]

1 School of Minerals Processing and Bioengineering, Central South University, Changsha 410083, China; huliang2018@csu.edu.cn
2 Key Laboratory of Biometallurgy of Ministry of Education, Central South University, Changsha 410083, China
3 Australian Centre for Water and Environmental Biotechnology, St Lucia, QLD 4072, Australia; zhigang.yu@uq.edu.au
* Correspondence: jiangluhua@csu.edu.cn

1. Introduction

This Editorial introduces the Special Issue titled "Integrated Remediation Processes toward Heavy Metal-Contaminated Environment". With the development of urban industrialization and the utilization of mineral resources, significant amounts of heavy metals have been released into aquatic ecosystems, agricultural soils, and the atmosphere, posing a serious threat to human health. Given the severe hazards associated with heavy metals, there is growing interest in exploring remediation technologies from a more comprehensive perspective [1]. Concurrently, under the national dual carbon target, researchers worldwide are striving to harness integrated technologies to treat contaminated water, soil, and even groundwater. It is expected that the dual carbon target and successful heavy metal remediation can be achieved concurrently by removing or stabilizing heavy metals in the environment, thereby reducing their ecological risks [2].

This Special Issue features nine articles, including seven research papers and two reviews. The research papers focus on comprehensive remediation strategies, underlying mechanisms, and impact assessments related to environmental heavy metal contamination. The reviews focus on recent advances, mechanisms of action, and future prospects of remediation agents for heavy metal pollution. These articles enhance our understanding of heavy metal pollution control and support its advancement toward greater efficiency, resource recovery, and ecological safety.

2. An Overview of Published Articles

The development of novel approaches for the efficient and safe remediation of heavy metal contamination is both imperative and challenging. Hence, Nie et al. (Contribution 1) provide a comprehensive analysis of recent advances in soil amendments for heavy metal-contaminated soil remediation, with a focused examination of natural, synthetic, natural–synthetic copolymer, and biological amendments. Through systematic comparison of their remediation mechanisms and effectiveness, this study offers a thorough evaluation of their impacts on soil physicochemical properties, leachable heavy metal concentrations, and microbial community structure.

This Special Issue presents five articles investigating heavy metal removal in polluted environments, three articles focusing on materials and their coupling technologies, and two articles concentrating on microbial approaches. Gertsen et al. (Contribution 2) synthesized organoclays based on bentonite using various amphoteric and nonionic surfactants and evaluated their performance as efficient sorbents for lead ions. Among them, the organoclay-based bentonite with alkyl polyglucoside exhibited the optimal performance, achieving a maximum adsorption capacity of 1.49 ± 0.05 mmol/g. These novel organoclays demonstrate significant potential for remediating lead-contaminated water and soil. Zhang et al. (Contribution 3) proposed a novel modified pulse electrochemical treatment (PECT) method that integrates biochar as a permeable reactive barrier into the PECT system, with acetate incorporation in the catholyte. This research revealed that the biochar-coupled PECT system with optimized pulse gradients achieved high lead removal efficiency while reducing energy consumption and treatment time. Notably, biochar has been widely applied for heavy metal remediation due to its exceptional adsorption capacity and persistent free radical (PFR) content. However, the environmental behavior and potential risks associated with both biochar and its associated PFRs warrant careful consideration. Alfei et al. (Contribution 4) conducted a systematic investigation into the application potential, mechanisms of action, and associated risks of biochar and its contained PFRs in environmental remediation. This review highlights that biochar—particularly the PFRs generated during its production via pyrolysis—may represent a double-edged sword in environmental applications. Consequently, comprehensive investigations into the ecological effects of biochar and the development of targeted mitigation strategies are critically needed.

For microbial approaches, Zhuang et al. (Contribution 5) analyzed the behavior and mechanism of sulfate-reducing bacteria *Desulfovibrio desulfuricans* in precipitating antimony from wastewater. The study demonstrated that SRB cells achieved antimony immobilization through a three-step process—adsorption, reduction, and sulfidation—on their surface, underscoring the coprecipitation role of phosphorus-containing groups. It also provided a theoretical foundation and technical parameters for SRB-based remediation of antimony-contaminated wastewater, particularly suitable for treating mining wastewater. Additionally, Hao et al. (Contribution 6) explored the impact of bioreactor scale-up cultivation on microbial succession in mixotrophic acidophiles from an industrial-scale perspective, alongside its application in remediating Cd-contaminated soil. Their study revealed that bioreactor scale-up drives pH reduction by altering bacterial communities (but not fungi), thereby indirectly enhancing cadmium (Cd) removal efficiency. They also identified 10 m^3 as the critical scale for microbial community and functional shifts.

Acid mine drainage is one of the major sources of heavy metals in contaminated environments. Therefore, the safe treatment and resource utilization of waste rock and tailings are effective methods for controlling heavy metal pollution. Zhang et al. (Contribution 7) developed a cooperative leaching system ($Fe_2(SO_4)_3$-O_3) to investigate the oxidative dissolution of the waste sulfides, achieving a zinc extraction efficiency of 97.8% under optimal conditions. Guo et al. (contribution 8) investigated the leaching behavior of As and Pb in lead–zinc mining waste rock under mine drainage and rainwater conditions. The study found that As and Pb in mine waste mainly exist in sulfides (e.g., arsenopyrite, pyrite, galena), with smaller amounts adsorbed onto clays. Acidic conditions enhanced As/Pb release, while alkaline conditions increased arsenate mobility. The sulfide content governs the waste's acid-generating potential and associated environmental risks. This study elucidated the mechanisms by which soil/water contamination occurs in lead–zinc mining areas, providing a scientific basis for pollution prevention and control. Moreover, in this Special Issue, Li et al. (contribution 9) highlighted the use of mathematical estimation of endogenous proline as a bioindicator for assessing the stress response of rice plants to trivalent chromium under varying nitrogenous conditions. The study showed that plants alleviate Cr(III) toxicity by regulating proline levels across nitrogen regimes. The proline biomarker system, when combined with MBMM, facilitates nitrogen source assessment and supports the selection of optimal remediation strategies.

3. Conclusions

Continuously exploring comprehensive remediation processes for heavy metal-contaminated environments under the dual carbon goals—including developing novel heavy metal removal methods, controlling heavy metal migration at the source, and mitigating ecological impacts—is a critical research direction requiring in-depth investigation. The articles in this Special Issue offer multidimensional insights into the mechanisms, application potential, and risk management of environmental heavy metal remediation technologies, offering scientific support for the development of an eco-friendly heavy metal pollution control system.

Data Availability Statement: Not applicable.

Acknowledgments: I would like to express my heartfelt thanks to the authors who have contributed their research to this Special Issue, the reviewers for their invaluable support.

Conflicts of Interest: The author declares no conflicts of interest.

List of Contributions:

1. Nie, X.; Huang, X.; Li, M.; Lu, Z.; Ling, X. Advances in Soil Amendments for Remediation of Heavy Metal-Contaminated Soils: Mechanisms, Impact, and Future Prospects. *Toxics* **2024**, *12*, 872.
2. Gertsen, M.; Perelomov, L.; Kharkova, A.; Burachevskaya, M.; Hemalatha, S.; Atroshchenko, Y. Removal of Lead Cations by Novel Organoclays Derived from Bentonite and Amphoteric and Nonionic Surfactants. *Toxics* **2024**, *12*, 713.
3. Zhang, Y.; Zang, L.; Zhao, Y.; Wei, Q.; Han, J. Removal of Pb from Contaminated Kaolin by Pulsed Electrochemical Treatment Coupled with a Permeable Reactive Barrier: Tuning Removal Efficiency and Energy Consumption. *Toxics* **2023**, *11*, 961.
4. Alfei, S.; Pandoli, O.G. Biochar-Derived Persistent Free Radicals: A Plethora of Environmental Applications in a Light and Shadows Scenario. *Toxics* **2024**, *12*, 245.
5. Zhuang, F.; Xiang, X.; Hu, J.; Xiong, J.; Zhang, T.; Zhou, L.; Jiang, G.; Zhang, M.; Liu, Z.; Yin, H., et al. Behavior and Mechanisms of Antimony Precipitation from Wastewater by Sulfate-Reducing Bacteria Desulfovibrio desulfuricans. *Toxics* **2025**, *13*, 17.
6. Hao, X.; Zhu, P.; Liu, X.; Jiang, L.; Jiang, H.; Liu, H.; Chen, Z. Bioreactor Expansion Affects Microbial Succession of Mixotrophic Acidophiles and Bioremediation of Cadmium-Contaminated Soils. *Toxics* **2024**, *12*, 362.
7. Zhang, M.; Zhao, H.; Zhang, Y.; Lv, X.; Zhang, L.; Shen, L.; Hu, L.; Wen, J.; Shen, L.; Luo, X. Oxidative Dissolution Process of Sphalerite in $Fe_2(SO_4)_3$-O_3 System: Implications for Heavy Metals Removal and Recovery. *Toxics* **2024**, *12*, 275.
8. Guo, Z.; Yang, J.; Li, K.; Shi, J.; Peng, Y.; Sarkodie, E.K.; Miao, B.; Liu, H.; Liu, X.; Jiang, L. Leaching Behavior of As and Pb in Lead–Zinc Mining Waste Rock under Mine Drainage and Rainwater. *Toxics* **2023**, *11*, 943.
9. Li, C.; Feng, Y.; Tian, P.; Yu, X. Mathematical Estimation of Endogenous Proline as a Bioindicator to Regulate the Stress of Trivalent Chromium on Rice Plants Grown in Different Nitrogenous Conditions. *Toxics* **2023**, *11*, 803.

References

1. Xu, D.; Fu, R.; Wang, J.; Shi, Y.; Guo, X. Chemical stabilization remediation for heavy metals in contaminated soils on the latest decade: Available stabilizing materials and associated evaluation methods—critical review. *J. Clean. Prod.* **2021**, *321*, 128730. [CrossRef]
2. Lin, H.; Zhou, M.; Li, B.; Dong, Y. Mechanisms, application advances and future perspectives of microbial-induced heavy metal precipitation: A review. *Int. Biodeter. Biodegr.* **2023**, *178*, 105544. [CrossRef]

toxics

MDPI

Article

Behavior and Mechanisms of Antimony Precipitation from Wastewater by Sulfate-Reducing Bacteria *Desulfovibrio desulfuricans*

Fan Zhuang [1], Xiaowu Xiang [2], Jin Hu [1], Jing Xiong [1,3], Teng Zhang [1,3], Lei Zhou [4], Guoping Jiang [4], Min Zhang [1], Zhenghua Liu [1], Huaqun Yin [1], Ling Xia [5], Ibrahim Ahmed Ibrahim Mahmoud [6] and Delong Meng [1,*]

1 Key Laboratory of Biometallurgy, Ministry of Education, School of Minerals Processing and Bioengineering, Central South University, Changsha 410083, China; fanzhuang@csu.edu.cn (F.Z.); 32210242@csu.edu.cn (J.H.); 19310085004@163.com (J.X.); zt18974215888@163.com (T.Z.); 205606002@csu.edu.cn (M.Z.); liuzhenghua2017csu@163.com (Z.L.); yinhuaqun_cs@sina.com (H.Y.)

2 Dongkou County Agricultural Bureau, Shaoyang 422300, China; m13278883049@163.com

3 Hunan Urban and Rural Environmental Construction Co., Ltd., Changsha 410118, China

4 Beijing Research Institute of Chemical Engineering and Metallurgy, Beijing 101148, China; lzhou0105@126.com (L.Z.); jiangguoping1984@163.com (G.J.)

5 Hubei Key Laboratory of Mineral Resources Processing and Environment, School of Resources and Environmental Engineering, Wuhan University of Technology, Luoshi Road 122, Wuhan 430070, China; xialing@whut.edu.cn

6 Central Metallurgical Research and Development Institute, Cairo 11421, Egypt; ibrahimcmrdi@yahoo.ca

* Correspondence: delong.meng@csu.edu.cn

Abstract: The development of the non-ferrous metal industry is generating increasingly large quantities of wastewater containing heavy metals (e.g., Sb). The precipitation of heavy metals by microorganisms involves complex mechanisms that require further investigation to optimize bioremediation technologies. In this study, we employed a sulfate-reducing bacteria (SRB) strain *Desulfovibrio desulfuricans* CSU_dl to treat the antimony (Sb)-containing wastewater; the behavior of Sb and mechanisms underlying precipitation were investigated by characterizing the precipitates. The results showed that the abiotic factors constraining SRB bacterial growth greatly affect Sb forms and precipitation. For instance, Sb precipitation maximumly occurred at pH 6 and 7, or C:N ratio of 10:1 and 40:3 for Sb(III) and Sb(V), respectively, resulting in a maximum Sb removal rate of 94%. Interestingly, we found that substantial antimonate and antimonite were adsorbed on the SRB cell surface, indicating that cell surface is a critical reaction site of Sb transformation and precipitation. Sb was adsorbed to the cell surface by C-C and C=O groups, and was further precipitated by forming Sb_2S_3 and Sb_2S_5 or was coprecipitated with the P-containing group. Partial Sb(V) reduction was also observed on the SRB cell surface. These results provided a deep insight into the Sb bio-transformation and were an advancement with respect to understanding bioremediation of Sb-contaminated wastewater.

Keywords: antimony pollution; sulfate-reducing bacteria; Sb; XPS; SEM-EDX

1. Introduction

In addition to being used in a wide variety of industries, such as electronics and metallurgy, antimony (Sb) is also known to cause water contamination due to improper handling [1]. Antimony-induced water pollution can be both natural and anthropogenic. The former sources include the weathering of rocks and minerals containing antimony, while the latter sources involve industrial activities, mining, and the discharge of wastewater containing antimony compounds. The non-ferrous metal industry, in particular, accounts for a significant portion in generating a substantial amount of heavy metal-containing wastewater, including antimony (Sb). Some studies have found that nearly 100 years of mining activities have caused severe soil contamination and water pollution in mining areas and these contaminations may alter the stability and functioning of ecosystems [2,3]. The unmitigated release of antimony mine drainage and wastewater has resulted in a persistent escalation of antimony pollutants [4]. It is known to all that the presence of antimony in water poses significant risks to both humans and ecosystems. Studies have shown that antimony (Sb) is listed as one of the major pollutants by the US EPA and the EU EPA due to its potential and proven carcinogenicity, immunotoxicity, genotoxicity, and reproductive toxicity like arsenic, and its toxicity Sb(metal) > Sb(III) > Sb(V), and the soluble compounds of antimony are more toxic than the insoluble compounds [5,6]. When the solubility of antimony in water reaches 3.5 mg/L, it becomes toxic to algae, and when it reaches 12 mg/L, it becomes toxic to fish and shrimps. Long-term ingestion of antimony-contaminated water may cause gastrointestinal disorders, skin irritation, respiratory problems, and may even lead to carcinogenic effects in humans. Antimony and its compounds have been consistently identified as priority contaminants due to their toxicity and biohazardous nature [1,4]. Antimony exhibits toxicity not only to higher organisms such as humans, algae, fish, and shrimps but also to microorganisms. The toxicity of Sb to bacteria, including sulfate-reducing bacteria (SRB), can significantly affect their metabolic activity and heavy metal-precipitation efficiency. Studies have shown that microbial toxicity assays, such as dehydrogenase activity tests and other redox reaction-based methods, provide valuable insights into the impact of toxic substances on microbial communities [7,8]). These methods could help elucidate the potential inhibitory effects of Sb on *Desulfovibrio* species and optimize bioremediation strategies.

The chemical speciation of antimony in water is highly dependent on the surrounding pH. Sb(III) undergoes significant changes in its chemical forms across a pH range of 1 to 12. At pH 1 to 4, the content of compounds in the form of positively charged $Sb(OH)_2{}^+$ decreases, while the content of compounds in the form of electrically neutral $Sb(OH)_3$ increases. Between pH 5 and 9, compounds mainly exist in the form of electrically neutral $Sb(OH)_3$. At pH 10 to 12, the content of electrically neutral $Sb(OH)_3$ decreases, and compounds in the form of negatively charged $Sb(OH)_4{}^-$ become more prevalent [9].

In order to mitigate the negative impacts of antimony pollution in water, it is essential to seek effective treatment methods. Numerous approaches have been explored in the field of antimony-removal technology, including physical processes (e.g., precipitation, filtration), chemical methods (e.g., flocculation, precipitation), and advanced techniques utilizing adsorption with materials like activated carbon or other specialized materials. Additionally, the use of sulfate-reducing bacteria (SRB) for mediating metal sulfide precipitation is seen as a promising approach for antimony removal [10,11]. The SRB bacteria put into application contain many species such as *Metallobacterium* [12], *Bacillus thiophilus*, and *Bacillus citriodora*. Extensive research has been carried out on the use of sulfate-reducing bacteria (SRB) for heavy metal remediation, and significant achievements have been made [13,14]. However, there are relatively few studies using SRB to treat Sb-containing wastewater. Wang demonstrated for the first time that SRB can convert $SO_4{}^{2-}$ to S^{2-} in mine drainage while converting Sb(V) to Sb(III), and verified that the ratio of $Sb(V)/SO_4{}^{2-}$ was an important parameter affecting the efficiency of antimony removal [11]. As previously reported, the addition of Fe^{2+} to the SRB system significantly increased the metabolic activity of SRB [15]. Complementarily, the addition of iron scrap and iron oxidizing bacteria (IOB) to the SRB system resulted in 99.98% Sb(V) removal for the Fe + IOB + SRB system [16]. Differences in carbon sources affected the efficiency of antimony removal in desulfurization processes, with the SRB system utilizing ethanol demonstrating a higher removal efficiency of 97.8% [17]. Temperature was also identified as a primary factor controlling microbial Sb reduction [17]. Characterization of the SRB antimony-removal system's precipitate indicated that Sb(V) was reduced to Sb_2S_3 [18,19]. Proteomic analysis showed that the extracellular protein functional groups of SRB were capable of adsorbing and immobilizing Sb(III), with a significant increase in extracellular proteins involved in electron transfer [20]. Despite many studies on Sb precipitation from wastewater by SRB, the effects of SRB on Sb(III)/Sb(V) precipitation in various habitats under anoxic conditions remain unclear, as do the magnitudes of the impacts of different environmental elements on precipitation efficiency. Furthermore, the precipitation mechanism of SRB on Sb(III)/Sb(V) under anoxic conditions requires further elucidation.

In this work, we analyzed the effect of different conditions (temperature, pH, C/N ratio, and SO_4^{2-} concentration) on the precipitation of Sb(III)/Sb(V) by SRB, and used scanning electron microscopy–energy dispersive X-ray (SEM-EDS), and X-ray photoelectron spectroscopy (XPS) to study the morphology, components, and surface chemistry of SRB before and after the immobilization of Sb(III)/Sb(V) by adsorption and the precipitation mechanism of Sb(III)/Sb(V) from water under oxygen conditions. The aim of this work is to reveal the mechanism related to the fixation of Sb(III) and Sb(V) by SRB. The work provided theoretical and technological foundations for the application of SRB in the remediation of antimony-polluted wastewater.

2. Materials and Methods

2.1. The SRB Strains and Culture Condition

The strain used in this study, *Desulfovibrio desulfuricans* subsp. CSU_dl, was isolated from underground water and preserved by the Key Laboratory of Biometallurgy of the Ministry of Education at Central South University, China. This sulfate-reducing bacterium is well known for its ability to reduce sulfate to sulfide and plays a crucial role in heavy metal precipitation. The culture conditions of the strain were as follows: anaerobic incubation for 3–5 days, incubation temperature of 30 °C. The lyophilised strain was activated, dissolved in 0.1–0.2 mL of culture medium or sterile water, and inoculated on 1–2 blood agar plates and incubated in an anaerobic incubator. After the colonies grew on the plates, the culture was enriched with modified Barr sulfate liquid medium. The strain was cultured in an anaerobic incubator. All those incubations and water treatment in this study were well controlled in anaerobic conditions by loading 97% N_2 and 3% H_2 to exclude oxygen.

2.2. Experiment Design

The initial concentrations of Sb(III) and Sb(V) were set at 15, 30, 45, and 60 mg/L. The logarithmic-phase SRB bacterial solution was inoculated into the medium containing Sb(III) and Sb(V) with a 10% inoculum volume. To optimize the precipitation conditions for Sb(III) and Sb(V), five experimental variables were investigated, with each condition performed in triplicate. Anaerobic flasks were incubated on a shaker at 180 rpm for 24 h. After the reaction, SRB cells were collected by centrifugation at 3500 rpm for 20 min. The experimental conditions included the following:

i. Temperature: Reaction temperatures were set at 20 °C, 25 °C, 30 °C, and 35 °C.
ii. Initial pH: The pH was adjusted to 4, 5, 6, 7, and 8.
iii. Carbon-to-Nitrogen (C/N) Ratio: C/N ratios were set at 20:1, 40:3, 10:1, 8:1, and 20:3.
iv. Initial Sulfate Concentration: SO_4^{2-} concentrations were set at 800, 1200, 1600, and 2000 mg/L.

2.3. Antimony Precipitation Efficiency

The concentration of SO_4^{2-} in the wastewater during treatment was determined using the barium chromate spectrophotometric method. The procedure was concise and well-organized, as follows: First, 1 mL of a 2.5 mol/L hydrochloric acid solution was added to both the water sample and the standard solution, and the mixture was boiled for approximately 5 min. Next, 25 mL of barium chromate suspension was added to each sample, and the mixture was boiled again for about 5 min. Afterward, the conical flask was removed, and the solution was allowed to cool slightly. Then, ammonia was added drop by drop until the solution turned lemon yellow, followed by the addition of two more drops. The solution underwent filtration using a slow qualitative filter paper, and the resulting filtrate was collected in a 50 mL cuvette. The conical flask and filter paper were rinsed three times with deionized water, and the filtered liquid was collected in a cuvette and was measured at 420 nm. Antimony concentration in wastewater was determined by ICP-AES.

Then, the best conditions for treating the Sb wastewater with SRB were determined. The subsequent experiment was set up as follows: 300 mL solution with 30 mg/L Sb(III) and 45 mg/L Sb(V), pH 7.1 $\pm$ 0.1, temperature 30 °C. The Sb-containing wastewater was treated with SRB (inoculation rate 10%) for 4 days. The bacterial cells, water, and precipitates were collected for further analysis.

2.4. Microbial Morphology

To detect the microbial morphology, the samples after 24 h treatment were collected by centrifuging. Scanning electron microscopy (SEM) was employed to examine the morphology of the sulphate-reducing bacteria. The specimens underwent double fixation with glutaraldehyde and osmium tetroxide ($H_2[OsO_4(OH)_2]$), followed by a gradient ethanol dehydration process lasting 15–25 min.

2.5. Morphological Structure and Chemical Composition of the Products

After washing and freeze-drying, two sample sets were obtained: untreated SRB organisms and SRB organisms adsorbed with Sb(III)/Sb(V). Scanning electron microscopy–energy dispersive X-ray (SEM-EDX) was used to analyze the morphology and composition of both untreated SRB bacteriophage samples and those with Sb(III)/Sb(V) adsorption. In addition, X-ray photoelectron spectroscopy (XPS) was employed to comprehensively characterize the elemental composition, including C, N, O, P, S, and Sb spectra, for further investigation of the chemical properties on the SRB surface.

3. Results and Discussion

3.1. Morphology and Growth Characteristics of SRB

The bacterial liquid was black in color, accompanied by hydrogen sulphide gas. Scanning electron microscope (SEM) observation showed that the SRB bacteria were in the shape of an arc (Figure A1). During the SRB growth, the pH value fluctuated between 7.2 and 7.5, the Eh redox potential was maintained between −55 mV and −70 mV, and the number of bacteria reached the maximum on the 4th day, and then began to decrease on the 5th day due to the accumulation of the product H_2S. After 7–9 days of incubation, the precipitation effect of SRB on SO_4^{2-} in the medium was higher than 91%, and a large desulphurization rate was achieved. It was shown that the sulphate-reduction rate of SRB reached a maximum in the pH range of 7.0–7.5, which justified the better desulphurization efficiency of the SRB bacteria used in this experiment [21,22].

3.2. Sb(III)/Sb(V) Precipitation by SRB Under Different Conditions

The pH, temperature, carbon to nitrogen ratio, and SO_4^{2-} concentration exert different impacts on the precipitation of Sb(III) and Sb(V) (Table 1).

Table 1. Precipitation of Sb(III)/Sb(V) by SRB under different conditions on day 7.

		Precipitation Rate of Sb(III) (%)	Precipitation Rate of Sb(V) (%)
pH	4	25.89 ± 4.20	62.07 ± 4.20
	5	45.44 ± 2.64	77.13 ± 3.23
	6	98.09 ± 5.35	87.17 ± 6.10
	7	91.57 ± 5.06	94.89 ± 4.37
	8	89.89 ± 3.45	91.21 ± 3.89
Temperature (°C)	20	88.97 ± 2.65	87.54 ± 5.06
	25	91.53 ± 5.56	91.56 ± 3.55
	30	91.57 ± 8.06	94.89 ± 4.37
	35	94.60 ± 2.65	97.53 ± 2.66
C/N	20:1	75.77 ± 4.20	88.36 ± 5.05
	40:3	90.26 ± 4.64	95.28 ± 6.55
	10:1	91.57 ± 4.06	94.89 ± 4.37
	8:1	86.66 ± 3.35	93.22 ± 6.39
	20:3	72.07 ± 4.45	87.46 ± 5.45
SO_4^{2-} (mg/L)	800	92.92 ± 2.65	96.27 ± 2.37
	1200	92.42 ± 5.56	95.97 ± 2.36
	1600	91.57 ± 8.06	94.89 ± 4.37
	2000	94.23 ± 4.25	94.16 ± 4.66

pH: For Sb(III), sulphate-reducing bacteria started to precipitate Sb(III) at partial neutrality (pH 6–8). The precipitating efficiencies of Sb(III) at pH 6, 7, and 8 on the seventh day were 98.09%, 91.57%, and 89.89%, while it was 25.89% at the weakly acidic condition (pH 4). Similar findings were observed for Sb(V). At pH 4, Sb(V) precipitated at a rate of 62.06%; at pH 7, it reached a maximum fixation rate of 94.89%; at pH 8, it decreased slightly to 91.21%. The pH values of 6 and 7, respectively, were optimal for the precipitation of Sb(III) and Sb(V) by SRB. The result is similar to a previous study [23], which showed that SRB were less active under acidic conditions and had a higher solution redox potential during growth, which reduced the efficiency of desulphurization and antimony precipitation.

The inhibitory effects of low pH on the desulphurization and antimony precipitation by SRB are varying and complicated. The effect of pH on SRB can be explained in two ways. Firstly, pH has a direct impact on SRB metabolism, disrupting its cellular homeostasis, destroying the pH gradient from the inside and outside of the cell, and ultimately causing energy loss. More protons diffuse through the cell membrane at lower pH levels than at neutral pH due to diffusion pressure across the membrane. To cope with low pH stress, SRB employ both active and passive mechanisms to maintain pH homeostasis. Active mechanisms include proton pumps, which actively expel excess H^+ ions from the cell using ATP, and amino acid decarboxylases, which produce neutralizing amines through decarboxylation reactions. Passive mechanisms involve changes in membrane lipid composition to reduce proton permeability and the expression of positively charged surface proteins that buffer external H^+ ions. These strategies help stabilize intracellular pH but divert energy from growth and metabolism, thereby impacting SRB performance in acidic environments. More energy from redox processes was needed to maintain pH homeostasis if the pH gradient in the extracellular and intracellular environments was too great [24]. Second, the pH changed the forms of some elements found in the environment, including organic acids, heavy metals, and sulfides. The forms of sulphur metabolites (H_2S, HS^- and S^{2-}) present in the solution of sulphate-reducing bacteria are dependent on the pH of the solution, and in acidic solutions (pH 4.5–5.27), the main form of sulfide substances present was molecular H_2S (~99%), which tended to volatilize easily [25], and therefore desulphurization for Sb(III) precipitation at a pH of 4.57 was less efficient. Under the condition of a neutral environment, the aqueous solution had a high concentration of sulfide, which increased the probability of Sb(III) binding with sulfide [26]. The SRB metabolism was stronger, so the neutral environment was more conducive to the desulfurization and antimony precipitation from wastewater by SRB. As shown in Figure 1a, as the initial pH increased, so did the removal of $SO4^{2-}$ and Sb(V). This might be the result of a higher pH, a drop in H^+ concentration, less competition, and a larger amount of heavy metal ion adsorption on the active sites, which would increase the removal of heavy metal ions [27].

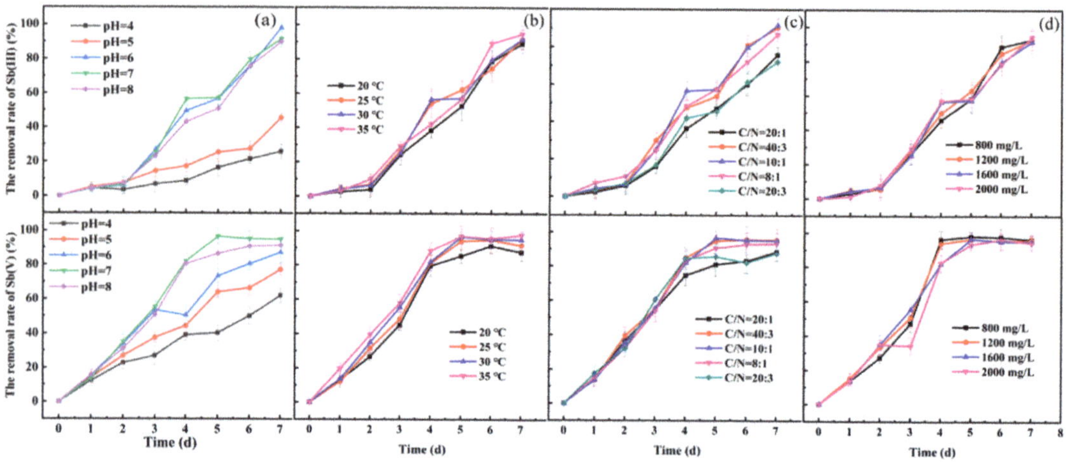

Figure 1. Effects of various parameters on the removal efficiency of Sb(III) and Sb(V) by sulfate-reducing bacteria (*Desulfovibrio desulfuricans*): (**a**) Initial pH (4, 5, 6, 7, and 8) at initial Sb concentration 30 mg/L, temperature 25 °C, C/N ratio 10:1, and sulfate concentration 1600 mg/L. (**b**) Temperature (20, 25, 30, and 35 °C) at initial Sb concentration 30 mg/L, pH 7, C/N ratio 10:1, and sulfate concentration 1600 mg/L. (**c**) Initial C/N ratio (20:1, 40:3, 10:1, 8:1, and 20:3) at initial Sb concentration 30 mg/L, temperature 25 °C, pH 7, and sulfate concentration 1600 mg/L. (**d**) Initial sulfate concentration (800, 1200, 1600, and 2000 mg/L) at initial Sb concentration 30 mg/L, temperature 25 °C, pH 7, and C/N ratio 10:1.

Temperature: Figure 1b shows the effect of different temperature conditions on antimony precipitation. On the 7th day, the precipitation efficiency of SRB for both Sb(III) and Sb(V) exceeded 90% at 25–35 °C. The precipitation efficiency of SRB for Sb(III) was 91.56%, 88.97%, and 94.60%, respectively. The precipitation efficiencies were 91.56%, 88.97%, and 94.60% for Sb(III) and 91.56%, 94.89%, and 97.53% for Sb(V). The combined findings demonstrated that 35 °C was the ideal temperature for both Sb(III) and Sb(V) precipitation by SRB in this investigation. A previous study concluded that SRB reduced thiosulfate at a growth temperature of 40–50 °C. However, when the growth temperature reached 30°C, virtually no H_2S was produced [28]. And the results showed that the ideal growth temperature range for SRB was 25 °C to 35 °C, which was consistent with this study.

There are two main ways that temperature affects Sb precipitation by SRB. On the one hand, temperature is an important influencing factor for the growth of SRB. Temperature modifies the microorganism's own enzyme activity, which in turn affects the efficiency of Sb(III) fixation. Certain enzymes in these cells will be inhibited by too high or too low a temperature, which will negatively impact cell growth and product synthesis, as well as alter cell morphology, metabolic function, and microbial toxicity, or even cause cell death [29]. Additionally, low temperature caused the lipids in the cell membrane to wax and the activity of membrane proteins to decrease, which limited the cell membrane's ability to transport electron donors and electron acceptors. This was the primary mechanism by which low temperature affected SRB metabolism [30]. On the other hand, the solubility of H_2S in wastewater was influenced by temperature; high temperatures make H_2S less soluble in wastewater, which lessens H_2S's inhibitory effect on SRB [31].

***C/N ratio*:** The C/N ratio represents the relative availability of carbon and nitrogen sources in the medium, which is a critical factor affecting bacterial growth and metabolism. Different C/N ratios in the substrate have certain effects on microbial growth. Reasonable adjustment of the C/N ratio in the substrate was one of the efficient measures to accelerate microbial growth and promote microbial action. The C/N ratio of Sb(III) and Sb(V) were shown in Figure 1c. Both high (20:1) and low (20:3) C/N ratio affected the growth of SRBs and their efficiency of desulphurization and antimony precipitation, and the immobilization rates of Sb(III) were 90.26%, 91.57%, and 86.66% for C/N ratios of 40:3, 10:1, and 8:1, respectively. When the C/N ratio was 40:3, the Sb(V) precipitation rate reached the highest value of 95.28%, which was the optimal C/N ratio.

At a low C/N ratio (e.g., 20:3), Sb(V)-removal efficiency decreased significantly due to limitations in microbial growth and metabolism. Microbial stoichiometry and metabolic theory state that when a substrate's low C/N ratio (or high N availability) satisfies microbial requirements or when the compounds are easily broken down by the microorganisms, increasing the production of enzymes and the breakdown of organic carbon, microbial activity is higher [32]. Therefore, the selection of an appropriate carbon-to-nitrogen ratio in microbial growth media can alleviate the metabolic constraints of microorganisms in heavy metal environments and enhance the efficiency of heavy metal fixation. Research has demonstrated that the growth and metabolism of SRB were restricted by a low C/N ratio [33]. In contrast, a high C/N ratio, insufficient nitrogen, low buffering capacity of the digestive solution, and easy decrease in pH caused the effectiveness of SRB's removal of Sb to decline. The selection of an appropriate carbon-to-nitrogen ratio is critical to alleviating metabolic constraints in microorganisms under heavy metal stress. While a low C/N ratio can inhibit Sb(V) removal by limiting SRB activity, excessively high C/N ratios can lead to nitrogen insufficiency, low buffering capacity, and pH reductions, all of which also reduce Sb(V) removal efficiency.

***SO_4^{2-} concentration*:** Increase in SO_4^{2-} concentration had little impact on the Sb(V) and Sb(III) immobilization efficiency of SRB; moreover, the Sb(V) immobilization efficiencies were higher than 94% in the range of sulphate concentration of 800–2000 mg/L. The results indicated that the immobilization efficiency of Sb(V) in the range of 800–2000 mg/L was higher than 94%. Meanwhile, the precipitation rate of Sb(III) under each SO_4^{2-} concentration condition was above 91.02%.

The SO_4^{2-} concentration in the solution directly reflects the equilibrium relationship between the substrate and microorganisms, and is also the main indicator of the system's ability to reduce sulfate ions. The production of hydrogen sulfide through heterogeneous sulfate reduction, or sulfide generation, is the fundamental metabolic characteristic of SRB. Sulfate and sulfite can be the electron acceptors for SRB during heterogeneous sulfate reduction [34]. When sulfate levels are sufficient, sulfate serves as the main electron acceptor for SRB metabolism; in situations where sulfate levels are insufficient, sulfite serves as the primary electron acceptor. It was demonstrated that SRBs can grow normally despite very low sulfate concentrations, and that microorganisms in the environment can propel this low-sulfate lake's sulfur recovery to high levels [35].

In this study, the SRB strain could grow well under the Sb(III) concentration of 30 mg/L. Under anaerobic conditions, when the Sb(III) concentration is 30 mg/L, the temperature is 30 °C, the pH value is 7.13, the C/N ratio is 10:1, and the SO_4^{2-} concentration is 1600 mg/L; the removal rate of Sb(III) by SRB could reach up to 91.02%. For Sb(V), the SRB strain could grow well under 60 mg/L of Sb(V). A lower toxicity of Sb(V) compared to Sb(III) would lead to the tolerance concentration of Sb(V) being higher than that of Sb(III). Under anaerobic conditions, when the Sb(V) concentration is 45 mg/L, the temperature is 35 °C, the pH value is 7.24, the C/N ratio is 40:3, the SO_4^{2-} concentration is 2000 mg/L, and the removal rate of Sb(V) by the SRB reaches 94%. Even though the precipitation rate was a bit lower than some previous studies which reached up to 98.7% removal rate [10], the initial Sb concentration was much higher in our study than other studies (45 mg/L vs. 20 mg/L). Much more Sb was removed in our study than previous studies. The results indicate that the *Desulfovibrio desulfuricans* CSU_dl strain had excellent Sb tolerance, as well as bioremediation potentials.

3.3. Characterization of Precipitates Generated by SRB

The morphology and chemical composition of Sb(III)/Sb(V) precipitates on the SRB cell surface were investigated using SEM-EDS (Figure 2). The SRB cells exhibited a smooth surface with significant pore cavities and thin pore walls. Chemical composition analysis of the SRB cell surface indicated that C, P, O, S, and N were the primary elements. After immobilization, Sb(III) adsorption on the SRB cell surface resulted in morphological changes, including a rougher surface, blocked pore cavities, thicker pore walls, and the formation of ellipsoidal insoluble material. The corresponding surface chemical analysis revealed that C, S, P, Sb, O, and N were the main elements in the reaction products. These observations suggest that Sb(III) precipitation occurred on the SRB cell surface, producing antimony trisulphide (Sb_2S_3) as the primary product. The mechanism driving Sb(III) precipitation is hypothesized to involve mineralization mediated by cell wall surface materials, as represented by the following reaction (Equation (1)):

$$3S^{2-} + 2Sb(OH)_3 + 6H^+ \rightarrow Sb_2S_3(s) + 6H_2O \quad (1)$$

The mechanism of Sb(V) precipitation by SRB organisms (Equations (1) and (2)) is inferred from the above results, where part of Sb(V) were reduced to Sb(III), and mineralization occurred under the action of cell wall surface substances to produce antimony trisulphide and antimony pentasulphide, resulting in the precipitation of Sb(V) from the wastewater by SRB.

$$5S^{2-} + 2Sb(OH)_6^- + 12H^+ \rightarrow Sb_2S_5(s) + 12H_2O \quad (2)$$

Cell morphology analysis by SEM indicated that the precipitation of Sb occurred mainly on the cell surface. In particular, abundant P elements were observed on the cell surface as indicated by EDS analysis. This could be due to P being the essential elements for the cell membrane. P could co-precipitate with heavy metals, which further promoted the precipitation of Sb. The results were further supported by the XPS and FTIR analysis.

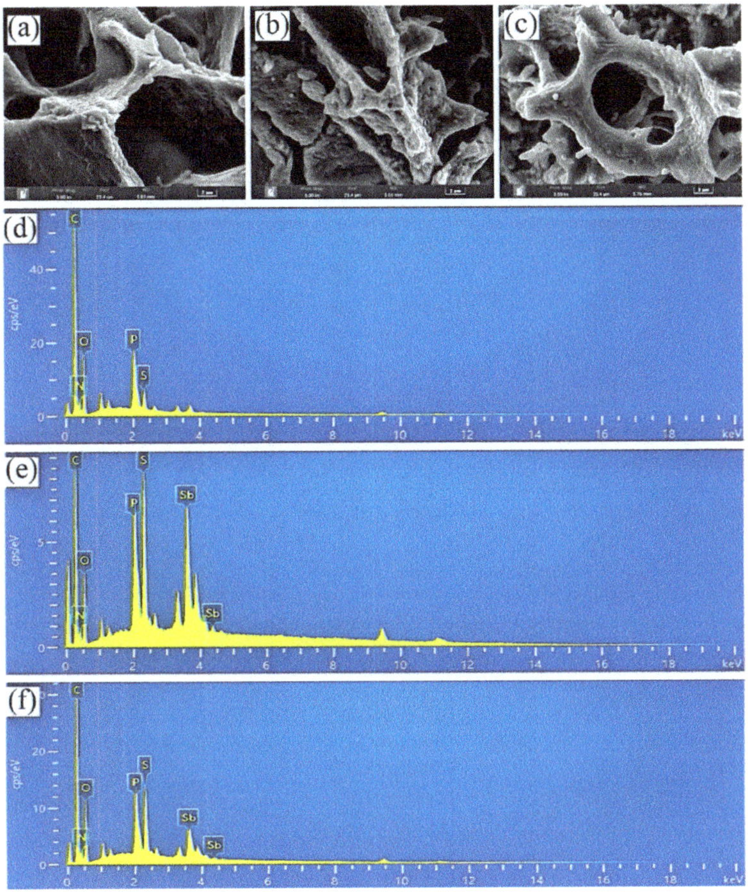

Figure 2. SEM images of freeze-dried SRB bacteria solution (**a**), SRB-Sb(III) immobilized products (**b**), SRB-Sb(V) immobilized products (**c**); EDM images of freeze-dried SRB bacteria solution (**e**), SRB-Sb(III) immobilized products (**d**), SRB-Sb(V) immobilized products (**f**).

3.4. Analysis of Surface Functional Groups Involved in the Adsorption and Immobilization of Sb(III)/Sb(V) by SRB

The chemical composition, elemental valence, and binding forms of SRB before and after immobilization of Sb(III)/Sb(V) by adsorption were investigated by XPS analysis. As shown in Figure 3, the SRB adsorption-immobilized Sb(III)/Sb(V) products clearly identified the peaks of P 2p, S 2p, C 1s, N 1s, O 1s, and Sb $3d_{5/2}$ in the scanning spectra, with the peaks of Sb $3d_{5/2}$ and O 1s being very close to each other at the binding energy of 528 eV–530.3 eV. The characteristic peak of Sb $3d_{3/2}$ was also identified in the product of SRB adsorption of immobilized Sb(III). The S 2p peak corresponds to a binding energy of 164.3 eV, and its increase indicated the production of sulphides in the product. This suggests that sulphate-reducing bacteria react with Sb(III) or Sb(V) in solution, allowing antimony to be adsorbed onto the product surface, which remains consistent with our previous predictions.

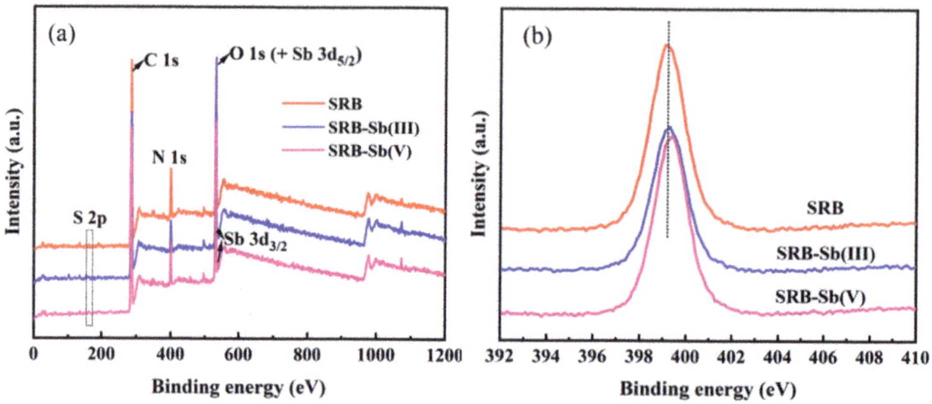

Figure 3. The XPS survey spectra of different SRB before and after Sb(III)/Sb(V) adsorption: (**a**) full spectrum; (**b**) N 1s orbital part spectrum. The red line represents the XPS analysis of the surface of normal SRB cells, the pink line represents the XPS analysis of SRB cell surfaces after Sb(V) adsorption, and the blue line represents the XPS analysis of SRB cell surfaces after Sb(III) adsorption.

On the basis of the shape of the XPS absorption peak in the N 1s orbital fractionation spectrum (Figure 3b), the binding energy of the N 1s peak is 399.08 eV in the SRB treatment, 399.10 eV in the SRB-Sb(III) treatment, and 399.38 eV in the SRB-Sb(V) treatment, suggesting that the binding energy of the N 1s peak in SRB to Sb(V) biosorption immobilization is shifted to the right, indicating a possible loss of electrons. This may be due to the metal ion chelating with the N atom in the amino and imine groups and the pair of lone electrons in the N atom being shared with the metal ion, resulting in a lower electron cloud density and a higher binding energy peak for the N atom [36].

Three peaks originally made up the C 1s orbital fractionation of the XPS absorption peaks (Figure 4a), which corresponded to C-C (284.08 eV), C-O (285.58 eV), and C=O (287.28 eV). The amount of C-O increases and the amount of C-C and C=O decreases following the adsorption immobilization of Sb(III) and Sb(V), as can be observed, suggesting the participation of carbonaceous groups in the Sb adsorption immobilization process.

The P 2p orbital spectra of the XPS absorption peaks (Figure 4b) initially consisted of two peaks corresponding to $O{=}P(OR)_3$ (133.18 eV) and P-C (132.28 eV), respectively. It can be seen that after the adsorption and immobilization of Sb(III)/Sb(V), the $O{=}P(OR)_3$ content decreased and the P-C content increased, and the P-C content exceeded the $O{=}P(OR)_3$ content, indicating that there is a relevant chemical reaction occurring between Sb(III)/Sb(V) and phosphorus-containing groups, and that phosphorus-containing functional groups play an important role in the adsorption and immobilization of Sb, which is consistent with the results of previous research [37].

The S 2p channel splitting spectrum of the XPS absorption peaks (Figure 4c) initially consisted of four peaks corresponding to S^{2-} (160.68 eV), S_2^{2-} (162.38 eV), S_n^{2-} (163.98 eV), and SO_4^{2-} (167.88 eV). Significant increases in S^{2-} and decreases in S_2^{2-} and S_n^{2-} were seen in Sb(III) and Sb(V); after Sb(III) was immobilized, SO_4^{2-} increased, and after Sb(V) was immobilized, SO_4^{2-} dropped. Sb $3d_{3/2}$, which does not overlap with the O 1s, was employed as a guide for the fitting of the Sb $3d_{5/2}$ peak because the Sb $3d_{5/2}$ peak overlaps with the O 1s. Therefore, the spin-orbit splitting and ratio of the Sb $3d_{3/2}$ peak dictate the strength and binding energy of the Sb $3d_{5/2}$ peak. The two characteristic peaks in Figure 4d for Sb(III) and Sb(V) at the binding site during Sb(V) adsorption show that Sb(V) was partially decreased during adsorption. The reduction reaction, however, is thought to occur during the adsorption of Sb(V) by SRB, with S_2^{2-} and S_n^{2-} in solution and their surface functional groups acting as electron donors, reducing Sb(V) to Sb(III) and immobilizing it by binding to functional groups to form mineralized substances. This is supported by the presence of the Sb(III) peak at 530.88 eV.

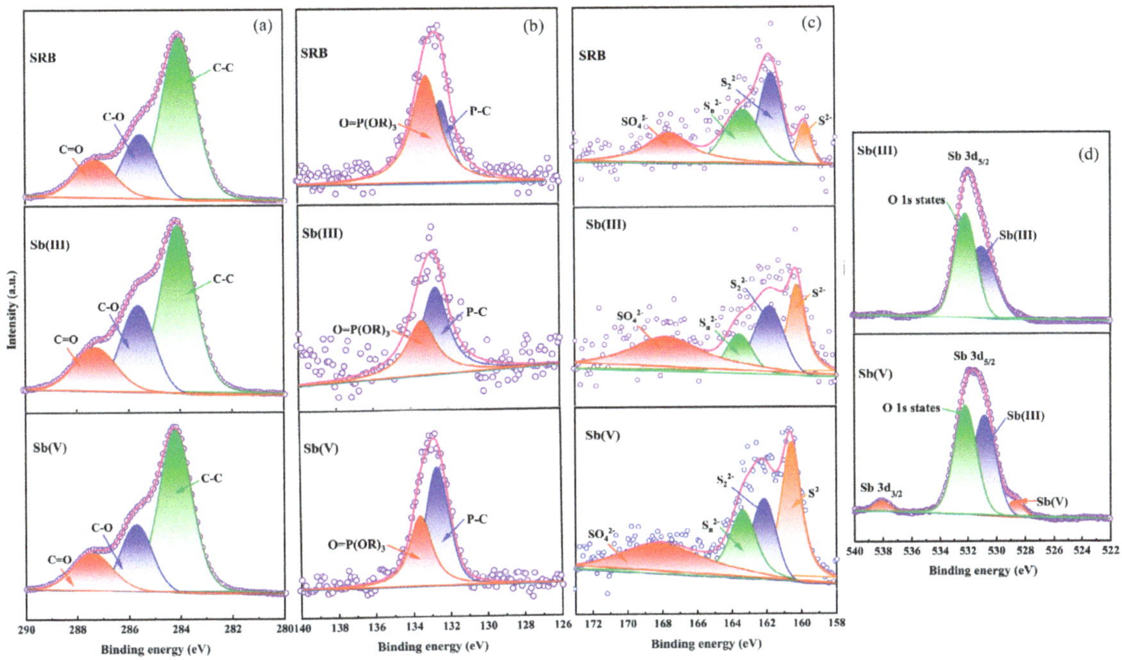

Figure 4. The XPS spectrum of different SRB before and after Sb(III)/Sb(V) adsorption (**a**) C 1s, (**b**) P 2p, (**c**) S 2p, (**d**) Sb. The pink line represents the total XPS spectra of each element, while the different filled lines represent the XPS spectra of the various chemical bonds of that element.

4. Conclusions

Due to the massive mining and utilization of antimony-mining resources, the surrounding water and soils are seriously polluted. The sulfate-reducing bacteria (SRB) method has the advantages of high efficiency, low cost, and no secondary pollution, and is highly advantageous in the treatment of antimony-contaminated water and soil. In this study, we find that under anoxic conditions, the precipitation rate of Sb(III) wastewater treated by SRB reached more than 91.02% when the Sb(III) concentration was 30 mg/L, temperature was 30 °C, pH was 7.13, C/N ratio was 10:1, and SO_4^{2-} concentration was 1600 mg/L. When the Sb(V) concentration was 45 mg/L, the temperature was 35 °C, the pH was 7.24, the C/N ratio was 40:3, and the SO_4^{2-} concentration was 2000 mg/L, the precipitation rate of Sb(V) from the wastewater by SRB was more than 94%. The morphology and precipitation of antimony are largely influenced by abiotic factors that restrict the growth of SRB bacteria. However, these factors do not work exactly the same for Sb(III) and Sb(V). The cell surface is the main activate site for Sb adsorption and precipitation. Sb was adsorbed to the SRB cell surface by the C-C and C=O group; the P-containing group also played a role in adsorption and precipitation of SRB. The adsorbed Sb was then precipitated by forming the sulfide precipitation. In this process, S_2^{2-} and S_n^{2-} and their surface functional groups acted as electron donors, reducing a part of Sb(V) to Sb(III) and combining with functional groups. Mineralization occurred on the surface of SRB cells, generating Sb_2S_3 and Sb_2S_5 solid precipitates, thus achieving the precipitation of Sb(III)/Sb(V) from wastewater by

SRB. However, this study used artificially generated wastewater, which may be different to real conditions. A further study using real wastewater with complex pollutants is also needed to promote the SRB remediation technology.

Author Contributions: F.Z.: Methodology, Visualization, Formal analysis, Investigation, Data curation, Writing—original draft, Writing—review & editing. X.X.: Visualization, Formal analysis, Investigation, Writing—review & editing. J.H.: Methodology, Visualization, Formal analysis, Investigation, Data curation, Writing—original draft, Writing—review & editing. J.X.: Methodology, Formal analysis, Visualization, Data curation, Writing—review & editing. T.Z.: Visualization Writing—review & editing. L.Z.: Conceptualization, Visualization, Writing—review & editing. G.J.: Conceptualization, Methodology, Visualization Writing—review & editing. M.Z.: Conceptualization, Methodology, Visualization Writing—review & editing. Z.L.: Visualization Writing—review & editing. H.Y.: Conceptualization, Methodology, Validation, Resources, Writing—review & editing, funding acquisition. L.X.: Visualization Writing—review & editing. I.A.I.M.: Visualization Writing—review & editing. D.M.: Supervision, Conceptualization, Methodology, Validation, Resources, Writing—review & editing, funding acquisition. All authors have read and agreed to the published version of the manuscript.

Funding: This research was supported by the National Key R&D Program of China: No. 2023YFE0114500; the National Natural Science Fundation of China: No. 42477134; the Natural Science Foundation of Hunan Province: No. 2023JJ30658.

Institutional Review Board Statement: Not applicable.

Informed Consent Statement: Not applicable.

Data Availability Statement: The data supporting the reported results are available from the corresponding author upon reasonable request.

Conflicts of Interest: Authors Jing Xiong, Teng Zhang were employed by the company Hunan Urban and Rural Environmental Construction Co., Ltd., Changsha 410118, China. The remaining authors declare that the research was conducted in the absence of any commercial or financial relationships that could be construed as a potential conflict of interest.

Appendix A

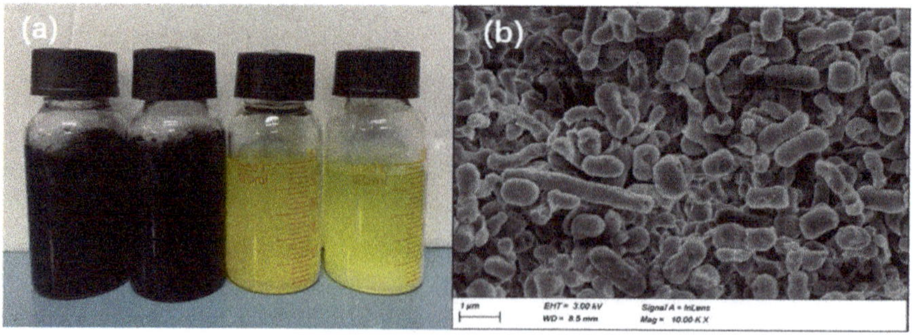

Figure A1. Sulfate-reducing bacteria solution (**a**) and scanning electron microscope image (**b**).

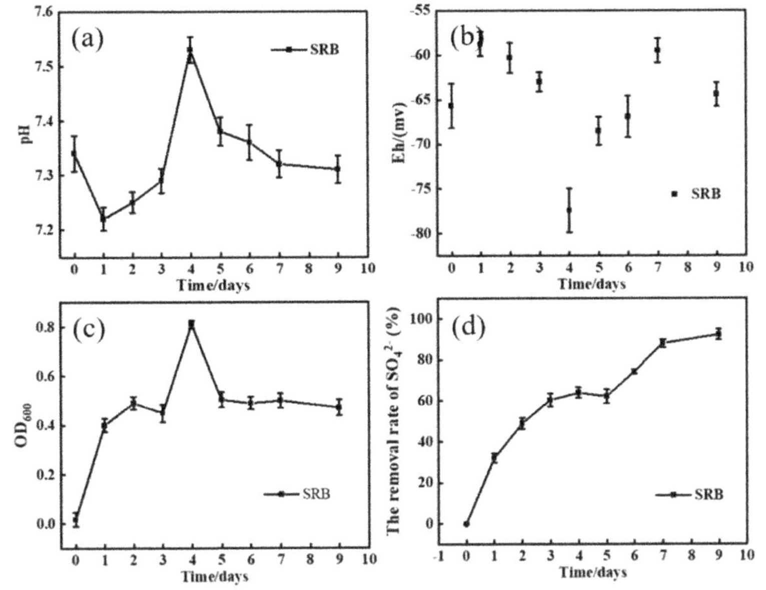

Figure A2. Parameter changes during the growth of sulfate-reducing bacteria (**a**) pH, (**b**) Eh, (**c**) OD_{600}, (**d**) SO_4^{2-} removal efficiency.

References

1. Filella, M.; Belzile, N.; Chen, Y.-W. Antimony in the environment: A review focused on natural waters: I. Occurrence. *Earth-Sci. Rev.* **2002**, *57*, 125–176. [CrossRef]
2. Okkenhaug, G.; Zhu, Y.G.; Luo, L.; Lei, M.; Li, X.; Mulder, J. Distribution, speciation and availability of antimony (Sb) in soils and terrestrial plants from an active Sb mining area. *Environ. Pollut.* **2011**, *159*, 2427–2434. [CrossRef] [PubMed]
3. Zhang, M.; Zhang, T.; Zhou, L.; Lou, W.; Zeng, W.; Liu, T.; Yin, H.; Liu, H.; Liu, X.; Mathivanan, K.; et al. Soil microbial community assembly model in response to heavy metal pollution. *Environ. Res.* **2022**, *213*, 113576. [CrossRef] [PubMed]
4. He, M.; Wang, X.; Wu, F.; Fu, Z. Antimony pollution in China. *Sci. Total Environ.* **2012**, *421–422*, 41–50. [CrossRef] [PubMed]
5. Arivalagan, P.; Singaraj, D.; Haridass, V.; Kaliannan, T. Removal of cadmium from aqueous solution by batch studies using Bacillus cereus. *Ecol. Eng.* **2014**, *71*, 728–735. [CrossRef]
6. Zhuang, F.; Huang, J.; Li, H.; Peng, X.; Xia, L.; Zhou, L.; Zhang, T.; Liu, Z.; He, Q.; Luo, F.; et al. Biogeochemical behavior and pollution control of arsenic in mining areas: A review. *Front. Microbiol.* **2023**, *14*, 1043024. [CrossRef]
7. Strotmann, U.; Zaremba, S.; Bias, W.R. Rapid toxicity tests for the determination of substance toxicity to activated sludge. *Acta Hydrochim. Hydrobiol.* **1992**, *20*, 136–142. [CrossRef]
8. Strotmann, U.; Durand, M.-J.; Thouand, G.; Eberlein, C.; Heipieper, H.J.; Gartiser, S.; Pagga, U. Microbiological toxicity tests using standardized ISO/OECD methods—Current state and outlook. *Appl. Microbiol. Biotechnol.* **2024**, *108*, 454. [CrossRef]
9. Xu, R.; Li, Q.; Nan, X.; Yang, Y.; Xu, B.; Li, K.; Wang, L.; Zhang, Y.; Jiang, T. Synthesis of nano-silica and biogenic iron (oxyhydr)oxides composites mediated by iron oxidizing bacteria to remove antimonite and antimonate from aqueous solution: Performance and mechanisms. *J. Hazard. Mater.* **2022**, *422*, 126821. [CrossRef]
10. Liu, F.; Zhang, G.; Liu, S.; Fu, Z.; Chen, J.; Ma, C. Bioremoval of arsenic and antimony from wastewater by a mixed culture of sulfate-reducing bacteria using lactate and ethanol as carbon sources. *Int. Biodeterior. Biodegrad.* **2018**, *126*, 152–159. [CrossRef]
11. Wang, H.; Chen, F.; Mu, S.; Zhang, D.; Pan, X.; Lee, D.J.; Chang, J.S. Removal of antimony (Sb(V)) from Sb mine drainage: Biological sulfate reduction and sulfide oxidation-precipitation. *Bioresour. Technol.* **2013**, *146*, 799–802. [CrossRef] [PubMed]
12. He, Q.; Liu, Y.; Wan, D.; Liu, Y.; Xiao, S.; Wang, Y.; Shi, Y. Enhanced biological antimony removal from water by combining elemental sulfur autotrophic reduction and disproportionation. *J. Hazard. Mater.* **2022**, *434*, 128926. [CrossRef] [PubMed]
13. Jong, T.; Parry, D.L. Removal of sulfate and heavy metals by sulfate reducing bacteria in short-term bench scale upflow anaerobic packed bed reactor runs. *Water Res.* **2003**, *37*, 3379–3389. [CrossRef] [PubMed]
14. Bai, H.; Kang, Y.; Quan, H.; Han, Y.; Sun, J.; Feng, Y. Bioremediation of copper-containing wastewater by sulfate reducing bacteria coupled with iron. *J. Environ. Manag.* **2013**, *129*, 350–356. [CrossRef]

15. Xi, Y.; Lan, S.; Li, X.; Wu, Y.; Yuan, X.; Zhang, C.; Liu, Y.; Huang, Y.; Quan, B.; Wu, S. Bioremediation of antimony from wastewater by sulfate-reducing bacteria: Effect of the coexisting ferrous ion. *Int. Biodeterior. Biodegrad.* **2020**, *148*, 104912. [CrossRef]
16. Li, Y.; Xu, Z.; Wu, J.; Mo, P. Efficiency and mechanisms of antimony removal from wastewater using mixed cultures of iron-oxidizing bacteria and sulfate-reducing bacteria based on scrap iron. *Sep. Purif. Technol.* **2020**, *246*, 116756. [CrossRef]
17. Kujala, K.; Laamanen, T.; Khan, U.A.; Besold, J.; Planer-Friedrich, B. Kinetics of arsenic and antimony reduction and oxidation in peatlands treating mining-affected waters: Effects of microbes, temperature, and carbon substrate. *Soil Biol. Biochem.* **2022**, *167*, 108598. [CrossRef]
18. Zhang, G.; Ouyang, X.; Li, H.; Fu, Z.; Chen, J. Bioremoval of antimony from contaminated waters by a mixed batch culture of sulfate-reducing bacteria. *Int. Biodeterior. Biodegrad.* **2016**, *115*, 148–155. [CrossRef]
19. Arulmani, S.R.B.; Dai, J.; Li, H.; Chen, Z.; Zhang, H.; Yan, J.; Xiao, T.; Sun, W. Efficient reduction of antimony by sulfate-reducer enriched bio-cathode with hydrogen production in a microbial electrolysis cell. *Sci. Total Environ.* **2021**, *774*, 145733. [CrossRef]
20. Yu, H.; Yan, X.; Weng, W.; Xu, S.; Xu, G.; Gu, T.; Guan, X.; Liu, S.; Chen, P.; Wu, Y.; et al. Extracellular proteins of *Desulfovibrio vulgaris* as adsorbents and redox shuttles promote biomineralization of antimony. *J. Hazard. Mater.* **2022**, *426*, 127795. [CrossRef]
21. Kikot, P.; Viera, M.; Mignone, C.; Donati, E. Study of the effect of pH and dissolved heavy metals on the growth of sulfate-reducing bacteria by a fractional factorial design. *Hydrometallurgy* **2010**, *104*, 494–500. [CrossRef]
22. Sharma, K.; Derlon, N.; Hu, S.; Yuan, Z. Modeling the pH effect on sulfidogenesis in anaerobic sewer biofilm. *Water Res.* **2014**, *49*, 175–185. [CrossRef] [PubMed]
23. Liu, J.L.; Yao, J.; Duran, R.; Mihucz, V.G.; Hudson-Edwards, K.A. Bacterial shifts during in-situ mineralization bio-treatment to non-ferrous metal(loid) tailings. *Environ. Pollut.* **2019**, *255*, 113165. [CrossRef] [PubMed]
24. Meier, J.; Piva, A.; Fortin, D. Enrichment of sulfate-reducing bacteria and resulting mineral formation in media mimicking pore water metal ion concentrations and pH conditions of acidic pit lakes. *FEMS Microbiol. Ecol.* **2012**, *79*, 69–84. [CrossRef]
25. Hsu, H.F.; Jhuo, Y.S.; Kumar, M.; Ma, Y.S.; Lin, J.G. Simultaneous sulfate reduction and copper removal by a PVA-immobilized sulfate reducing bacterial culture. *Bioresour. Technol.* **2010**, *101*, 4354–4361. [CrossRef]
26. Mitsunobu, S.; Harada, T.; Takahashi, Y. Comparison of antimony behavior with that of arsenic under various soil redox conditions. *Environ. Sci. Technol.* **2006**, *40*, 7270–7276. [CrossRef]
27. Wang, C.; Yang, R.; Wang, H. Synthesis of ZIF-8/Fly Ash Composite for Adsorption of Cu^{2+}, Zn^{2+} and Ni^{2+} from Aqueous Solutions. *Materials* **2020**, *13*, 214. [CrossRef]
28. Zhang, S.H.; Kuang, F.; Zhang, J.; Zhou, X.; Lan, G.; Zou, C. Influences of Thermophilic Sulfate-Reducing Bacteria on the Corrosion of Carbon Steel at Different Temperature. *Adv. Mater. Res.* **2011**, *396–398*, 1963–1968. [CrossRef]
29. Huang, J.; Ou, Y.; Zhang, D.; Zhang, G.; Pan, Y. Optimization of the culture condition of Bacillus mucilaginous using Agaricus bisporus industrial wastewater by Plackett-Burman combined with Box-Behnken response surface method. *AMB Express* **2018**, *8*, 141. [CrossRef]
30. Nevatalo, L.M.; Bijmans, M.F.M.; Lens, P.N.L.; Kaksonen, A.H.; Puhakka, J.A. The effect of sub-optimal temperature on specific sulfidogenic activity of mesophilic SRB in an H2-fed membrane bioreactor. *Process Biochem.* **2010**, *45*, 363–368. [CrossRef]
31. Kaksonen, A.; Puhakka, J. Sulfate Reduction Based Bioprocesses for the Treatment of Acid Mine Drainage and the Recovery of Metals. *Eng. Life Sci.* **2007**, *7*, 541–564. [CrossRef]
32. Li, X.; Li, Y.; Wu, T.; Qu, C.; Ning, P.; Shi, J.; Tian, X. Potassium fertilization combined with crop straw incorporation alters soil potassium fractions and availability in northwest China: An incubation study. *PLoS ONE* **2020**, *15*, e0236634. [CrossRef] [PubMed]
33. Robinson-Lora, M.A.; Brennan, R.A. Efficient metal removal and neutralization of acid mine drainage by crab-shell chitin under batch and continuous-flow conditions. *Bioresour. Technol.* **2009**, *100*, 5063–5071. [CrossRef]
34. Kushkevych, I.; Abdulina, D.; Kovac, J.; Dordevic, D.; Vitezova, M.; Iutynska, G.; Rittmann, S.K.R. Adenosine-5′-Phosphosulfate- and Sulfite Reductases Activities of Sulfate-Reducing Bacteria from Various Environments. *Biomolecules* **2020**, *10*, 921. [CrossRef]
35. Berg, J.S.; Jezequel, D.; Duverger, A.; Lamy, D.; Laberty-Robert, C.; Miot, J. Microbial diversity involved in iron and cryptic sulfur cycling in the ferruginous, low-sulfate waters of Lake Pavin. *PLoS ONE* **2019**, *14*, e0212787. [CrossRef]
36. Zhang, M.; Hou, S.; Li, Y.; Hou, Y.; Yang, P. Single evaluation and selection of functional groups containing N or O atoms to heavy metal adsorption: Law of electric neutrality. *Chemosphere* **2022**, *287*, 132207. [CrossRef]
37. Li, Y.; Song, S.; Xia, L.; Yin, H.; García Meza, J.V.; Ju, W. Enhanced Pb(II) removal by algal-based biosorbent cultivated in high-phosphorus cultures. *Chem. Eng. J.* **2019**, *361*, 167–179. [CrossRef]

toxics MDPI

Article

Removal of Lead Cations by Novel Organoclays Derived from Bentonite and Amphoteric and Nonionic Surfactants

Maria Gertsen [1,*], Leonid Perelomov [1,2], Anna Kharkova [3], Marina Burachevskaya [1], S. Hemalatha [4] and Yury Atroshchenko [1]

[1] Laboratory of Soil Chemistry and Ecology, Faculty of Natural Sciences, Tula State Lev Tolstoy Pedagogical University (Tolstoy University), Lenin Avenue, 125, 300026 Tula, Russia; perelomov@rambler.ru (L.P.); marina.0911@mail.ru (M.B.); reaktiv@tsput.ru (Y.A.)

[2] Laboratory of Biogeochemistry, Faculty of Natural Sciences, Tula State Lev Tolstoy Pedagogical University (Tolstoy University), Lenin Avenue, 125, 300026 Tula, Russia

[3] Natural Sciences Institute, Tula State University, Lenin Avenue, 92, 300012 Tula, Russia; anyuta_zaytseva@mail.ru

[4] School of Life Sciences, B.S. Abdur Rahman Crescent Institute of Science and Technology, Chennai 600048, India; hemalatha.sls@crescent.education

* Correspondence: gertsen@tsput.ru

Abstract: For many decades, natural and modified clay minerals have been used as adsorbents to clean up aquatic and soil ecosystems contaminated with organic and inorganic pollutants. In this study, organoclays based on bentonite and various amphoteric and nonionic surfactants were synthesized and tested as effective sorbents for lead ions. The maximum values of R were obtained when describing the sorption processes using the Langmuir model, which ranged from 0.97 to 0.99. The adsorption of lead ions by these organoclays was investigated using different sorption models including the Langmuir, Freundlich, and BET. It was found that, according to the values of limiting adsorption to the Langmuir equation, the synthesized organoclays formed an increasing series: organoclay with cocamide diethanolamine < bentonite < organoclay with lauramine oxide < organoclay with sodium cocoiminodipropionate < organoclay with disodium cocoamphodiacetate < organoclay with alkyl polyglucoside. The Gibbs energy for all of the analyzed samples was calculated and found to be negative, indicating the spontaneity of the cation adsorption process in the forward direction. The maximum value of the adsorption capacity of lead cations on organoclay-based bentonite with alkyl polyglucoside was 1.49 ± 0.05 mmol/g according to the Langmuir model, and 0.523 ± 0.003 mmol/g as determined by the BET model. In the process of modifying bentonite, there was an increase in negative values of the zeta potential for organoclays compared to the initial mineral, which clearly enhanced their electrostatic interactions with the positively charged lead ions. It was hypothesized, based on the physicochemical principles, that exchange adsorption is the main mechanism for lead absorption. Based on chemical approaches, organoclays based on amphoteric surfactants absorb lead mainly through the mechanisms of electrostatic attraction, ion exchange, and complexation as well as the formation of insoluble precipitates. Organoclays based on nonionic surfactants, on the other hand, absorb lead through mechanisms of complexation (including chelation) and the formation of insoluble chemical precipitates. The comparison of isotherms from different models allows us to find the most accurate match between the model and the experimental data, and to better understand the nature of the processes involved.

Keywords: clays; surfactants; modification; adsorption; heavy metals; ecosystem remediation

1. Introduction

The widespread heavy metal pollution of environmental objects is one of the most significant environmental issues today [1]. Metals do not decompose biologically or chemically, but instead undergo transformations of their compounds, leading to their long-term presence in soil, ground, and sediment, which can become sources of secondary pollution over time. Numerous studies have demonstrated that the accumulation of heavy metals in soil can pose risks to soil fertility and quality, disrupting its ecological functions [2–4]. In soil environments, processes such as sorption, desorption, chemical complexation with inorganic and organic ligands, and oxidation–reduction are crucial for monitoring the availability, mobility, and toxicity of heavy metals. These reactions are influenced by various factors including pH, the composition of the soil, the presence of inorganic and organic ligands such as humic and fulvic acids, and microbial metabolites and root exudates [5,6].

Currently, a significant challenge lies in the search for and development of efficient, environmentally friendly, and cost-effective remediation techniques for soils contaminated with heavy metals. These techniques should reduce the mobility and biological activity of these pollutants in order to improve the overall health of the environment. Several approaches have been explored including the use of graphene oxide modified with chitosan to adsorb lead(II) ions [7], polymer adsorbents based on ethylenediamine, diethylenetriamine, and tetraethylenepentamine [8], carbon nanotubes [9], humic acids [10,11], biochar [12], and natural and modified clay minerals [13,14]. These materials have shown promise in reducing the concentration of heavy metals in contaminated soils. Natural and modified clay minerals with expanding structural cells have also been considered as potential remediation materials due to their ability to interact with and remove heavy metals from soil. These materials hold great potential for future research and development efforts in this area.

Natural montmorillonite has traditionally been used to absorb and immobilize various cationic pollutants such as heavy metals, dyes, and some pesticides [15]. Bentonite clay pastes and suspensions have been used to decontaminate clothing, equipment, and building materials during the clean-up of the Chernobyl disaster as well as prevent the movement of radionuclides through water [16]. It has been shown that natural clays are highly effective at absorbing Cs^+ from radioactive waste [17].

Various physical and chemical methods for modifying clay can significantly improve its sorption properties and expand its application for new substances. Currently, there is active research underway on the sorption properties of clay minerals modified with organic substances, also known as organoclays. The sorption efficiency and strength of pollutant binding as well as the methods for synthesizing and applying organoclays are all being optimized [18–20].

The interaction between organic matter and clay minerals is a complex process that is influenced by the chemical nature and structural characteristics of the minerals as well as the physical and chemical properties of the organic matter. It is also influenced by the environmental conditions under which the interaction occurs. Organic matter has a significant impact on clay minerals, altering their surface through the formation of chemical bonds and intermolecular aggregates. These interactions are facilitated by hydrogen bonds and Van der Waals forces as well as physical contact. As a result, the resulting organoclay products exhibit properties that differ significantly from those of the individual starting materials. At present, the mechanisms of interaction between heavy metals and the complex organoclays have not been thoroughly investigated. The study of these mechanisms, using individual organic matter with known properties, also contributes to our understanding of the interactions between specific and nonspecific soil organic matter and the mineral components of soil in complex natural systems.

The most common way to describe adsorption processes and the properties of adsorbents and adsorbates is through the use of adsorption isotherms, which are constructed based on experimental data. These isotherms are equations that describe the equilibrium between the concentration of retained metal ions at the surface of the adsorbing material and the concentration in the surrounding solution. Equilibrium data from experiments are fitted into these equations to create adsorption isotherm models [21]. An adsorption isotherm can be used to visualize the dynamic equilibrium between these two concentrations at a given temperature and initial adsorbate concentration. It can also help to determine the type of adsorption (monolayer or multilayer) and the nature of the surface (homogenous or heterogenous). Additionally, it can distinguish between physical adsorption and chemisorption [22]. Adsorption isotherms can be used to determine the thermodynamic parameters that characterize the adsorbed layer such as the heat of adsorption, changes in entropy, and heat capacity associated with the adsorption process. In the case of physical adsorption isotherms, they can also be used to calculate the surface area of the adsorbed layer [23]. The equations for adsorption isotherms provide valuable information about the capacity of the adsorbent and the strength of the interaction between the adsorbate and adsorbent [24].

Experimental data obtained from the equilibrium distribution of the adsorbate provide specific and individual characteristics for each system. Therefore, modeling the adsorption isotherms is essential for assessing the applicability of adsorbents and designing an optimal adsorbent–adsorbate system for the desired separations [25]. Isothermal models, such as Langmuir, Freundlich, Dubinin–Radushkevich, Brunauer–Emmett–Teller (BET), Temkin, and others, are used to predict the equilibrium behavior of metal ions during adsorption on clay adsorbents [26]. Although there are many isotherm models available in the literature, it is important to select those that describe the physical and chemical properties of the adsorbing material most closely. Comparing isotherms from different models allows for a more accurate match with the experimental data and a better understanding of the underlying processes.

The aim of this study was to investigate the sorption characteristics of organomineral complexes formed by bentonite clay and amphoteric or nonionic surfactants in relation to lead ions by using adsorption isotherm models.

2. Materials and Methods

2.1. The Composition and Properties of the Initial Clays

Bentonite from the Sarigyukh deposit in Armenia (Bento Group Minerals, Moscow, Russia) was used to create the organoclays. The manufacturer states that it has the following oxide composition: SiO_2—58.3%, Al_2O_3—14.3%, Fe_2O_3—4.4%, MgO—3.6%, Na_2O—2.3%, K_2O—1.2%, and CaO—2.1%. The mineral's cation exchange capacity is 105 mg eq/100 g.

2.2. X-ray Diffraction Analysis

A X-ray diffractometric study of a clay mineral was conducted at the Institute of Physicochemical and Biological Problems of Soil Science of the Russian Academy of Sciences using a DRON-3 diffractometer (SPE Burevestnik, St. Petersburg, Russia). The experiment was performed under normal conditions with CuKα radiation. The survey was conducted in the angle range of $2° < 2\theta < 20°$, with a scanning step of 0.05°, and an exposure time of $\tau = 1$ s. Diffractograms were taken for qualitative and semi-quantitative determinations from oriented preparations prepared by the natural orientation of clay particles on a glass substrate at normal pressure and room temperature. Three preparations with the same suspension density and thickness were studied on the substrate [27].

2.3. Surfactants Used for Modification

For the synthesis of organoclays based on bentonite, commercially available amphoteric and nonionic preparations of surfactants were used (ETS, Moscow, Russia) (Table 1).

These surfactants are used as detergents.

Table 1. Surfactants used for the synthesis of organoclays.

Name	Formula
Amphoteric surfactants	
Disodium cocoamphodiacetate	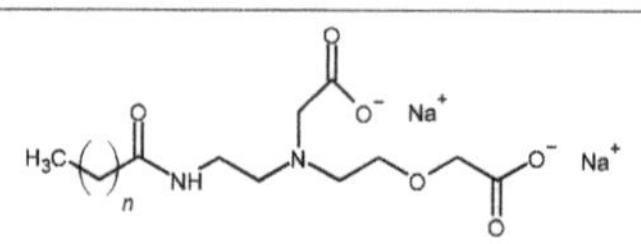
Sodium cocoiminodipropionate	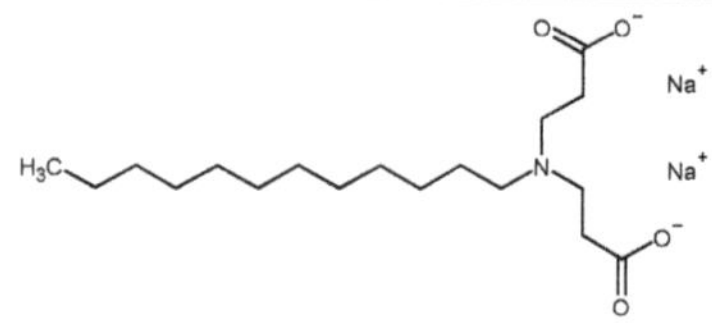
Nonionic surfactants	
Lauramine oxide	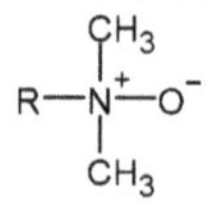
Cocamide diethanolamine	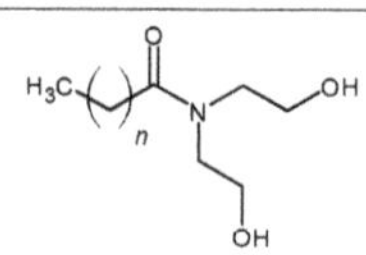
Alkyl polyglucoside	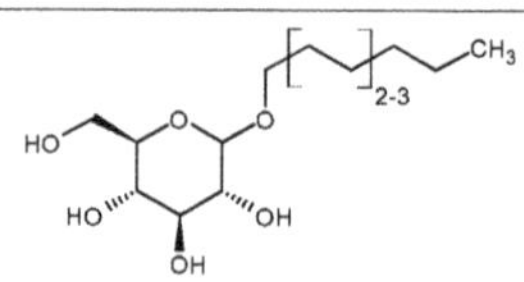

2.4. Organoclay Synthesis

The monoionic sodium form of bentonite was used to synthesize organoclays. To prepare the monoionic form of the mineral, 0.2 N NaCl was added to a purified natural phyllosilicate that had been crushed to particles smaller than 75 μm twice—first for 6 h, and then for another 18 h at a ratio of 1:50 (weight:volume). After the procedure was completed, the mineral was washed with deionized water several times until a negative chloride ion test with $AgNO_3$ was obtained. The solid phase was separated from the liquid by centrifugation at 10,000 rpm for 10 min, and the mineral was then dried at 60 ± 2 °C until it reached a constant weight. Finally, the dried mineral was crushed into a fine powder.

A total of 1 g of the surfactant was dissolved in 95 mL of deionized water and added to 5 g of the mineral in a conical flask. The mixture was thoroughly stirred with a glass rod and then placed on a digital orbital shaker S-3M.A10 (SIA "ELMI", Riga, Latvia) in a closed flask at a frequency of 180 strokes per minute for 24 h at a temperature of 22–23 °C. Afterward, the organoclay and the liquid phase were separated by centrifugation using a laboratory bench top centrifuge CLn-16 (Xiangzhi Centrifuge, Changsha, China) at 10,000 rpm per minute for 10 min, followed by decantation of the supernate.

The sediment was washed three times with 40 mL of deionized water in centrifuge tubes, and thoroughly stirred for 60 s with a glass rod. The liquid phase was then separated by centrifugation at 10,000 rpm for 10 min, and the supernatant was decanted. After combining the sediment in a Petri dish, the organoclay was dried at 60 ± 2 °C until it reached constant weight. This temperature was chosen to prevent the decomposition and transformation of the sorbed organic matter. After grinding the organoclay in an EM-100 laboratory mill (Vetinstrument, Rodninki, Russia), the resulting material was stored in hermetically sealed weighing bottles with a ground-in lid and placed in a desiccator.

2.5. Adsorption Experiment

The study of heavy metal adsorption by the synthesized organoclays was conducted through static sorption experiments. Lead cations, in the form of nitrate, were chosen as the model heavy metal. A solution of metal nitrate with varying concentrations (ranging from 0.4 mM to 4 mM) was added to 0.2 g of organoclay in a 100 mL conical flask. The experiments were performed in an electrolyte solution of 0.01 M KNO_3 with a pH of 5 to ensure equalization of the ionic strength of the solution. The solid-to-liquid ratio was 0.2:25, and the suspension was thoroughly mixed. After mixing, the flasks were shaken on an orbital shaker at a frequency of 180 movements per minute, for 3 h, at a temperature of 22–23 ° C. After this, the organoclay and liquid phases were separated by centrifugation using a laboratory bench top centrifuge CLn-16 (Xiangzhi Centrifuge, Changsha, China) at 10,000 rpm for 10 min. The supernatant liquid was then decanted and filtered through a syringe filter with a pore size of 0.45 μm.

The content of lead ions in the liquid phase was determined using flame atomic absorption spectrometry on an Analytik Jena ContrAA® 800 F spectrometer (Analytik Jena AG, Jena, Germany). The amount of metal adsorbed was calculated by comparing its concentration in the initial solution with its concentration in equilibrium. To calculate the sorption capacity of Pb^{2+} ions in the system, Equation (1) was used.

$$Q = \frac{C_0 - C}{m} \cdot V \tag{1}$$

where Q is the amount of absorbed cations, mmol/g; C is the equilibrium concentration of the metal with the sorbent, mmol/L; C_0 is the analytical concentration of the metal, mmol/L; V is the volume of the solution, l; m is the weight of the adsorbent, g.

The adsorption of Pb^{2+} ions on the surface of the initial bentonite and organoclays based on amphoteric and nonionic surfactants was studied using the Langmuir, Freundlich, and BET theoretical models.

2.6. Determination of pH of the Salt Extract

The exchange acidity of the initial bentonite and organoclays was determined using a 1 M KCl solution at a bentonite:solution ratio of 1:5 and potentiometric measurement of the solution pH with the pH meter/ionometer ITAN (LLC NPP "Tomyaanalit", Tomsk, Russia) [28].

2.7. FTIR Spectroscopy

The infrared spectra of the samples were obtained using an FSM-2202 infrared Fourier spectrometer (INFRASPEC, St. Petersburg, Russia). The spectra were collected in the range of 4000–400 cm^{-1}, in transmission mode, with a resolution of 4 cm^{-1} and a number of scans of 4. The Norton–Beer average apodization technique was used. Before obtaining the spectrum for each sample, a comparison sample of pure KBr was removed [29].

2.8. Determination of Zeta Potential by Electrokinetic Method

Determination of the zeta potential was carried out in the Department of Fundamental Chemistry at The Novomoskovsk Institute of the D.I. Mendeleev University of Chemical Technology. The zeta potential of samples was measured using a Stabino II (Colloid metrix, Meerbusch, Germany) equipped with a microprocessor unit. The unit automatically calculates the electrophoresis mobility of the particles and converts it to the zeta potential using the Smoluchowski equation [30]. A sample of 1 g in 100 mL distilled water was added to a thermostatic shaker bath (BIOBASE, Jinan, China) and rinsed for 24 h at 20 ± 2 °C. The samples were allowed to tend for 5 min to let larger particles settle. An aliquot taken from the supernatant was used to measure the zeta potential. The average of five measurements was taken to represent the measured potential. The applied voltage during the measurement was generally varied in the range of 50–150 mV.

3. Results and Discussion

3.1. X-ray Diffraction Analysis

The results of the X-ray diffraction analysis of the bentonite used are shown in Figure 1. The analytical data indicate a high content of montmorillonite (12.49 Å) and minor impurities of quartz (3.32 Å) and feldspar (3.20 Å).

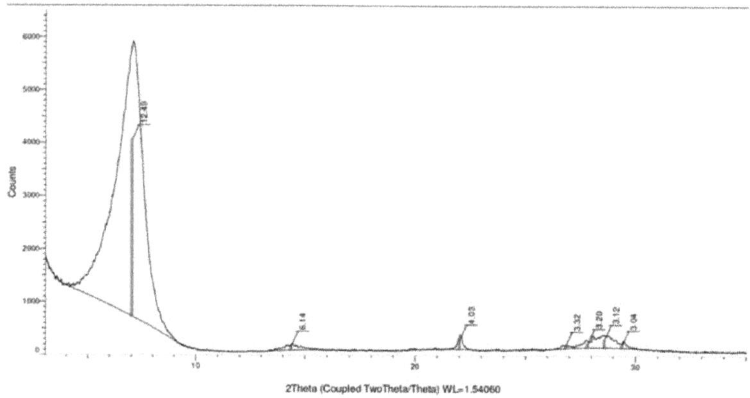

Figure 1. The results of the X-ray diffraction analysis of bentonite used in the synthesis of organoclays.

When the initial bentonite is converted to the monoionic sodium form, the distance between particles remains practically unchanged (d001 = 12.45 Å). Therefore, the physical distance between particles with a silicate plate thickness of 9.6 Å in the monoionic form is approximately 2.85 Å.

3.2. IR Spectroscopy

According to the IR spectroscopy data (Figure 2), several characteristic absorption bands were observed in the spectrum of the initial bentonite. These included bands at 978 and 1197 cm^{-1}, which corresponded to free and associated forms of the Si–OH groups. The absorption band at 1638 cm^{-1} was related to the stretching and deformation vibrations of the OH group. The bands in the region of 798–838 cm^{-1} were caused by the vibrations of the Si-O bond in tetrahedral molecules, while the bands in the range 426–439 cm^{-1} corresponded to the vibrations of this bond within the tetrahedron [31]. The bands between 3000 and 3700 cm^{-1} represented the stretching and deformation of OH groups in free or bound water molecules, while the absorption at 2512 cm^{-1} corresponded to the formation of chelate H-bridges with OH groups. Additionally, characteristic absorption bands for Al-O groups in Al_2O_3 (1111 cm^{-1}) and Si-O-Al stretches (645–669 cm^{-1}) were also observed as well as those related to O-C-O vibrations in CO_3^{2-} ions (1197 cm^{-1}) [32].

The IR spectra of bentonite samples treated with amphoteric surfactants showed absorption bands that were characteristic of the original mineral form. However, there was a decrease in the number of free -OH groups, which was likely due to the distribution of surfactant molecules within the interlayer spaces of bentonite and the displacement of water that was previously located there. The absorption bands at 1465, 2853–2927, and 1378 cm^{-1} corresponded to vibrations of the CH_2 bond and CH_3, respectively, indicating the presence of organic compounds within the structure. Additionally, the appearance of absorption bands at 1542, 1320, and 1583 cm^{-1} was related to amino groups and carboxylate acid residues (Figure 2).

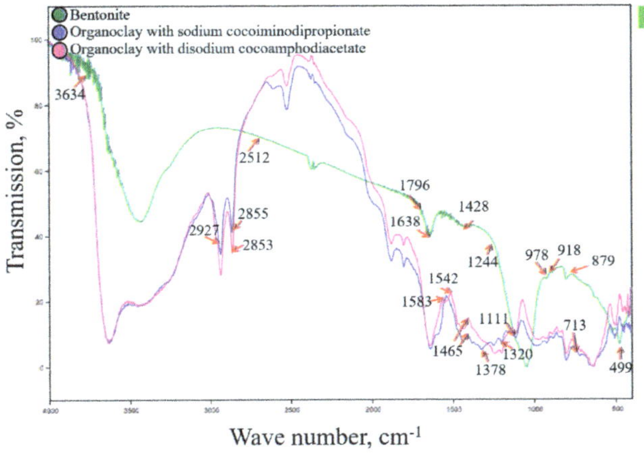

Wave number, cm^{-1}	Structural fragment
3634	ν free –OH groups
2927	ν_{as} CH$_2$
2855	ν_s CH$_2$
2853	ν_s CH$_2$
2553	ν_s CH$_2$
2512	chelate H–bridges with –OH groups
1796	CO_3^{2-}
1638	δ –OH
1583	ν_{as} –C–O–
1542	δ –N–H
1465	δ CH$_2$– scissor
1428	ν_{as} O–C–O CO_3^{2-}
1378	δ_s –CH$_3$
1320	ν_s –C–O–
1244	ν Si–O in layered silicates
1111	ν Al–O in Al_2O_3
978	δ –OH в (O–Si) –OH
918	δ Al–O–H
879	δ_{as} CO_3^{2-}
713	δ CH$_2$– pendulum
499	δ Si–O in O–Si–O tetrahedra

Figure 2. IR spectra of the synthesized organoclays using amphoteric surfactants.

For organoclays synthesized with the participation of nonionic surfactants, a similar situation was observed as with the modification by amphoteric surfactants (Figure 3).

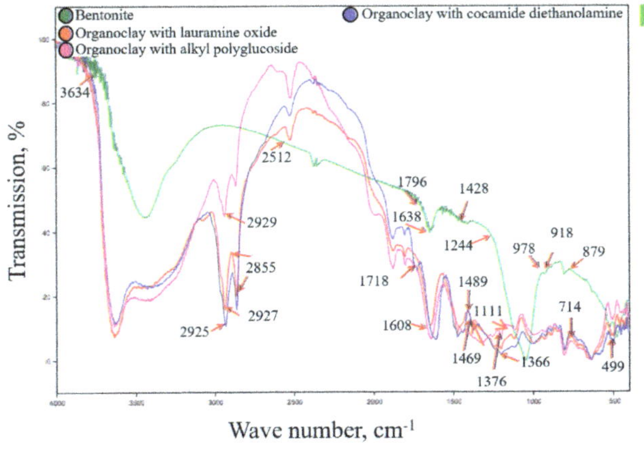

Wave number, cm^{-1}	Structural fragment
3634	ν free –OH groups
2929	ν_{as} CH$_2$
2927	ν_{as} CH$_2$
2925	ν_{as} CH$_2$
2855	ν_s CH$_2$
2512	chelate H-bridges with –OH groups
1796	CO_3^{2-}
1718	ν_s –C=O
1638	δ –OH
1608	shift to the long-wave region δ –OH
1489	δ -CH
1469	δ CH$_2$– scissor
1428	ν_{as} O–C–O CO_3^{2-}
1376	δ_s –CH$_3$
1366	ν_s –C–N–
1244	ν Si–O in layered silicates
1111	ν Al–O in Al_2O_3
978	δ –OH в (O–Si) –OH
918	δ Al–O–H
879	δ_{as} CO_3^{2-}
714	δ CH$_2$– pendulum
499	δ Si–O in O–Si–O tetrahedra

Figure 3. IR spectra of the synthesized organoclays using nonionic surfactants.

In addition, there was a shift to the long-wave region of hydroxyls (1608 cm^{-1}), carbonyls (-C=O 1718 cm^{-1}), and absorption bands 2855–2929 cm^{-1}, belonging to asymmetric and symmetric stretching vibrations of the CH_{2-} group of the adsorbed surfactant.

3.3. The Acid–Base Properties of the Original and Modified Bentonite

When characterizing the cation exchange capacity of bentonite and the organoclays, the acid–base properties of the samples were determined. The most significant factor for the cation exchange process is the exchange acidity, which is determined in a salt extract. H^+ and Al^{3+} ions from this type of acidity are the first to enter into exchange reactions and compete with exchange cations. The results of the acid–base property determination are presented in Table 2.

Table 2. The acid–base properties of the original and modified bentonite.

Samples	pH of the Aqueous Extract	pH of the Salt Extract
Bentonite	8.37 ± 0.05	9.79 ± 0.05
Organoclay with sodium cocoiminodipropionate	8.38 ± 0.04	9.78 ± 0.02
Organoclay with lauramine oxide	8.19 ± 0.02	9.81 ± 0.02
Organoclay with cocamide diethanolamine	8.62 ± 0.05	10.92 ± 0.04
Organoclay with disodium cocoamphodiacetate	8.41 ± 0.02	9.78 ± 0.01
Organoclay with alkyl polyglucoside	8.63 ± 0.01	10.10 ± 0.01

The pH values of the aqueous and salt suspensions of the original and modified forms of bentonite were alkaline, ranging from 8.19 ± 0.02 to 10.92 ± 0.04. The original form of bentonite had a slightly alkaline environment (aqueous pH of 8.38 ± 0.04). The salt extract had a pH of 9.78 ± 0.01, indicating the presence of some exchange acidity due to the presence of mobile exchange ions in the material.

The modification of bentonite with different types of surfactants (amphoteric and nonionic) affected the dehydration of the surface functional groups of the bentonite. The pH of the solutions of the aqueous and salt extracts from the modified samples shifted toward the alkaline region to varying degrees

3.4. Isotherm Studies

An accurate representation of the dynamic adsorption process of a solute on an adsorbent depends on the description of its equilibrium distribution between the two phases. Equilibrium is reached when the amount of solute adsorbed on the adsorbent is equal to the amount disturbed in the solution. The equilibrium concentration of the solution remains constant. The graph displaying the dependence of the concentration of the adsorbed solute in the solid phase and the concentration in the liquid phase shows the equilibrium adsorption isotherm [33–36].

Sorption equilibrium isotherms are an important tool for analyzing and designing adsorption systems and provide a way to assess and interpret thermodynamic parameters. While isotherms may accurately match the experimental data under certain conditions, they can completely disagree under other conditions. This paper presents the results of fitting adsorption isotherms using the Langmuir, Freundlich, and BET models as they allowed us to find the best possible fit between the experimental and theoretical data.

3.4.1. The Langmuir Isotherm

The Langmuir adsorption isotherm theory assumes that a monolayer of the adsorbate covers a uniform surface of the adsorbent. This is represented graphically by a plateau in the isotherm. The model states that at equilibrium, the adsorption reaches a saturation point where no further absorption can occur. Equation (2) was used to analyze the experimental data on adsorption processes within the context of the Langmuir model.

$$Q = Q_{\infty} \frac{K_L C_{eq}}{(1 + K_L C_{eq})} \quad (2)$$

where Q is the amount of absorbed cations, mg/g; Q_{∞} is the maximum adsorption value of an element, mmol/g; K_L is the Langmuir constant, L/mmol; C_{eq} is the concentration of an element in an equilibrium solution, mmol/L.

The obtained sorption isotherms are shown in Figure 4.

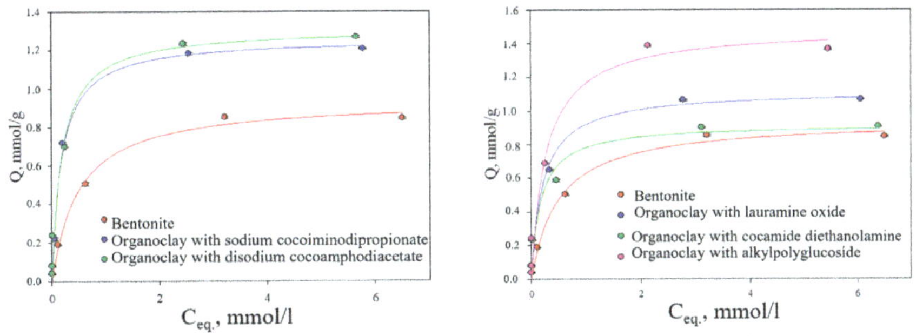

(a) (b)

Figure 4. Adsorption isotherms of Pb^{2+} ions on the initial and surfactant-modified bentonite forms: (**a**) organoclays modified with amphoteric surfactants; (**b**) organoclays modified with nonionic surfactants.

The Langmuir adsorption isotherms for Pb^{2+} ions on bentonite and its modified forms with the participation of amphoteric and nonionic surfactants followed the L-type according to the Giles classification [37]. The shape of the isotherm at low concentrations indicates that the binding strength to the sorbent is relatively strong. At the beginning of the process, there is an excess of available functional groups for binding and a high concentration of metal in the solution. As equilibrium is approached, functional groups become occupied by metal ions, and filling the adsorption surface becomes more difficult due to repulsive forces between the bound metal ions and those still present in solution [38]. The parameters of the Langmuir isotherm can be found in Table 3.

Table 3. Sorption process parameters based on the Langmuir equation.

Sorbent	R	Q_∞, mmol/g	K_L, L/mmol
Bentonite	0.99	0.95 ± 0.03	2.03 ± 0.04
With the use of amphoteric nonionic surfactants			
Organoclay with sodium cocoiminodipropionate	0.99	1.26 ± 0.07	5.67 ± 0.01
Organoclay with disodium cocoamphodiacetate	0.99	1.31 ± 0.04	5.41 ± 0.09
With the use of nonionic surfactants			
Organoclay with lauramine oxide	0.97	1.12 ± 0.04	4.67 ± 0.02
Organoclay with cocamide diethanolamine	0.98	0.92 ± 0.04	5.81 ± 0.06
Organoclay with alkyl polyglucoside	0.98	1.49 ± 0.05	3.61 ± 0.05

It was found that high correlation coefficients (above 0.97) indicated the applicability of the Langmuir model for describing the adsorption of lead ions on the surface of the initial bentonite and organoclays. Based on the value of maximum adsorption (A_∞), the sorbents studied with the use of amphoteric surfactants can be arranged in the following order: bentonite, organoclay containing sodium cocoamphodiacetate, and organoclay containing disodium cocoamphodiacetate. The parameter K_L, which describes the strength of the bond between metal ions and functional centers on the sorptive surface, increased in the series: bentonite < organoclay with disodium cocoamphodiacetate < organoclay with sodium cocoiminodipropionate.

In the case of modification with nonionic surfactants, the sorbents analyzed can be arranged in order of increasing A_∞ value as follows: organoclay with cocamide diethanolamine < bentonite < organoclay with lauramine oxide < organoclay with alkyl polyglucoside. Although the maximum calculated adsorption capacity of the organomineral complex involving diethanolamine cocamide may have been lower than that of bentonite, it was still absorbed quite effectively by this organoclay at relatively low concentrations of metal (Figure 4b). Based on the K_L value, the order is: bentonite < organoclay with alkyl polyglycoside < organoclay with lauramine oxide < organoclay with cocamide diethanolamine.

In terms of maximum adsorption, organoclays synthesized using both amphoteric and nonionic surfactants formed a series in ascending order: organoclay with cocamide diethanolamine (nonionic surfactant) < bentonite < organoclay with lauramine oxide (nonionic surfactant) < organoclay with sodium cocoiminodipropionate (amphoteric surfactant) < organoclay with disodium cocoamphodiacetate (amphoteric surfactant) < organoclay with alkyl polyglucosides (nonionic surfactant).

The Langmuir constant varied in the following order for both types of surfactants: bentonite < organoclay with alkyl polyglucoside (nonionic surfactant) < organoclay with lauramine oxide (nonionic surfactant) < organoclay with disodium cocoamphodiacetate (amphoteric surfactant) < organoclay with sodium cocoiminodipropionate (amphoteric surfactant) < organoclay with diethanolamine cocamide (nonionic surfactant). Therefore, based on the K_L value, both amphoteric and nonionic surfactants improved the bond strength between bluestone and the organometallic complex when compared to the original bentonite.

The data obtained on the adsorption of Pb^{2+} ions by the initial form of bentonite and the synthesized organoclays from aqueous solutions can be compared with the literature data for existing analogs (Table 4):

Table 4. Comparison of the adsorption capacity of the synthesized organoclays for Pb^{2+} ions with the best sorption capacities of different adsorbents.

Sorbent	Q_∞, mg/g
Bentonite (this work)	194.5
Bentonite (Saudi Arabia)	51.19 [39]
Magnetically modified zeolite	123 [40]
Green AC/HKUST-1 Nanocomposite	249.4 [41]
Organoclay with disodium cocoamphodiacetate	271.2
Organoclay with alkyl polyglucoside	380.8

Comparison of the obtained data with analogous materials (within the studied range of concentrations and pH) led to the conclusion that the synthesized organoclay materials could be used as efficient sorbents for Pb^{2+} ions.

3.4.2. The Freundlich Isotherm

The Freundlich model is used to describe adsorption behavior in heterogeneous systems. The surface of natural sorbents is believed to be non-uniform, with several different sorption sites characterized by different energy levels [42]. The main limitation of the Freundlich model lies in the fact that the adsorption capacity is not fully saturated, leading to a continuous increase in adsorption. Consequently, the parameters of the Freundlich equation can only provide information about the degree of surface non-uniformity, and indirectly, the number of adsorption sites [43].

The two-parameter Freundlich sorption model is described by the exponential Equation (3):

$$Q = K_F C_{eq}^{1/n} \quad (3)$$

where Q is the amount of absorbed cations, mmol/g; K_F is the Freundlich constant, l/mmol; C_{eq} is the concentration of an element in an equilibrium solution, mmol/L; 1/n is the empirical exponent.

This expression is characterized by a heterogeneity factor of 1/n, which allows the Freundlich isotherm to be used to describe heterogeneous systems. Theoretically, an infinite amount of adsorption can occur using this expression [34].

The Freundlich isotherms and parameters for this model are presented in Figure 5 and Table 5.

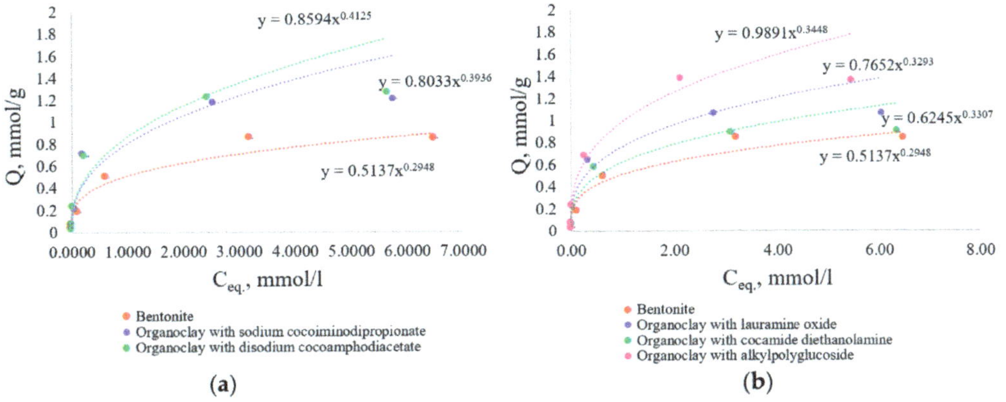

Figure 5. Adsorption isotherms of Pb^{2+} ions on the initial and surfactant-modified forms of bentonite: (**a**) organoclays modified with amphoteric surfactants; (**b**) organoclays modified with nonionic surfactants.

Table 5. Sorption process parameters based on the Freundlich equation.

Sorbent	R	K_F, L/mmol	1/n
Bentonite	0.99	0.514 ± 0.004	0.295 ± 0.003
With the use of amphoteric nonionic surfactants			
Organoclay with sodium cocoiminodipropionate	0.97	0.803 ± 0.006	0.394 ± 0.003
Organoclay with disodium cocoamphodiacetate	0.97	0.859 ± 0.008	0.413 ± 0.005
With the use of nonionic surfactants			
Organoclay with lauramine oxide	0.98	0.765 ± 0.003	0.329 ± 0.002
Organoclay with cocamide diethanolamine	0.98	0.625 ± 0.008	0.331 ± 0.004
Organoclay with alkyl polyglucoside	0.98	0.989 ± 0.004	0.345 ± 0.003

Based on the correlation coefficients, the Freundlich model is less suitable for describing lead ion adsorption processes on organoclay surfaces than the Langmuir model. The coefficient K_F, which measures absorption capacity, increased in the following order when the organoclays were modified with amphoteric surfactants: bentonite < organoclay with sodium cocoiminodipropionate < organoclay with disodium cocoamphodiacetate. In the case of nonionic surfactant modification, the sorbents analyzed can be arranged in order of increasing K_F value as follows: bentonite < organoclay with cocamide diethanolamine < organoclay with lauramine oxide < organoclay with alkyl polyglucoside. The dimensionless parameter 1/n can be used to identify the energy heterogeneity of reaction centers on sorption surfaces, and its value can vary between 0 and 1. When 1/n is close to 0, there is high heterogeneity among the sorption sites, while when it is closer to 1, the sites become more homogeneous. The specified parameter for all analyzed sorbents varied within the range from 0.295 ± 0.003 to 0.413 ± 0.005, indicating an increase in the heterogeneity of the sorption layer.

The high correlation coefficients and 1/n values indicate that the Pb^{2+} ions were more effectively adsorbed onto the organoclays compared to the original bentonite form. This was likely due to a redistribution of electron density within the adsorbent atoms, leading to the formation of active adsorption sites on the surface.

3.4.3. The BET Isotherm

Both the Langmuir and Freundlich models have drawbacks: the data on sorption equilibrium at different concentrations cannot be accurately described by a single set of parameters. The Langmuir model allows us to calculate the limiting adsorption capacity, which is useful in systems where all adsorption sites have the same energy level and ability to bind heavy metals. However, the Freundlich model is better suited for describing processes occurring on heterogeneous surfaces with multilayer adsorption. It cannot calculate the maximum adsorption capacity, but it can provide information about the heterogeneity of the surface. The BET model addresses the limitations of both models by considering energy heterogeneity on porous surfaces. Adsorbed molecules often interact with each other, leading to a violation of stoichiometry and limiting adsorption to the formation of multiple layers. The molecules of the first layer adhere to the surface of the adsorbent due to the intermolecular interaction between the adsorbent and adsorbate. Each adsorbed molecule in the first adsorption layer can act as a center for adsorption of molecules in the second layer, and so on. This process leads to the formation of subsequent sorption layers, for which physical sorption is possible. Therefore, the value of A_∞ becomes more accurate compared to the Langmuir model.

To describe polymolecular adsorption, the Brunauer–Emmett–Teller (BET) theory was used. The isotherm equation can be expressed as follows:

$$Q = \frac{Q_\infty \cdot c \cdot (C/C_0)}{(1 - C/C_0)\cdot[1 + (c-1)(C/C_0)]} \tag{4}$$

where Q is the adsorption value, mmol/g; Q_∞ is the maximum adsorption capacity of monolayer, mmol/g; C is a constant for a given adsorption system, directly related to the heat and entropy of adsorption; C, C_0 is the equilibrium and initial concentration of lead (Pb^{2+}) ions, mmol/L.

The BET adsorption equation in linear form is shown in Equation (5):

$$\frac{C/C_0}{Q(1 - C/C_0)} = \frac{1}{Q_\infty \cdot c} + \frac{c-1}{Q_\infty \cdot c}\cdot C/C_0 \tag{5}$$

The results of the experiment on the adsorption of lead ions onto clay particles were analyzed using the BET equation, which was applied in a linear form (Figure 6).

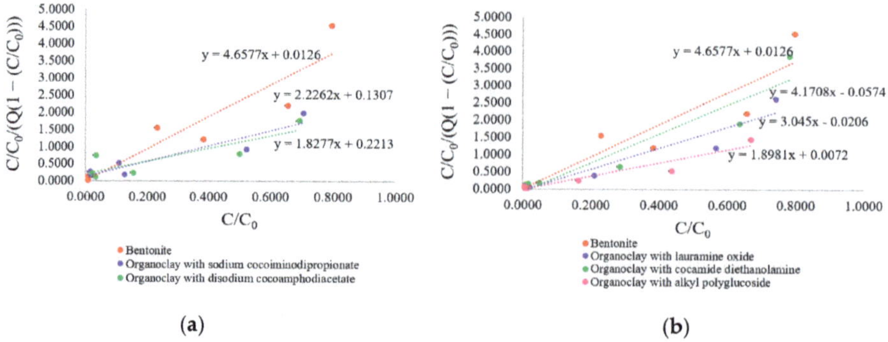

Figure 6. Adsorption isotherms of Pb^{2+} ions on the original and surfactant-modified forms of bentonite: (**a**) organoclays modified with amphoteric surfactants; (**b**) organoclays modified with nonionic surfactants.

From the linear dependencies shown in Figure 6, the values of the maximum adsorption (maximum adsorption capacity of the monolayer) (A_∞) and the constant (c) were determined. This was conducted by using the tangent of the angle of inclination of the straight lines obtained, and the size of the segments cut off on the y-axis (Table 6).

Table 6. Sorption parameters of the process according to the BET equation.

Sorbent	R	Q_∞, mmol/g	$c \cdot 10^{-3}$
Bentonite	0.96	0.214 ± 0.002	0.37 ± 0.01
With the use of amphoteric surfactants			
Organoclay with sodium cocoiminodipropionate	0.96	0.422 ± 0.004	0.018 ± 0.002
Organoclay with disodium cocoamphodiacetate	0.93	0.491 ± 0.003	0.091 ± 0.002
With the use of nonionic surfactants			
Organoclay with lauramine oxide	0.97	0.328 ± 0.002	−0.146 ± 0.004
Organoclay with cocamide diethanolamine	0.97	0.241 ± 0.002	−0.071 ± 0.001
Organoclay with alkyl polyglucoside	0.98	0.523 ± 0.003	0.264 ± 0.006

It has been established that the correlation coefficients for describing the sorption process using the BET equation are lower than those obtained using the Freundlich and Langmuir models. This is due to the fact that the BET model is only applicable in the region of high concentrations, and therefore is not suitable for describing sorption processes at low concentrations. Based on the value of the maximum adsorption (A_∞), the order of the studied sorbent materials can be arranged as follows: bentonite < organoclay with cocamide diethanolamine < organoclay with lauramine oxide < organoclay with sodium cocoiminodipropionate < organoclay with disodium cocoamphodiacetate < organoclay with alkyl polyglucoside. The constant of the BET equation—c, which characterizes the ratio of the energy of intermolecular interactions between the sorbent (initial and modified clays) and Pb^{2+} ions, increased in the following order: organoclay with lauramine oxide < organoclay with cocamide diethanolamine < organoclay with sodium cocoiminodipropionate < organoclay with disodium cocoamphodiacetate < organoclay with alkyl polyglucoside < bentonite. Additionally, for organoclays derived from lauramine oxide and cocamide diethanolamine, the values of c were less than 1, at -0.146 ± 0.004 and -0.071 ± 0.001, respectively. When c is less than 1, even before the formation of a monolayer of adsorbed molecules is complete, the formation of a polymolecular layer begins.

Since the sorption process is best described by the Langmuir equation (with the highest coefficients of determination), the adsorption equilibrium constants obtained were used to calculate the standard Gibbs free energy (Formula (6), Table 7).

$$\Delta G = -RT\ln K \tag{6}$$

where ΔG is the Gibbs energy, kJ/mol; R is the universal gas constant, J/(mol K); T is the temperature, 298 K, kJ/mol, K is the Langmuir constant.

Table 7. Gibbs energy (Langmuir model).

Sorbent	ΔG^0, kJ/mol
Bentonite	−1.7
Organoclay with sodium cocoiminodipropionate	−4.2
Organoclay with disodium cocoamphodiacetate	−4.1
Organoclay with lauramine oxide	−3.8
Organoclay with cocamide diethanolamine	−4.3
Organoclay with alkyl polyglucoside	−3.1

The results obtained indicate that the adsorption of Pb^{2+} ions on both the original bentonite form and on the synthesized organoclays occurred spontaneously in the forward direction.

In our case, we observed the highest correlation coefficients when using the Langmuir model to describe the adsorption processes. This model assumes monomolecular adsorption, heterogeneity of the adsorbent surface, and predominant interaction due to chemical forces. A number of studies have also shown that the adsorption of Cu(II) onto organoclays with zwitterionic surfactants is better described by the Langmuir model compared to the Freundlich model for adsorption isotherms [44]. At the same time, [45] suggested that the adsorption of Pb^{2+} onto montmorillonite modified with different chain length betaines could be well-described by the Freundlich adsorption isotherm. The Langmuir and Freundlich R^2 calculated from Cu^{2+}, Pb^{2+}, Ni^{2+}, Cd^{2+}, Fe^{2+}, and Zn^{2+} isotherms at the metal adsorption on natural zeolite (clinoptilolite) modified by nonionic surfactant Triton X-100 showed that for Pb, Cd, and Fe, the adsorption data were fitted equally well by both the Langmuir equation and the Freundlich equation. Adsorption data for Cu, Ni, and Zn were fitted better by the Freundlich adsorption isotherm [46].

3.5. Determination of the Zeta Potential of Bentonite and Organoclays

The electrokinetic or zeta potential (ζ potential) is the voltage on the slip plane, which is the difference in charge between the bulk fluid and the fluid layer containing ions that are attached to the surface of a particle. For molecules and nanoparticles, a higher ζ potential means greater stability, as the solution or suspension will be less likely to aggregate. Conversely, when the ζ potential is lower, the attraction between particles exceeds the repulsive force, disrupting the stability of the suspension. Therefore, colloids with high ζ potentials are electrically stable, while those with low ζ potentials tend to coagulate or form flocs [47]. For stability, it does not matter whether the variance has a positive or negative sign; however, the sign of the variance can have a huge impact on the application of the variance.

It is now recognized that a zeta potential of at least 30 mV (optimum > 60 mV) is required for complete electrostatic stabilization, while potentials between 5 and 15 mV are in the region of limited flocculation, and those between 5 and 3 mV correspond to maximum flocculation [48]. It is important to note that the magnitude of the charge on the surface of the nanoparticles depends on the pH of the solution.

Table 8 presents the change in the zeta potential of the original bentonite and its modified versions, which were obtained using amphoteric and nonionic surfactants.

Table 8. The zeta potential of the original bentonite and the synthesized organoclays.

Sorbent	ζ, mV
Bentonite	−37 ± 1
With the use of amphoteric surfactants	
Organoclay with sodium cocoiminodipropionate	−68 ± 2
Organoclay with disodium cocoamphodiacetate	−73 ± 1
With the use of nonionic surfactants	
Organoclay with lauramine oxide	−49 ± 2
Organoclay with cocamide diethanolamine	−64 ± 2
Organoclay with alkyl polyglucoside	−70 ± 3

The measured zeta potential of the initial bentonite sample was −37 ± 1 mV, indicating resistance to aggregation of its negatively charged particles. When the bentonite was modified with amphoteric and nonionic surfactants, the negative values of the zeta potential increased, ranging from −49 ± 2 to −73 ± 1 mV. This indicates an increase in the stability of the bentonite suspension and a greater possibility of electrostatic interaction between the negatively charged bentonite particles and positively charged lead ions.

3.6. Possible Mechanisms of Lead Cation Sorption on Synthesized Organoclays

Modern concepts of the structure of the interface between solid mineral particles and the solution as well as the mechanisms of interaction between the solid and liquid phases are based on two main methodological approaches [49,50]. The first of these uses the principles and concepts of colloid chemistry, which underlie the theory of the electrical double layer [51]. The second approach is based on the theoretical principles of coordination compound chemistry. When using this approach, the reactions of interaction between solid soil particles and solution components are considered as processes of the formation of surface complexes [52].

According to the first approach, the more efficient absorption of lead cations by the colloidal particles of organoclays compared to the colloidal particles of the original clay mineral can be explained by the increase in the negative charge of the colloidal particles. The zeta potential, which is not equal to the adsorption potential or surface potential in the double electric layer, is often used to assess the properties of this layer. An increase in the zeta potential indicates an increase in colloid charge, which is associated with specific adsorption. In this type of adsorption, ions are adsorbed on the surface of the solid phase without releasing an equivalent number of ions into the solution, leading to the acquisition of an electric charge by the solid phase. This leads to the fact that, near the surface, due to the action of electrostatic forces of attraction, an equal number of ions with opposite charges are grouped, forming a double electric layer.

According to the Paneth–Fajans–Hahn adsorption rule, ions that are capable of completing their crystal lattice or forming a poorly soluble compound with one of the ions present in the crystal are specifically adsorbed onto the surface of a crystalline solid from an electrolyte solution. Lead is unlikely to be incorporated into the crystal structure of modified bentonite, but the formation of precipitates of lead hydroxides or salts is possible (see below). The observed acidification of the equilibrium solution during lead adsorption, compared to the organoclay solution, may indicate a predominant exchange nature of metal adsorption by organoclays. In general, the addition of salts containing higher-valence cations (less hydrated) can displace hydrogen and other metal ions from the surface of the micelles [53].

In bentonite, the permanent negative charge is quite high, accounting for 90–95% of the total charge. This high negative charge density is due to a significant degree of isomorphic substitution, where Fe^{2+} or Mg^{2+} replace the Al^{3+} element in the octahedral layer, and some Al^{3+} ions replace Si^{4+} ions in the tetrahedral layer. These permanent charges are mainly located on the basal surface and account for most of the surface charge density of bentonite. Positive charges are mainly located at the edges of the clay structures and only account for 5–10% of the total charge [54]. According to the second approach, positively charged fragments of amphoteric surfactants interact with negatively charged groups on the surface of bentonite, while their own anionic fragments remain unoccupied and can interact with lead cations. In this scenario, each molecule of the amphoteric surfactant contains two carboxyl groups.

During the synthesis of organoclays, surfactant molecules are added and adsorbed on the clay surface as well as penetrate (at least partially) into the interlayer space. The authors in [54] investigated the effect of the length of the carbon chain in the amphoteric surfactant, betaine, on the interlayer spacing of clays. They found that the increase in basal spacing was most significant when betaine molecules were first introduced into the galleries, but then changed only slightly as the carbon chain length increased. Therefore, an increase in betaine molecule size forced them to reorganize within the existing space rather than expanding the galleries further. The adsorption of amphoteric surfactants into interlayer spaces occurs in a monolayer, and it should be noted that we refer to small carbon chains within the molecule in this case.

The authors in [55], based on the results of infrared (IR) and X-ray photoelectron spectroscopy (XPS) studies and sorption experiments, proposed several mechanisms of interaction between Cd^{2+} and bentonite or organoclay including electrostatic interactions, ion exchange, surface complexation, and chelation. Chelation is also responsible for the adsorption of Cd^{2+} onto organoclays that contain amphoteric surfactants.

Theoretically, electrostatic interactions between molecules of nonionic surfactants and clay minerals should be absent. However, the synthesis of organoclays using nonionic surfactants is a widely used process [20,56,57]. Compared to ionic surfactants, clays modified with nonionic compounds have a hydrophobic surface without changing the degree of surface charge [56]. Organoclays synthesized with nonionic surfactants are a type of organomineral complex whose formation is not based on ion exchange. Instead, the authors suggest that the nonionic surfactant molecules interact with the silicate through their functional groups and with water molecules associated with exchange cations in clay minerals through ion-dipole and hydrogen bonds [58]. It was also noted that nonionic surfactants adsorb onto clay surfaces due to the combination of polar attractions, Van der Waals forces, and weak C-H. . .O (clay) bonds [59]. This can lead to the displacement of water molecules and the formation of organic layers within the interlayer space. X-ray diffraction analysis of systems containing nonionic surfactants and bentonite has shown that these surfactants can cause a slight increase in the spacing between the layers of smectite [58].

In our opinion, the main mechanism of interaction between organoclays based on nonionic surfactants and lead cations involves the formation of complex compounds including chelate complexes between the structures of surfactant molecules adsorbed on the clay and the cations. Hypothetical modeling of Pb^{2+} adsorption onto synthesized organoclays can be represented as follows (Figure 7).

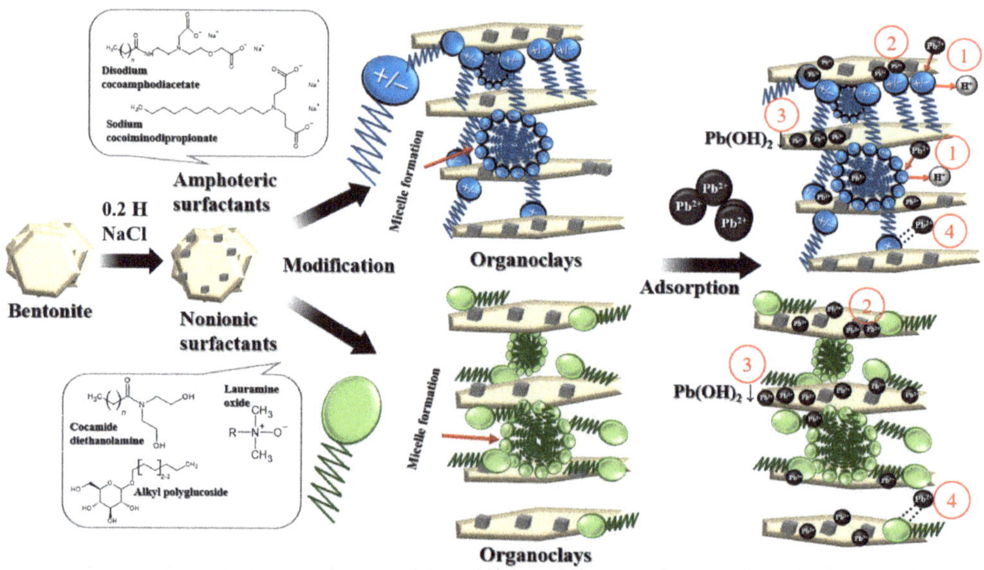

Figure 7. Schematic illustration of the possible mechanism for the adsorption of Pb^{2+} onto organoclays modified with amphoteric and nonionic surfactants: 1—ion exchange; 2—complexation; 3—precipitation; 4—long-range forces (physical adsorption).

During the sorption experiment, a decrease in the pH of the equilibrium solutions was observed as a result of the interaction between the initial working solutions of lead nitrate and the mineral or organic clays based on both amphoteric and nonionic surfactants. The pH of the solutions decreased from 8.51 ± 0.05 to 5.11 ± 0.08, depending on the type of surfactant used for modification. This was compared to the pH values of the salt extracts, which are shown in Table 2. It is known that at pH > 6, it is possible for lead cations to form the $Pb(OH)^+$ complex or the $Pb(OH)_2$ precipitate, as shown in Figure 7(2) and Figure 7(3), respectively [60,61]. The XRD patterns also indicated the formation of lead carbonate hydroxide (hydrocerussite) as a result of Pb adsorption on Na-bentonite under similar experimental conditions [62].

4. Conclusions

The structural sorption properties of organoclays based on bentonite and various amphoteric (sodium cocoiminodipropionate, disodium cocoamphodiacetate) and nonionic (lauramine oxide, cocamide diethanolamine, alkyl polyglucoside) surfactants with respect to lead ions were synthesized and investigated. According to the Fourier transform infrared spectroscopy (FTIR) data, it was found that when bentonite is modified with these amphoteric and nonionic surfactants, there is no replacement of the existing absorption bands in the original bentonite structure.

A decrease in absorption in the long-wave region of hydroxyl groups was observed due to the distribution of surfactant molecules in the interlayer space of bentonite and the displacement of water located there.

The sorption properties of bentonite and the synthesized organoclays were investigated using various sorption models such as the Langmuir, Freundlich, and BET. It was found that the maximum adsorption capacity for lead cations in organoclays with different surfactants was observed, with the highest value being 1.49 ± 0.05 mmol/g according to the Langmuir model, and 0.523 ± 0.003 mmol/g according to the BET. The Langmuir constants, which characterize the strength of the binding between lead ions and the organoclays, varied in the following order: bentonite < organoclay with alkyl polyglucoside < organoclay with lauramine oxide < organoclay with disodium cocoamphodiacetate < organoclay with sodium cocoiminodipropionate < organoclay with cocamide diethanolamine. For the analyzed sorbents, the value of the $1/n$ parameter (the Freundlich equation) varied between 0.295 ± 0.003 and 0.413 ± 0.005, which indicates an increase in the heterogeneity of the sorption layer.

The negative values of the zeta potential of the synthesized organoclays increased in modulus compared to the initial bentonite (-49 ± 2 to -73 ± 1 mV) and were related to the limiting adsorption values, indicating an increase in suspension stability and an increase in the possibility of electrostatic interaction between organoclays and positively charged lead ions.

Based on the zeta potential values, structural formulas used for the synthesis of organoclays and the literature data, we made assumptions about the possible mechanisms for increasing the maximum adsorption of lead by organoclays based on amphoteric and nonionic surfactants.

Author Contributions: Conceptualization, Y.A. and S.H.; Methodology, M.G. and A.K.; Formal analysis, L.P. and M.G.; Investigation, M.G. and A.K; Resources, L.P. and M.B.; Writing—original draft preparation, M.G. and A.K.; Writing—review and editing, L.P. and M.G.; Visualization, M.G.; Supervision, L.P. and M.G.; Project administration, L.P. and M.G.; Funding acquisition, L.P. All authors have read and agreed to the published version of the manuscript.

Funding: This paper was prepared as part of the state assignment on the topic: "Immobilization of heavy metals by products of interactions of layered silicates with soil organic matter and microorganisms" (Agreement No. 073-00033-24-01 with the Ministry of Education of Russia from 09.02.2024).

Institutional Review Board Statement: Not applicable.

Informed Consent Statement: Not applicable.

Data Availability Statement: All data are included in the manuscript.

Acknowledgments: X-ray diffractometric study of the clay mineral was conducted by T.V. Alekseeva at the Institute of Physicochemical and Biological Problems of Soil Science of the Russian Academy of Sciences. We would like to express our sincere gratitude to T.V. Alekseeva. Determination of the zeta potentials was carried out at the Department of Fundamental Chemistry in The Novomoskovsk Institute of the D.I. Mendeleev University of Chemical Technology. We would like to express our gratitude to N.F. Kizim for providing us with the opportunity to carry out this work. We would like to express our gratitude to D.L. Pinsky from the Institute of Physicochemical and Biological Problems of Soil Science at the Russian Academy of Sciences for his valuable consultations.

Conflicts of Interest: The authors declare no conflicts of interest.

References

1. Singh, O.; Singh, S.; Shahi, U.P.; Singh, R.; Singh, P.K.; Grigoryan, K.K.; Khachatryan, H.; Rajput, P.; Perelomov, L.; Kozmenko, S.; et al. Nanotechnological Innovations for Soil Pollution Remediation and Environmental Conservation. In *Harnessing NanoOmics and Nanozymes for Sustainable Agriculture*; Rajput, V., Singh, A., Ghazaryan, K., Alexiou, A., Said Al-Tawaha, A., Eds.; IGI Global: Hershey, PA, USA, 2024; pp. 237–255. [CrossRef]
2. Motuzova, G.V.; Minkina, T.M.; Karpova, E.A.; Barsova, N.U.; Mandzhieva, S.S. Soil contamination with heavy metals as a potential and real risk to the environment. *J. Geochem. Explor.* **2024**, *144*, 241–246. [CrossRef]
3. Kolesnikov, S.; Minnikova, T.; Kazeev, K.; Akimenko, Y.; Evstegneeva, N. Assessment of the Ecotoxicity of Pollution by Potentially Toxic Elements by Biological Indicators of Haplic Chernozem of Southern Russia (Rostov region). *Water Air Soil. Pollut.* **2022**, *233*, 18. [CrossRef] [PubMed]
4. Gui, H.; Yang, Q.; Lu, X.; Wang, H.; Gu, Q.; Martin, J.D. Spatial distribution, contamination characteristics and ecological-health risk assessment of toxic heavy metals in soils near a smelting area. *Environ. Res.* **2023**, *222*, 115328. [CrossRef]
5. Violante, A.; Cozzolino, V.; Perelomov, L.; Caporale, A.G.; Pigna, M. Mobility and bioavailability of heavy metals and metalloids in soil environments. *J. Soil Sci. Plant Nutr.* **2010**, *10*, 268–292. [CrossRef]
6. Pinskii, D.L.; Shary, P.A.; Mandzhieva, S.S.; Minkina, T.M.; Perelomov, L.V.; Maltseva, A.N.; Dudnikova, T.S. Effect of Composition and Properties of Soils and Soil-Sand Substrates Contaminated with Copper on Morphometric Parameters of Barley Plants. *Eurasian Soil. Sci.* **2023**, *56*, 352–362. [CrossRef]
7. Mkrtchyan, E.S.; Anan'eva, O.A.; Burakova, I.V.; Burakov, A.E.; Tkachev, A.G. Graphene nanocomposites modified with organic polymers as effective heavy metal sorbents in aqueous media. *J. Adv. Mater. Technol.* **2022**, *7*, 228–237. [CrossRef]
8. Erdogan, S.; Baysal, A.; Akba, O.; Hamamci, C. Interaction of metals with humic acid isolated from oxidized coal. *Pol. J. Environ. Stud.* **2007**, *16*, 671–675.
9. Perelomov, L.; Sarkar, B.; Pinsky, D.; Atroshchenko, Y.; Perelomova, I.; Mukhtorov, L.; Mazur, A. Trace elements adsorption by natural and chemically modified humic acids. *Environ. Geochem. Health* **2021**, *43*, 127–138. [CrossRef]
10. Gertsen, M.M.; Arlyapov, V.A.; Perelomov, L.V.; Kharkova, A.S.; Golysheva, A.N.; Atroshchenko, Y.M.; Cardinale, A.M.; Reverberi, A.P. Environmental Implications of Energy Sources: A Review on Technologies for Cleaning Oil-Contaminated Ecosystems. *Energies* **2024**, *17*, 3561. [CrossRef]
11. Gertsen, M.; Dmitrieva, E. The sorption ability of humic acids of black alder fen peat on montmorillonite containing clays in the presence of heavy metal ions. *Moroc. J. Chem.* **2019**, *7*, 474–481. [CrossRef]
12. Qiu, B.; Tao, X.; Wang, H.; Li, W.; Ding, X.; Chu, H. Biochar as a low-cost adsorbent for aqueous heavy metal removal: A review. *J. Anal. Appl. Pyrolysis* **2021**, *155*, 105081. [CrossRef]
13. Sarkar, B.; Rusmin, R.; Ugochukwu, U.C.; Mukhopadhyay, R.; Manjaiah, K.M. Modified clay minerals for environmental applications. In *Modified Clay and Zeolite Nanocomposite Materials: Environmental and Pharmaceutical Applications*; Mercurio, M., Sarkar, B., Langella, A., Eds.; Elsevier: Amsterdam, The Netherlands, 2019; pp. 113–127. [CrossRef]
14. Perelomov, L.; Mandzhieva, S.; Minkina, T.; Atroshchenko, Y.; Perelomova, I.; Bauer, T.; Pinsky, D.; Barakhov, A. The Synthesis of Organoclays Based on Clay Minerals with Different Structural Expansion Capacities. *Minerals* **2021**, *11*, 707. [CrossRef]
15. Barakan, S.; Aghazadeh, V. The advantages of clay mineral modification methods for enhancing adsorption efficiency in wastewater treatment: A review. *Environ. Sci. Pollut. Res.* **2021**, *28*, 2572–2599. [CrossRef] [PubMed]
16. Kretinin, A.A. Optimizaciya Parametrov Ekraniruyushchego Sloya v Pripoverhnostnyh Hranilishchah Radioaktivnyh Othodov. Ph.D. Thesis, MGRI, Moscow, Russia, 2004. (In Russian).
17. Yıldız, B.; Erten, H.N.; Kış, M. The sorption behavior of Cs + ion on clay minerals and zeolite in radioactive waste management: Sorption kinetics and thermodynamics. *J. Radioanal. Nucl. Chem.* **2011**, *288*, 475–483. [CrossRef]
18. Jiménez-Castañeda, M.E.; Medina, D.I. Use of Surfactant-Modified Zeolites and Clays for the Removal of Heavy Metals from Water. *Water* **2017**, *9*, 235. [CrossRef]
19. Sarkar, B.; Xi, Y.; Megharaj, M.; Krishnamurti, G.S.R.; Bowman, M.; Rose, H.; Naidu, R. Bioreactive organoclay: A new technology for environmental remediation. *Crit. Rev. Environ. Sci. Technol.* **2012**, *42*, 435–488. [CrossRef]
20. Perelomov, L.; Gertsen, M.; Burachevskaya, M.; Hemalatha, S.; Vijayalakshmi, A.; Perelomova, I.; Atroshchenko, Y. Organoclays Based on Bentonite and Various Types of Surfactants as Heavy Metal Remediants. *Sustainability* **2024**, *16*, 4804. [CrossRef]
21. Muliwa, A.M.; Oyewo, O.A.; Maity, A. Recent progress on the removal of aqueous mercury by carbon-based adsorbents: A review. *Inorg. Chem. Commun.* **2023**, *156*, 111207. [CrossRef]
22. Wang, J.; Guo, X. Adsorption isotherm models: Classification, physical meaning, application and solving method. *Chemosphere* **2020**, *258*, 127279. [CrossRef]

23. Simon, R.; Bare, G.A. Somorjai, Surface Chemistry. In *Encyclopedia of Physical Science and Technology*, 3rd ed.; Meyers, R.A., Ed.; Academic Press: Cambridge, MA, USA, 2003; pp. 373–421. [CrossRef]
24. Saleh, T.A. Chapter 4—Isotherm models of adsorption processes on adsorbents and nanoadsorbents. In *Interface Science and Technology*; Saleh, T.A., Ed.; Elsevier: Amsterdam, The Netherlands, 2022; Volume 34, pp. 99–126. [CrossRef]
25. Molina-Calderón, L.; Basualto-Flores, C.; Paredes-García, V.; Venegas-Yazigi, D. Advances of magnetic nanohydrometallurgy using superparamagnetic nanomaterials as rare earth ions adsorbents: A grand opportunity for sustainable rare earth recovery. *Sep. Purif. Technol.* **2022**, *299*, 121708. [CrossRef]
26. Limousin, G.; Gaudet, J.P.; Charlet, L.; Szenknect, S.; Barthes, V.; Krimissa, M. Sorption isotherms: A review on physical bases, modeling and measurement. *Appl. Geochem.* **2007**, *22*, 249–275. [CrossRef]
27. Liu, Z.; Uddin, M.A.; Sun, Z. FT-IR and XRD analysis of natural Na-bentonite and Cu (II)-loaded Na-bentonite. *Spectrochim. Acta Part A Mol. Biomol. Spectrosc.* **2011**, *79*, 1013–1016. [CrossRef]
28. Jo, H.Y.; Benson, C.H.; Edil, T.B. Hydraulic Conductivity and Cation Exchange in Non-prehydrated And Prehydrated Bentonite Permeated with Weak Inorganic Salt Solutions. *Clays Clay Miner.* **2004**, *52*, 661–679. [CrossRef]
29. Parolo, M.E.; Pettinari, G.R.; Musso, T.B.; Sánchez-Izquierdo, M.P.; Fernández, L.G. Characterization of organo-modified bentonite sorbents: The effect of modification conditions on adsorption performance. *Appl. Surf. Sci.* **2014**, *320*, 356–363. [CrossRef]
30. Akbulut, S.; Albayrak, Z.N.K.; Arasan, S. Electrokinetic properties of surfactant modified clays. *Int. J. Civil. Struct. Eng.* **2010**, *1*, 354–361.
31. Tabak, A.; Yilmaz, N.; Eren, E.; Caglar, B.; Afsin, B.; Sarihan, A. Structural analysis of naproxen-intercalated bentonite (Unye). *Chem. Eng. J.* **2011**, *174*, 281–288. [CrossRef]
32. Kumar, A.; Lingfa, P. Sodium bentonite and kaolin clays: Comparative study of their FT-IX, XFR and XRD. *Mater. Proc.* **2020**, *22*, 737–742. [CrossRef]
33. Kalam, S.; Abu-Khamsin, S.A.; Kamal, M.S.; Patil, S. Surfactant adsorption isotherms: A review. *ACS Omega* **2021**, *6*, 32342–32348. [CrossRef]
34. Lima, É.C.; Dehghani, M.H.; Guleria, A.; Sher, F.; Karri, R.R.; Dotto, G.L.; Tran, H.N. Adsorption: Fundamental aspects and applications of adsorption for effluent treatment. In *Green Technologies for the Defluoridation of Water*; Dehghani, M.H., Karri, R.R., Lima, É.C., Eds.; Elsevier: Amsterdam, The Netherlands, 2021; pp. 41–88. [CrossRef]
35. Tahiruddin, N.S.M.; Aziz, R.A.; Ali, R.; Taib, N.I. Potential of using jackfruit peel (*Artocarpus heterophyllus*) as green solution for removal of copper (II) and zinc (II) from aqueous solution: Adsorption kinetics, isotherm and thermodynamic studies. *J. Environ. Chem. Eng.* **2023**, *11*, 109953. [CrossRef]
36. Majd, M.M.; Kordzadeh-Kermani, V.; Ghalandari, V.; Askari, A.; Sillanpää, M. Adsorption isotherm models: A comprehensive and systematic review (2010–2020). *Sci. Total Environ.* **2022**, *812*, 151334. [CrossRef] [PubMed]
37. Al-Ghouti, M.A.; Da'ana, D.A. Guidelines for the use and interpretation of adsorption isotherm models: A review. *J. Hazard. Mater.* **2020**, *393*, 122383. [CrossRef] [PubMed]
38. Futalan, C.M.; Kan, C.C.; Dalida, M.L.; Hsien, K.J.; Pascua, C.; Wan, M.W. Comparative and competitive adsorption of copper, lead, and nickel using chitosan immobilized on bentonite. *Carbohydr. Polym.* **2011**, *83*, 528–536. [CrossRef]
39. Al-Jlil, S.A. Kinetic of Adsorption of Chromium and Lead Ions on Bentonite Clay Using Novel Internal Parallel Model. *Res. J. Environ. Toxicol.* **2015**, *9*, 1–16. [CrossRef]
40. Nah, I.W.; Hwang, K.-Y.; Jeon, C.; Choi, H.B. Removal of Pb ion from water by magnetically modified zeolite. *Miner. Eng.* **2006**, *19*, 1452–1455. [CrossRef]
41. Abhari, P.S.; Manteghi, F.; Tehrani, Z. Adsorption of Lead Ions by a Green AC/HKUST-1 Nanocomposite. *Nanomaterials* **2020**, *10*, 1647. [CrossRef] [PubMed]
42. Pandey, S. A comprehensive review on recent developments in bentonite-based materials used as adsorbents for wastewater treatment. *J. Mol. Liq.* **2017**, *241*, 1091–1113. [CrossRef]
43. Qasem, N.A.A.; Mohammed, R.H.; Lawal, D.U. Removal of heavy metal ions from wastewater: A comprehensive and critical review. *npj Clean Water* **2021**, *4*, 36. Available online: https://www.nature.com/articles/s41545-021-00127-0 (accessed on 1 September 2024). [CrossRef]
44. Ma, L.; Chen, Q.; Zhu, J.; Xi, Y.; He, H.; Zhu, R.; Tao, Q.; Ayoko, G.A. Adsorption of phenol and Cu(II) onto cationic and zwitterionic surfactant modified montmorillonite in single and binary systems. *Chem. Eng. J.* **2016**, *283*, 880–888. [CrossRef]
45. Liu, C.; Wu, P.; Tran, L.; Zhu, N.; Dang, Z. Organo-montmorillonites for efficient and rapid water remediation: Sequential and simultaneous adsorption of lead and bisphenol A. *Environ. Chem.* **2018**, *15*, 286–295. [CrossRef]
46. Mirbaloochzehi, M.R.; Rezvani, A.; Samimi, A.; Shayesteh, M. Application of a Novel Surfactant-Modified Natural Nano-Zeolite for Removal of Heavy Metals from Drinking Water. *Adv. J. Chem. A* **2020**, *3*, 612–620. [CrossRef]
47. Heurtault, B.; Saulnier, P.; Pech, B.; Proust, J.-E.; Benoit, J.-P. Physico-chemical stability of colloidal lipid particles. *Biomaterials* **2003**, *24*, 4283–4300. [CrossRef]
48. Onugwu, A.L.; Nwagwu, C.S.; Onugwu, O.S.; Echezona, A.C.; Agbo, C.P.; Ihim, S.A.; Emeh, P.; Nnamani, P.O.; Attama, A.A.; Khutoryanskiy, V.V. Nanotechnology based drug delivery systems for the treatment of anterior segment eye diseases. *J. Control. Release* **2023**, *354*, 465–488. [CrossRef] [PubMed]
49. Sokolova, T.A.; Trofimov, S.Y. *Sorbcionnye Svojstva Pochv. Adsorbciya. Kationnyj Obmen: Uch. Posob. po Nekotorym Glavam Himii Pochv*; Grif i K: Tula, Russia, 2009; 172p. (In Russian)

50. Perelomov, L.V.; Chilachava, K.B.; Shvykin, A.Y.; Atroschenko, Y.M. Vliyanie organicheskih veshchestv gumusa na pogloshchenie tyazhelyh metallov glinistymi mineralami. *Agrochemistry* **2017**, *2*, 99–107. (In Russian)
51. Sposito, G. *The Environmental Chemistry of Aluminum*; CRC Press, Inc.: Boca Raton, FL, USA, 1996; 480p. [CrossRef]
52. Stumm, W. *Chemistry of the Solid-Water Interface: Processes at the Mineral-Water and Particle-Water Interface in Natural Systems*; John Wiley & Sons, Inc.: New York, NY, USA, 1992; 448p.
53. Sofroniou, C.; Chazapi, I.; Leontidis, E. Binding of lanthanide salts to zwitterionic phospholipid micelles. *J. Colloid. Interface Sci.* **2019**, *557*, 568–579. [CrossRef]
54. Sarkar, P.; Ghimire, S.; Vlasov, S.; Mukhopadhyay, K. Effect of clay-zwitterionic interactions in controlling the viscoelastic properties in organomodified clays. *iScience* **2023**, *26*, 108388. [CrossRef]
55. Ren, S.; Meng, Z.; Sun, X.; Lu, H.; Zhang, M.; Lahori, A.H.; Bu, S. Comparison of Cd^{2+} adsorption onto amphoteric, amphotericcationic and amphoteric-anionic modified magnetic bentonites. *Chemosphere* **2019**, *239*, 124840. [CrossRef] [PubMed]
56. Guégan, R. Intercalation of a Nonionic Surfactant (C10E3) Bilayer into a Na- Montmorillonite Clay. *Langmuir* **2010**, *26*, 19175–19180. [CrossRef]
57. Fu, M.; Zhang, Z.P.; Wu, L.M.; Zhuang, G.Z.; Zhang, S.; Yuan, J.Y.; Liao, L.B. Investigation on the co-modification process of montmorillonite by anionic and cationic surfactants. *Appl. Clay Sci.* **2016**, *132–133*, 694–701. [CrossRef]
58. Andrunik, M.; Bajda, T. Modification of Bentonite with Cationic and Nonionic Surfactants: Structural and Textural Features. *Materials* **2019**, *12*, 3772. [CrossRef]
59. Park, Y.; Ayoko, G.A.; Frost, R.L. Application of organoclays for the adsorption of recalcitrant organic molecules from aqueous media. *J. Colloid Interface Sci.* **2011**, *354*, 292–305. [CrossRef]
60. Zyrin, N.G.; Sadovnikova, L.K.; Malinina, M.S.; Motuzova, G.V.; Gorbatov, V.S.; Karpova, E.A.; Nadezhkin, V.V.; Reshetnikov, S.I.; Yarilova, L.S. *Himiya Tyazhelyh Metallov, Mysh'yaka i Molibdena v Pochvah*; MSU: Moscow, Russia, 1985; pp. 104–127. (In Russian)
61. Mineev, V.G.; Alekseev, A.A.; Trishina, T.A. *Tyazhelye Metally i Okruzhayushchaya Sreda v Usloviyah Sovremennoj Intensivnoj Himizacii*; MSU: Moscow, Russia, 2005; pp. 248–268. (In Russian)
62. Perelomov, L.; Sarkar, B.; Rahman, M.M.; Goryacheva, A.; Naidu, R. Uptake of lead by Na-exchanged and Al-pillared bentonite in the presence of organic acids with different functional groups. *Appl. Clay Sci.* **2016**, *119*, 417–423. [CrossRef]

toxics

MDPI

Article

Bioreactor Expansion Affects Microbial Succession of Mixotrophic Acidophiles and Bioremediation of Cadmium-Contaminated Soils

Xiaodong Hao [1,2], Ping Zhu [1,2], Xueduan Liu [2], Luhua Jiang [2], Huidan Jiang [3], Hongwei Liu [2,*] and Zhiqun Chen [4,*]

1 Shandong Provincial Key Laboratory of Water and Soil Conservation and Environmental Protection, College of Resources and Environment, Linyi University, Linyi 276000, China; haoxiaodong@lyu.edu.cn (X.H.); zhuping199508@163.com (P.Z.)
2 School of Minerals Processing and Bioengineering, Central South University, Changsha 410083, China; xueduanliucsu@163.com (X.L.); jiangluhua@csu.edu.cn (L.J.)
3 Biotechnology Research Institute, Hunan Academy of Agricultural Sciences, Changsha 410125, China; huidanjiangcsu@163.com
4 College of Life Science, Linyi University, Linyi 276000, China
* Correspondence: hongweiliu@csu.edu.cn (H.L.); chenzhiqun@lyu.edu.cn (Z.C.)

Abstract: Microbial scale-up cultivation is the first step to bioremediating cadmium (Cd)-contaminated soils at the industrial scale. However, the changes in the microbial community as the bioreactor volume expands and their associations with soil Cd removal remain unclear. Herein, a six-stage scale-up cultivation process of mixotrophic acidophiles was conducted, scaling from 0.1 L to 10 m^3, to remediate Cd-contaminated soils. The findings showed that bioreactor expansion led to a delay in sulfur and glucose oxidations, resulting in a reduced decline in solution pH and cell density. There were minimal differences observed in bacterial alpha-diversity and community structure as the bioreactor volume increased, except for the 10 m^3 scale. However, bioreactor expansion decreased fungal alpha-diversity, changed the community structure, and simplified fungal community compositions. At the family level, *Acidithiobacillaceae* and *Debaryomycetaceae* dominated the bacterial and fungal communities throughout the scale-up process, respectively. Correlation analysis indicated that the indirect effect of mixotrophic acidophiles played a significant role in soil Cd removal. Bacterial community shifts, driven by changes in bioreactor volume, decreased the pH value through sulfur oxidation, thereby indirectly enhancing Cd removal efficiency. This study will contribute to the potential industrial application of mixotrophic acidophiles in bioremediating Cd-contaminated soils.

Keywords: scale-up cultivation; mixotrophic acidophiles; microbial community dynamics; Cd removal; indirect effect

1. Introduction

Soil cadmium (Cd) contamination has become a global environmental concern resulting from industrialization, urbanization, and unsustainable farming practices [1–3]. Even at low concentration, Cd is highly toxic and carcinogenic [4]. While Cd is resistant to degradation by organisms, it is easily accumulated in soils and crops, and subsequently enters the human food chain [5]. This poses a significant threat to environmental ecology and human health [6]. Therefore, it is imperative to remediate Cd-contaminated soils to mitigate the risks associated with Cd exposure and ensure food security.

The soil Cd mobilization strategy, compared to the Cd stabilization technique, focuses on minimizing Cd pollution by transforming Cd speciation into dissolved fractions and/or directly extracting Cd from soils [7]. Phytoextraction is an economic and ecological approach for removing Cd from soils [8]. However, the extensive application of phytoremediation is limited by factors such as a long soil restoration time, slow plant growth, and variable soil and weather conditions. Microbial extraction (bioleaching) is a promising approach for the efficient extraction of heavy metals from solid materials. It involves the direct biocorrosion of microbial metabolism or the indirect solubilization of microbial metabolites [9]. Bioleaching technology has been extensively studied due to its cost-effectiveness, eco-friendliness, and effective resource recovery. It has been successfully applied for in situ processing of copper sulfide ores at an industrial scale and for recovering valuable metals from various secondary sources at a pilot scale [10,11]. The employment of microbial extraction technology in large-scale remediation efforts for Cd-contaminated soils worldwide holds significant promise for enhancing the sustainability and efficiency of the soil treatment process. However, the application of bioleaching for Cd bio-extraction from heavy metal-contaminated soils is limited to laboratory-level studies, mainly due to factors such as low microbial functional diversity and slow dissolution kinetics [12]. Consequently, the selection of appropriate microbial functional flora and the provision of supporting hardware facility and cultivation condition are crucial for ensuring the feasibility and functionality of the scale-up cultivation of microbes.

The microbial composition is the most important factor influencing the extraction rate of heavy metals in bioleaching systems. Autotrophic bacteria and heterotrophic fungi have demonstrated their capacity to dissolve Cd complexes in soils through processes such as displacement of Cd ion by hydrion or the formation of chelate through the microbial production of inorganic/organic acids, leaching agents, biosurfactants, and other metabolites [13,14]. Cd extraction from soil matrices using pure species or monotrophic cultures typically involves the solubilization of Cd fractions in specific areas [15]. However, the bioleaching efficiency, when using mixed cultures, is higher for the same Cd fraction compared to that of a single population. Recently, mixotrophic acidophiles consisting of autotrophic and heterotrophic communities with supernally diverse phylogenetic and functional characteristics have been found to exhibit high Cd removal capability from soils [16]. In cooperative bioleaching systems involving the symbiotic co-culture of bacterial and fungal consortia, mixotrophic acidophiles exhibit significant potential in enhancing the extraction efficiency of soil Cd [17].

The successful industrial application of bioleaching relies on the scale-up cultivation of microorganisms. Bioreactor cultivation offers numerous advantages over traditional culture tanks, including better control of physicochemical properties in the solution and the provision of suitable growth conditions [18,19]. The microbial community structure in the scale-up cultivation of mixotrophic acidophiles is closely linked to key biochemical functions involved in the activation and extraction of Cd from soils [20]. As the bioreactor volume increases, microbial growth necessitates adaptation to larger spaces and environmental fluctuations, which can affect their resistance to environmental stress. Larger bioreactors require more time to achieve uniform distribution of nutrients and energy substrates, influencing microbial growth rates and distribution and thereby impacting the microbial community structure. Concurrently, with increased volume, dissolved oxygen concentrations may decline, impacting the growth of aerobic and oxygen-demanding microorganisms. The expansion of the bioreactor also affects the production of microbial metabolites, as larger spaces enable microorganisms to produce more products at different stages of metabolism, thereby influencing the microbial community's functionality [21]. Presently, research on the effects of bioreactor expansion on changes in microbial community structure has primarily focused on the liquid fermentation process [22]. Limited work has been conducted to examine microbial community successions in the context of increasing bioreactor volume, especially regarding the complex mixotrophic community at different scale-up cultivation stages. The investigation of the broad applicability of the functional microbial population across diverse geographical locations and climatic environments is important for the implementation of microbial expansion technology in the soil remediation field.

To address this gap, the present study aims to investigate the following questions: (i) to measure the effects of bioreactor expansion on the microbial community successions of mixotrophic acidophiles; and (ii) to evaluate the effects of changes in solution properties and microbial structure on the soil Cd removal efficiency. Furthermore, the contribution of bioreactor parameters to soil Cd extraction, considering the direct and indirect actions of microbes and solution properties, also requires further analysis, which will aid in adjusting bioreactor parameters to maintain a stable microbial community structure and ensure efficient Cd bioleaching.

2. Materials and Methods

2.1. Initial Inoculum for the Scale-Up Cultivation of Mixotrophic Acidophiles

Leachate and Cd-contaminated soils were utilized to enrich the initial inoculum for the scale-up cultivation of mixotrophic acidophiles. The leachate sample was collected from a bioleaching heap of sulfide minerals in Dexing City, Jiangxi Province, China (29°04′ N, 117°71′ E). The Cd-contaminated soil was obtained from a polluted rice-growing region in Xiangtan County, Hunan Province, China (27°77′ N, 112°88′ E). The leachate (10 mL) and fresh soil (10 g) samples were combined and added to a 250 mL conical flask containing 100 mL of basal medium. The basal medium composition included 3 g/L $(NH_4)_2SO_4$, 0.1 g/L KCl, 0.5 g/L K_2HPO_4, 0.5 g/L $MgSO_4{\cdot}7H_2O$, 0.01 g/L $Ca(NO_3)_2$, 1 g/L elemental sulfur, 0.7 g/L glucose, and 0.3 g/L yeast extract. The solution's pH value was adjusted to 3.5 using sulfuric acid, and then incubated in an oscillating incubator at 32 °C and 175 rpm. When the solution's pH dropped below 2.5, the supernatant microbial fluid was transferred to a new basal medium (10%, vol/vol) and cultured under the same conditions. This process of uniform subculturing was repeated 10 times, and the microbial solution from the tenth culture was used as the initial inoculum for the scale-up cultivation of mixotrophic acidophiles.

2.2. Scale-Up Cultivation Process of Mixotrophic Acidophiles

A six-stage scale-up cultivation system, including the lab-scale (0.1 L and 2 L), pilot-scale (10 L and 100 L), and commercial-scale (1 m^3 and 10 m^3) cultivation stages, was designed for the scale-up cultivation of mixotrophic acidophiles. These cultivation processes were conducted in vessels of varying sizes: a 250 mL glass conical flask; a 2.5 L glass beaker; and 12 L, 120 L, 1.5 m^3 and 12 m^3 cylindroid polyethylene reactors. The details of the bioreactor parameters are shown in Figure 1. The initial inoculum served as the seed liquid of the 0.1 L cultivation system. The scale-up process of mixotrophic acidophiles was continuous, where the microbial solution from a smaller cultivation stage acted as the seed liquid for the next stage once the solution pH dropped below 2.5 or the cell density exceeded 1×10^9 cells/mL. The seed liquid dose was 10% (vol/vol). The cultivation media used in each stage were identical to the basal medium used for the initial inoculum, maintaining a constant temperature of 32 °C and an initial pH of 3.5. Each cultivation stage lasted approximately four to five days. Finally, the microbial solutions from each of the six scale-up cultivation stages were collected separately for microbial community analysis and soil Cd removal experiments.

2.3. Bioremediation of Cd-Contaminated Soils

The Cd-contaminated soils used in this study had been exposed to industrial sewage for approximately ten years. The soil type was clay. Top soil samples (0~20 cm) were collected using a T-sampler. Fresh soils were utilized to enrich the initial inoculum, as described in Section 2.1. The remaining soil samples were air-dried and passed through a 2 mm plastic sieve for the soil bioremediation experiment and the measurement of soil chemical properties (Table 1). The soil bioremediation experiment was conducted in 250 mL conical flasks. A soil sample (10 g) and microbial solution (100 mL) from the six scale-up cultivation stages were added to the conical flasks. To account for the potential impact of pH on soil Cd removal efficiency, a control treatment (CK) using the basal medium with a pH of 2.5 was established. The flasks were incubated in an oscillating incubator at a consistent temperature of 32 °C and a rotational speed of 175 rpm for 7 days. Each treatment was replicated three times. At the end of the bioremediation experiment, soil residue samples were obtained by centrifuging at 3000× g for 10 min. The supernatants were discarded, and the solid residues were air-dried. Total Cd and DTPA-Cd contents in the soil residues were measured to calculate the Cd removal efficiencies.

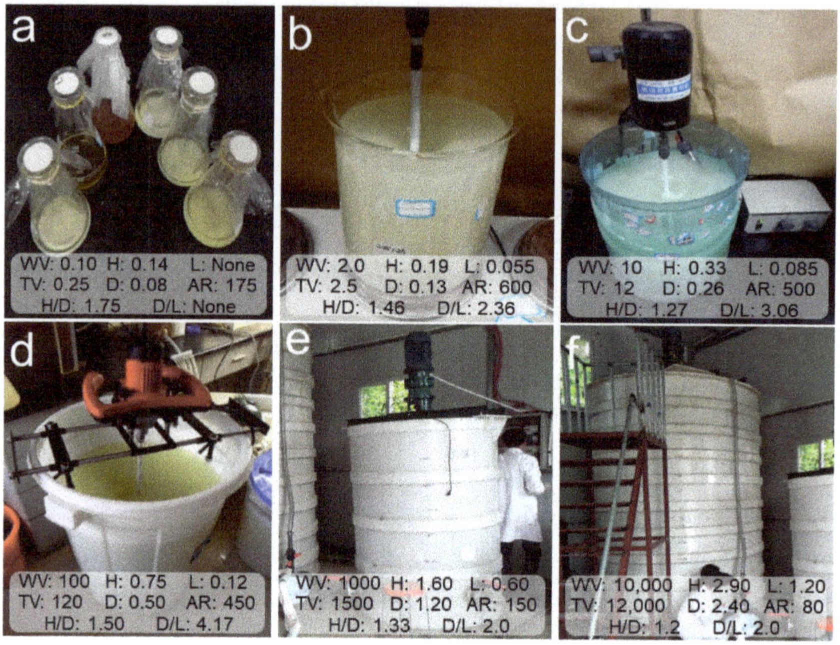

Figure 1. Bioreactor parameters used for the scale-up cultivation process of mixotrophic acidophiles. (**a–f**) Scale-up cultivation stages of 0.1 L, 2 L, 10 L, 100 L, 1 m^3, and 10 m^3, respectively. WV, work volume (L); TV, total volume (L); H, bioreactor height (m); D, bioreactor diameter (m); L, paddle length (m); AR, agitator agitation rate (rpm).

Table 1. Chemical properties of Cd-contaminated soils used in this study (n = 3).

Item	Mean Value ± SD
pH	6.23 ± 0.06
Organic matter (g/kg)	17.3 ± 0.5
Total N (g/kg)	1.74 ± 0.12
Total P (g/kg)	0.736 ± 0.056
Total K (g/kg)	12.7 ± 0.3
Cu (mg/kg)	21.2 ± 0.1
Cd (mg/kg)	13.2 ± 0.2
DTPA-Cd (mg/kg)	7.41 ± 0.83
Pb (mg/kg)	27.8 ± 2.2
Cr (mg/kg)	125 ± 22
Mn (mg/kg)	304 ± 11

2.4. Chemical Analysis

The pH values of microbial solutions in each scale-up cultivation system were monitored daily using a pH meter (BPH-220, Bell Instrument, Dalian, China). The sulfate radical concentration and glucose concentration in the microbial solution were analyzed at the beginning and end of each scale-up cultivation performance. The sulfate radical concentration was measured using the barium sulfate precipitation method [23]. The glucose concentration was determined using the Agilent 1200 HPLC (Agilent Technologies Co. Ltd., USA). The increased sulfate radical concentration (ISC) was calculated as the difference between the sulfate radical concentrations after and before the cultivation performance, while the decreased glucose concentration (DGC) was calculated as the difference between the glucose concentrations before and after the cultivation performance. The cell density (CD) of the microbial solution was assessed at the end of each scale-up cultivation performance using a hemocytometer with an optical microscope (BX41, Olympus Instruments, Tokyo, Japan).

The total Cd content in the soil was determined using the acid digestion method, which involved the use of a mixture of HNO_3, HF, and $HClO_4$ (10:5:1, vol/vol). The Cd ion concentration was then measured using an inductively coupled plasma–optical emission spectrometer (ICP–OES, Optima 5300DV, PerkinElmer, Shelton, CT, USA). The diethylenetriamine pentaacetic acid-soluble Cd (DTPA-Cd) content in the soils was evaluated using a soil-to-liquid mass ratio of 1:5 with a diethylenetriamine pentaacetic acid (DTPA) solution mixture. The mixture contained DTPA (0.005 M), triethanolamine (0.1 M), and $CaCl_2$ (0.01 M) [24]. The soil removal efficiencies of total Cd and DTPA-Cd were calculated using the formula: removal efficiency (%) = $(M_1 - M_2)/M_1 \times 100$. Here, M_1 represents the Cd quantity in the soils before the bioremediation treatment, and M_2 represents the Cd quantity in the soils after the bioremediation treatment.

2.5. Microbial Community Analysis

The microbes from microbial solutions (50 mL) in each scale-up cultivation stage were centrifuged at 12,000× g for 15 min. Total genomic DNA was extracted using the E.Z.N.A. Water DNA kit (Omega BioTek Inc., Norcross, GA, USA). The quality and quantity of the extracted DNA were assessed using a NanoDrop 2000 Spectrophotometer (Bio-Rad Laboratories Inc., Hercules, CA, USA). For bacterial analysis, the V4~V5 region of the bacterial 16S rRNA gene was amplified using the 515F (5′−GTGCCAGCMGCCGCGGTAA−3′) and 907R (5′−CCGTCAATTCCTTTGAGTTT−3′) primers [25]. For fungal analysis, the ITS region of the fungal rRNA gene was amplified using the ITS1F (5′−CTTGGTCATTTAGAGGA AGTAA−3′) and ITS2R (5′−GCTGCGTTCTTCATCGATGC−3′) primers [26]. The PCR reaction systems and amplification procedure followed a previously published protocol [27]. The PCR products from three replicates of the same sample were pooled and purified using the E.Z.N.A.TM Gel Extraction Kit (Omega BioTek Inc., USA). The purified amplicons from all samples were then sent to E-Gene Biotech Co., Ltd. (Shenzhen, China) for PE300 sequencing on the Illumina HiSeq platform.

The raw sequencing data were merged using fast length adjustment of short reads (FLASH) [28] and assigned to each sample based on the specific barcodes. Low-quality reads were filtered and removed using QIIME's quality filters to eliminate chimeras [29]. The remaining high-quality sequences were binned into operational taxonomic units (OTUs) at a 97% similarity level using Uparse [30]. The representative sequence for each OTU was selected as the most frequent sequence. Taxonomic information for bacteria and fungi was annotated using the Ribosomal Database Project's classifier [31] against the SILVA 132 database [32], with a confidence threshold of 0.8~1.0. The assigned sequences for bacteria ranged from 23,731 to 53,194, and for fungi, from 30,372 to 59,959. To normalize the read counts in the OTU table across all samples, the minimum sequence numbers were selected.

2.6. Statistical Analysis

One-way ANOVA was conducted to analyze the significance of differences between means, with a significance level of $p < 0.05$, following the least significant difference test (LSD). Error bars accompanied by lowercase letters were employed to indicate statistically significant variations in mean values across distinct groups. Bacterial and fungal alpha-diversities were evaluated using the abundance-based coverage estimator (ACE) and Shannon indices. The microbial beta-diversity was assessed using principal coordinate analysis (PCoA) based on the Bray-Curtis distance metrics. To estimate the contributions of the bioreactor parameter and solution property to the microbial community, variation partitioning analysis (VPA) was performed. Pearson correlation and the Spearman test were used to determine correlations between microbial community data, the bioreactor parameter, and the solution property. Partial least square path modeling (PLSPM) was employed to analyze the effects of bioreactor expansion on the microbial community, solution property, and soil Cd removal. These statistical analyses were carried out using R software version 3.5.1, with the packages vegan, amap, shape, and diagram.

3. Results

3.1. Properties of Microbial Solution in Scale-Up Cultivation Process of Mixotrophic Acidophiles

During the scale-up cultivation of mixotrophic acidophiles, the properties of microbial solutions underwent significant changes, especially in the 10 m^3 stage (Figure 2). When the bioreactor volume increased from 0.1 L to 1 m^3, the solution pH decreased by 1.1~1.3 units. However, in the 10 m^3 stage, the solution pH only decreased by 0.6 units within a five-day period (Figure 2a). The bioreactor expansion from the 2 L stage to the 1 m^3 stage consistently increased the concentration of sulfate radicals (ISC) (Figure 2b). However, in the 10 m^3 stage, the ISC was significantly lower ($p < 0.05$) compared to the 10 L stage, with a decrease of 65.9% in five days. While no significant ($p > 0.05$) variations in decreased glucose concentration (DGC) were observed for the small system (from 0.1 L stage to 10 L stage) or the large system (from 100 L stage to 10 m^3 stage), the large system exhibited a significant ($p < 0.05$) decrease in DGC compared to the small system (Figure 2c). The effect of bioreactor expansion on cell density (CD) was consistent between the 0.1 L stage and the 10 L stage ($p > 0.05$) (Figure 2d), but a significant ($p < 0.05$) downward trend in CD was observed with an increase in bioreactor volume.

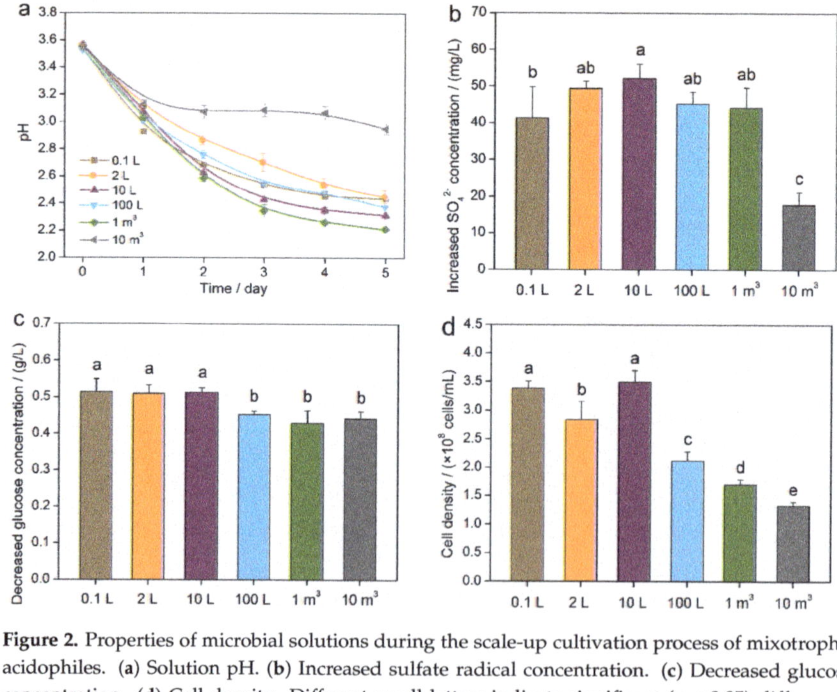

Figure 2. Properties of microbial solutions during the scale-up cultivation process of mixotrophic acidophiles. **(a)** Solution pH. **(b)** Increased sulfate radical concentration. **(c)** Decreased glucose concentration. **(d)** Cell density. Different small letters indicate significant ($p < 0.05$) differences according to the LSD test.

3.2. Microbial Community Diversities and Compositions in Scale-Up Cultivation Process of Mixotrophic Acidophiles

The ACE indices did not show significant differences ($p < 0.05$) in the scale-up cultivation process for the bacterial community (Figure 3a). However, the Shannon indices in the

1 m^3 stage and 10 m^3 stage were significantly ($p < 0.05$) lower and higher, respectively, than those of the other cultivation stages (Figure 3b). The richness and diversity of microbial OTUs in the fungal community were reduced with bioreactor expansion (Figure 3a,b). The Shannon indices displayed a clear downward trend with the expansion of the cultivation system. Principal coordinate analysis (PCoA) showed inverse changes between the bacterial and fungal communities for the 0.1 L, 2 L, and 10 L stages. The bacterial community structures were not altered in the 0.1 L, 2 L, 10 L, or 1 m^3 stages based on the PCoA 1 axis (explaining 85.8% of the total variation), but they were different from the bacterial communities in the 10 m^3 stage (Figure 3c). Bioreactor expansion significantly changed the solution's fungal community structures for the 0.1 L, 2 L, 10 L, and 10 m^3 stages (Figure 3d), but it had no effect on the fungal community structures in the 100 L or 1 m^3 stages (PCoA 1 axis, explaining 75.9% of the total variation).

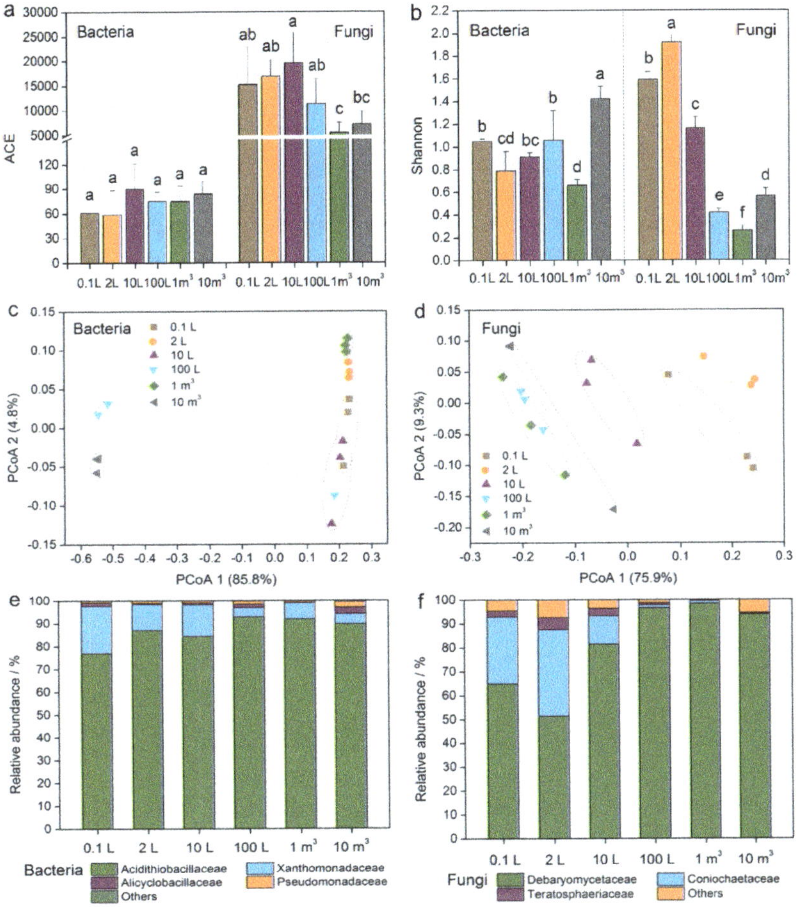

Figure 3. Microbial community diversities and compositions during the scale-up cultivation process of mixotrophic acidophiles. (**a**,**b**) Bacterial and fungal community alpha-diversity, indicated by ACE and Shannon indices. (**c**,**d**) Bacterial and fungal community beta-diversity, indicated by principal coordinate analysis (PCoA) based on the Bray-Curtis distance metrics. (**e**,**f**) Dominant bacteria and fungi at the family level. Different small letters indicate significant ($p < 0.05$) differences according to the LSD test.

The relative abundances of dominant fungi at the family level were more sensitive to the expansion of the cultivation system (Figure 3e,f). Within the scale-up cultivation process, the relative abundance of *Acidithiobacillaceae* increased from 77.0% (0.1 L stage) to 89.8% (10 m^3 stage), while *Xanthomonadaceae* decreased from 20.7% (0.1 L stage) to 4.4% (10 m^3 stage) (Figure 3e). Additionally, the relative abundances of *Alicyclobacillaceae* and *Pseudomonadaceae* remained stable, ranging from 0.9% to 5.2% in all cultivation systems. For fungal community composition, family-level fungi including *Debaryomycetaceae*, *Coniochaetaceae*, and *Teratosphaeriaceae* were the dominant taxa in the small stages of 0.1 L, 2 L, and 10 L (Figure 3f). However, bioreactor expansion remarkably enriched the family *Debaryomycetaceae*, which accounted for more than 93% of the relative abundance in the large stages of 100 L, 1 m^3, and 10 m^3.

3.3. Soil Removal Efficiencies of Total Cd and DTPA-Cd Using Mixotrophic Acidophiles in Different Scale-Up Cultivation Process

The removal efficiencies of total Cd from the 0.1 L group to the 1 m^3 group exceeded 92.8%, and were significantly ($p < 0.05$) higher than those of the CK group (76.9%) (Figure 4a). The removal efficiency of total Cd in the 10 m^3 group (85.3%) was also significantly ($p < 0.05$) higher than that in the CK group, but it was the lowest among all the biological groups. The removal efficiency of soil DTPA-Cd followed a similar trend to total Cd extraction, reaching 89.0% (0.1 L group), 90.8% (2 L group), 89.0% (10 L group), 88.3% (100 L group), and 88.9% (1 m^3 group), respectively (Figure 4b). These values were significantly ($p < 0.05$) higher than those in the CK group (70.2%). The removal efficiency of DTPA-Cd in the 10 m^3 group was 81.3%, which was also significantly ($p < 0.05$) higher than that in the CK group.

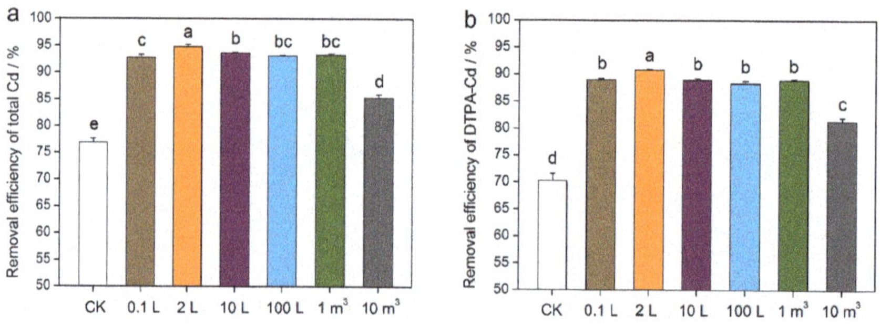

Figure 4. Soil Cd removal efficiencies using microbial solutions of mixotrophic acidophiles during different scale-up cultivation process. (**a**) Total Cd. (**b**) Diethylenetriamine pentaacetic acid-soluble Cd (DTPA-Cd). Different small letters indicate significant ($p < 0.05$) differences according to the LSD test.

3.4. Links between Microbial Solution Properties, Solution Microbes, and Cd Removal Efficiencies

The results of the Spearman test revealed that the solution pH had a significant ($p < 0.01$) positive correlation with bioreactor expansion, including the WV, L, H, and D of the bioreactor (Figure 5a). The solution DGC, CD, and ISC all displayed significant ($p < 0.05$) positive associations with the AR of the paddle. Furthermore, ISC showed a negative correlation with the solution pH ($p < 0.001$), but a positive correlation with DGC ($p < 0.05$). Both ISC and DGC were significantly ($p < 0.01$) positively correlated with the solution CD. Variation partitioning analysis (VPA) indicated that the bioreactor parameter and solution property accounted for 27% and 4% of the variation in the bacterial community, respectively, while they accounted for 33% and 1% of the variation in the fungal community (Figure 5a). Specifically, bioreactor expansion (WV, L, H, and D) showed significant ($p < 0.05$) correlations with the bacterial community, as revealed by PCoA 1. Additionally, the bacterial community was significantly ($p < 0.05$) associated with solution pH, CD, and DGC. As for the fungal community, parameters such as the L, H, D, and H/D of the bioreactor, as well as DGC and CD of the solution, displayed stronger relationships with variations in fungal composition.

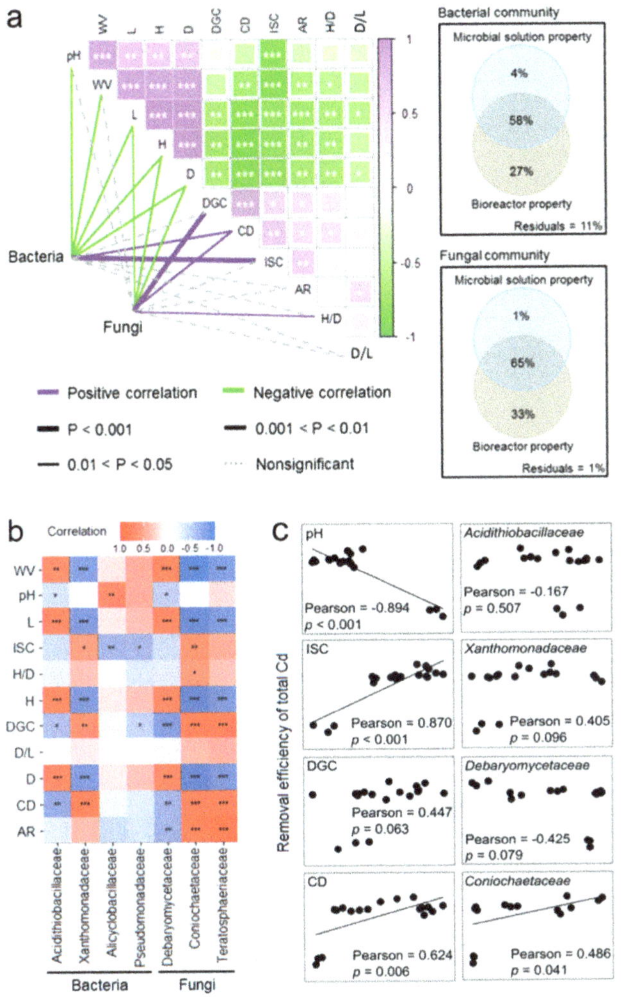

Figure 5. Correlation analysis between microbial community data, bioreactor parameter, and solution property. (**a**) Microbial community structure, bioreactor parameter, and solution property. (**b**) Dominant microbe, bioreactor parameter, and solution property. (**c**) Soil Cd removal, solution property, and dominant microbe. WV, work volume; TV, total volume; H, bioreactor height; D, bioreactor diameter; L, paddle length; AR, agitator agitation rate; ISC, increased sulfate radical concentration; DGC, decreased glucose concentration; CD, cell density. Significances of these correlations are indicated by *** when $p < 0.001$, ** when $p < 0.01$, and * when $p < 0.05$.

The WV, L, H, and D of the bioreactor parameters showed significant ($p < 0.01$) correlations with these dominant taxa (Figure 5b). The solution pH showed a significant ($p < 0.05$) negative relationship with the relative abundances of *Acidithiobacillaceae* and *Debaryomycetaceae*, but a significant ($p < 0.01$) positive relationship with *Alicyclobacillaceae*. Solution DGC and CD were significantly ($p < 0.05$) associated with the relative abundance of most dominant family-level microbes. The bioleaching efficiency of total Cd showed significant positive correlations with the solution ISC ($p < 0.001$) and CD ($p = 0.006$), but a significant negative correlation with the solution pH ($p < 0.001$) (Figure 5c). However, dominant taxa such as *Acidithiobacillaceae*, *Xanthomonadaceae*, and *Debaryomycetaceae* did not show significant correlations with the total Cd removal. Only the relative abundance of *Coniochaetaceae* showed a significantly ($p < 0.05$) positive correlation with the total Cd removal.

The results of partial least square path modeling (PLSPM) showed that the increase in bioreactor volume had a significant ($p < 0.001$) impact on the bacterial community structure with the scale-up cultivation of mixotrophic acidophiles (Figure 6). Additionally, there were significant ($p < 0.01$) links between the solution ISC and CD (Figure 6a). The solution ISC and DGC had significant ($p < 0.05$) direct correlations with the solution pH, which in turn significantly ($p < 0.01$) affected the removal efficiency of the total soil Cd. The differences in fungal community structure were significantly ($p < 0.01$) correlated with the solution DGC and CD (Figure 6b). The solution ISC directly affected the solution pH ($p < 0.001$) and indirectly increased the efficiency of total soil Cd extraction ($p < 0.001$).

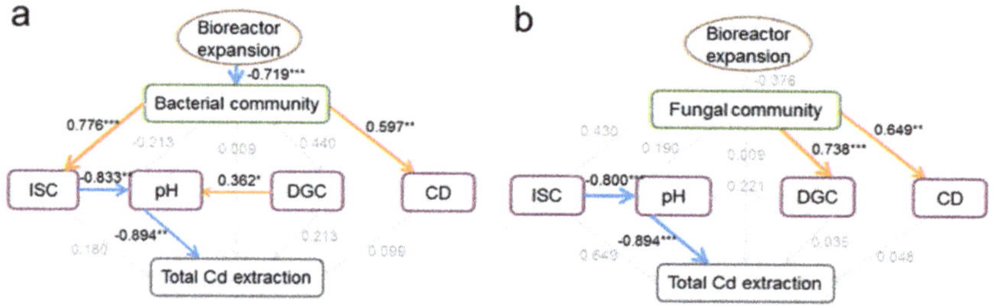

Figure 6. Effects of bioreactor expansion on microbial community and soil Cd extraction, indicated by partial least square path models. (**a**) Bacteria. (**b**) Fungi. Goodness of fit of the modeling was 0.4072. Red and blue arrows represent the significant positive and negative path coefficients. Gray dashed indicate the correlations at a non-significant level. ISC, increased sulfate radical concentration; DGC, decreased glucose concentration; CD, cell density. The significances of these correlations are indicated by *** when $p < 0.001$, ** when $p < 0.01$, and * when $p < 0.05$.

According to Figures 5c and 6, the solution pH was found to play a more significant role in total soil Cd removal. However, the highest solution pH value was obtained in the 10 m^3 cultivation stage (Figure 2a). To address this issue, modifications were made to the bioreactor parameter for the 10 m^3 volume by lengthening the agitator blade from 1.2 m to 1.8 m and increasing the agitation rate from 80 rpm to 150 rpm. Mixotrophic acidophiles were then cultured again in the modified bioreactors with a volume of 10 m^3. The results showed that both modifications, i.e., lengthening the agitator blade and increasing the agitation rate, led to decreases in the solution pH values to 2.4 within five days (Figure 7). However, the modification involving lengthening the agitator blade resulted in a faster reduction rate of solution pH compared to the modification involving increasing the agitation rate.

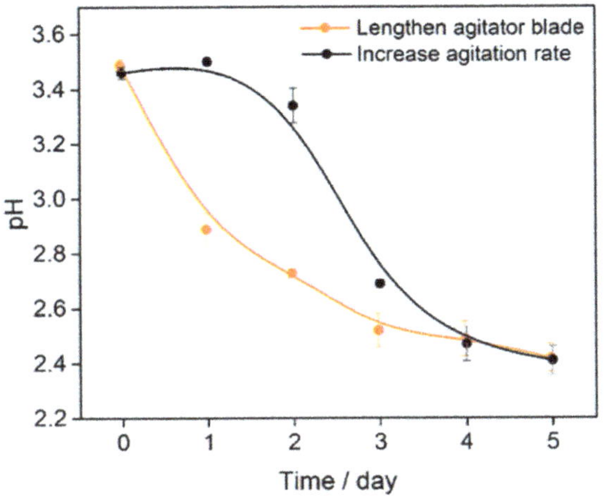

Figure 7. Solution pH changes in the cultivation system of 10 m^3 after modifying the bioreactor parameters. Red and black lines represent the modifications of lengthening the agitator blade and increasing the agitation rate.

4. Discussion

This study systematically examined the effects of bioreactor expansion on the microbial community succession in mixotrophic acidophiles and the associated changes in solution properties. Additionally, we explored the roles of these factors in Cd removal. The results indicated that bioreactor expansion led to a decrease in microbial growth and biomass accumulation in mixotrophic acidophiles (Figure 2d), hindering the utilization of energy substrates (Figure 2b,c) [33]. Autotrophic microbes were capable of oxidizing elemental sulfur to sulfuric acid and inorganic sulfur compounds [34], while heterotrophic microbes could convert glucose and yeast extract into organic acids and organic substances [35]. Interestingly, we observed a substantial rise in solution pH accompanied by a decrease in solution ISC during the scale-up cultivation process (Figure 5a), suggesting that the decline in solution pH was primarily attributed to the production of sulfuric acid through the oxidation of elemental sulfur.

The alpha-diversity of the fungal community decreased with increasing cultivation volume (Figure 3a,b), suggesting a decline in the variations in fungal species. The similarity in the fungal community structure observed in a large cultivation system (Figure 3d) was potentially due to the simplification of fungal composition (Figure 3f). However, the relationships between the solution property and dominant microbe were particularly noteworthy. The relative abundances of *Acidithiobacillaceae* and *Debaryomycetaceae* showed negative correlations with the solution pH, indicating that their growth and metabolism of these microbes could lead to a reduction in solution pH [36,37]. Given the greater impact of sulfuric acid production on pH variation (Figure 5a), *Acidithiobacillaceae* played a more significant role in reducing solution acidity [38].

The scale-up cultivation process also resulted in changes in the bacterial community, particularly in the 10 m^3 cultivation stage (Figure 3c). The observed disparity can be traced back to the solubility properties of organic (elemental sulfur) and inorganic (glucose and yeast extract) energy sources, as well as the distinct growth characteristics of bacteria and fungi. Bacterial growth, particularly autotrophic strains, exhibits lower oxygen demand compared to fungal growth [39,40]. For instance, *Acidithiobacillus ferrooxidans* and *Acidithiobacillus thiooxidans*, members of the *Acidithiobacillaceae* family, utilize carbon dioxide as a carbon source [41]. In the bioreactor, the agitation mechanism facilitates bacterial growth by suspending insoluble powdered sulfur, enabling its utilization. However, in the larger 10 m^3 system, the sulfur powder tended to settle, limiting its accessibility to sulfur-oxidizing bacteria, thus raising the solution pH. On the contrary, organic energy sources dissolve easily, fueling fungal growth, which necessitates substantial oxygen. As the bioreactor volume expanded, the dissolved oxygen concentration declined, thereby decelerating fungal biomass accumulation. In industrial production, measures such as reactor aeration and increasing the rotational speed can be taken to ensure fungal growth. Furthermore, the changes in the bacterial community were correlated with the solution ISC, indicating a limitation in the utilization of elemental sulfur by bacteria (Figure 5a). Meanwhile, the fungal community was correlated with the solution DGC, suggesting a restriction on the utilization of glucose by fungi. These correlations further highlight the impact of bioreactor expansion on the utilization of energy sources by bacteria and fungi.

The PLSPM analysis revealed that bacterial community's indirect effects on soil total Cd removal were more substantial (Figure 6). Sulfur-oxidizing bacteria in mixotrophic acidophiles convert elemental sulfur to sulfuric acid, lowering the solution pH [42], thus enhancing total Cd extraction efficiency. A low pH of the leaching solution facilitates increased Cd mobility and its release from soils into a stable solution phase. This study observed a decline in microbial solution pH from the 0.1 L to 1 m^3 stages. Solution pH fluctuations influenced the soil pH, altering Cd adsorption sites, adsorption interface stability, and Cd speciation distribution. As the soil pH decreased, the contents of carbonate-bound Cd, reducible Cd, and residual Cd diminished. These Cd forms were more easily converted to available Cd speciation at a low pH [43,44]. Acid-soluble Cd could be released by the dissolution of solid materials with a pH 5 reagent. The reducible Cd was bound to hydrous oxides of Fe, Mn, and Al in soils. Hydroxylamine hydrochloride ($E^0 = -1.87$ V and pH 2.0) could extract this fraction. Residual Cd was encapsulated in a crystalline lattice of silicate and primary and secondary minerals, requiring strong acid digestion for release. This study employed low-pH microbial solutions (pH 2.5) for the bioremediation of Cd-contaminated soils. Consequently, Cd compounds were transformed into Cd^{2+} in the solution and then separated from the soils, reducing the total soil Cd and DTPA-Cd levels.

The solution pH exhibited a significant negative correlation with the removal efficiency of total soil Cd (Figures 5c and 6). Therefore, it was necessary to lower the solution pH during the scale-up cultivation process, particularly at the 10 m^3 stage. The solution pH was positively correlated with the agitator blade's length. The high solution pH observed in the 10 m^3 stage was due to the inadequate blade length to suspend sulfur powder effectively [45]. Although the solution pH was not significantly correlated with AR, AR showed a positive correlation with solution DGC and ISC. Consequently, the bioreactor parameters in the 10 m^3 stage were readjusted by extending the agitator blade and boosting the agitation rate. As a result, the solution pH values were reduced in both modified bioreactors, with the longer blade resulting in a more rapid pH decline. However, considering energy conservation, increasing the agitation rate could further increase the electrical energy and strain stirring machine. Therefore, it was advised to prioritize extending the agitator blade length as the preferred modification strategy for the 10 m^3 bioreactor.

The study's results demonstrate potential for practical implementation, as industrial-scale cultivation of functional strains or microflora has been successfully realized. Mixotrophic acidophiles exhibit a promising capacity to meet remediation demands at the field level. A microbial cultivation facility can strategically be situated near contaminated rice paddies, with the microbial solutions disseminated through irrigation channels or carriers. Irrigation water serves as a medium for cultivating these microorganisms. To replace the flask mixing used in this study, conventional tillage methods liking plowing can be employed to blend the microbial solution with contaminated soil, followed by natural sedimentation. It is crucial to avoid deep soil disruption during mixing process, necessitating the development of automated machinery for consistent-depth plowing for solid–liquid mixing. Supernatants can be drained, separated, and collected in a reservoir lined with impermeable membranes to prevent leakage. Incorporating heavy-metal adsorbents into the reservoir aids in capturing Cd^{2+} from the supernatant. The purified liquid can be repurposed for microbial consortium cultivation, facilitating the transition of mixotrophic acidophilic remediation from laboratory to field-scale implementation.

5. Conclusions

This study examined the effects of bioreactor expansion on the succession of mixotrophic acidophiles and its implications for soil Cd remediation. Bioreactor expansion resulted in a delay in sulfur and glucose oxidations and reduced a less pronounced decrease in solution pH and cell density. Bioreactor expansion significantly decreased fungal alpha-diversity, altered the community structure, and simplified the fungal community's composition. Bacterial diversity and community structure displayed more substantial variations between the 10 m^3 stage and other cultivation scales. The family-level microbes of *Acidithiobacillaceae* and *Debaryomycetaceae* dominated the bacterial and fungal communities throughout the scale-up process. The indirect effects of mixotrophic acidophiles played a significant role in soil Cd removal. Bacterial community shifts with the changes in bioreactor volume decreased the solution pH, indirectly enhancing Cd removal efficiency. To optimize the 10 m^3 bioreactor for efficient Cd remediation, it was suggested to increase the agitator blade length. This study offers new insights into the relationships between microbial community dynamics and Cd removal, shedding light on the bioremediation of Cd-contaminated soils in agricultural systems.

Author Contributions: X.H.: conceptualization, investigation, data curation, methodology, visualization, writing—review and editing, supervision, and funding acquisition. P.Z.: investigation, methodology, and software. X.L.: conceptualization and supervision. L.J.: software and visualization. H.J.: methodology. H.L.: methodology, software, and visualization. Z.C.: supervision and funding acquisition. All authors have read and agreed to the published version of the manuscript.

Funding: This research was funded by the Key Research and Development Program of Linyi City (2022020), the Shandong Provincial Natural Science Foundation (ZR2020QD120), and the Special Project of Central Government Guiding Local Scientific and Technological Development (YDZX2021070).

Institutional Review Board Statement: Not applicable.

Informed Consent Statement: Not applicable.

Data Availability Statement: The raw data supporting the conclusions of this article will be made available by the authors upon request.

Conflicts of Interest: The authors declare no conflict of interest.

References

1. Ren, S.; Song, C.; Ye, S.; Cheng, C.; Gao, C. The spatiotemporal variation in heavy metals in China's farmland soil over the past 20 years: A meta-analysis. *Sci. Total Environ.* **2022**, *806*, 150322. [CrossRef] [PubMed]
2. Liu, X.; Chen, S.; Yan, X.; Liang, T.; Yang, X.; El-Naggar, A.; Liu, J.; Chen, H. Evaluation of potential ecological risks in potential toxic elements contaminated agricultural soils: Correlations between soil contamination and polymetallic mining activity. *J. Environ. Manag.* **2021**, *300*, 113679. [CrossRef] [PubMed]
3. Bind, A.; Kushwaha, A.; Devi, G.; Goswami, S.; Sen, B.; Prakash, V. Biosorption valorization of floating and submerged macrophytes for heavy-metal removal in a multi-component system. *Appl. Water Sci.* **2019**, *9*, 95. [CrossRef]
4. Shahid, M.; Dumat, C.; Khalid, S.; Niazi, N.; Antunes, P. Cadmium bioavailability, uptake, toxicity and detoxification in soil-plant system. *Rev. Environ. Contam. Toxicol.* **2017**, *241*, 73–137. [PubMed]
5. Dutta, M.; Kushwaha, A.; Kalita, S.; Devi, G.; Bhuyan, M. Assessment of bioaccumulation and detoxification of cadmium in soil-plant-insect food chain. *Bioresour. Technol. Rep.* **2019**, *7*, 100242. [CrossRef]
6. Li, Y.; Sun, J.; Qian, J.; Huang, T.; Su, F. Study on the remediation of cadmium/mercury contaminated soil by leaching: Effectiveness, conditions, and ecological risks. *Water Air Soil Pollut.* **2023**, *234*, 23. [CrossRef]
7. Fonti, V.; Dell'Anno, A.; Beolchini, F. Does bioleaching represent a biotechnological strategy for remediation of contaminated sediments? *Sci. Total Environ.* **2016**, *563–564*, 302–319. [CrossRef] [PubMed]
8. Wang, L.; Zhang, Q.Y.; Liao, X.Y.; Li, X.H.; Zheng, S.N.; Zhao, F.H. Phytoexclusion of heavy metals using low heavy metal accumulating cultivars: A green technology. *J. Hazard. Mater.* **2021**, *413*, 125427. [CrossRef] [PubMed]

9. Verma, S.; Bhatt, P.; Verma, A.; Mudila, H.; Prasher, P.; Rene, E. Microbial technologies for heavy metal remediation: Effect of process conditions and current practices. *Clean Technol. Environ. Policy* **2023**, *25*, 1485–1507. [CrossRef]
10. Srichandan, H.; Mohapatra, R.K.; Parhi, P.K.; Mishra, S. Bioleaching approach for extraction of metal values from secondary solid wastes: A critical review. *Hydrometallurgy* **2019**, *189*, 105122. [CrossRef]
11. Potysz, A.; Van Hullebusch, E.D.; Kierczak, J. Perspectives regarding the use of metallurgical slags as secondary metal resources—A review of bioleaching approaches. *J. Environ. Manag.* **2018**, *219*, 138–152. [CrossRef]
12. Raffa, C.; Chiampo, F.; Shanthakumar, S. Remediation of metal/metalloid-polluted soils: A short review. *Appl. Sci.* **2021**, *11*, 4134. [CrossRef]
13. Bosecker, K. Bioleaching: Metal solubilization by microorganisms. *FEMS Microbiol. Rev.* **1997**, *20*, 591–604. [CrossRef]
14. Brandl, H. Microbial leaching of metals. *Biotechnology* **2001**, *10*, 191–224.
15. Acharya, C.; Kar, R.; Sukla, L. Short Communication: Leaching of chromite overburden with various native bacterial strains. *World J. Microbiol. Biotechnol.* **1998**, *14*, 769–771. [CrossRef]
16. Zhu, J.; Zhang, J.; Li, Q.; Han, T.; Hu, Y.; Liu, X.; Qin, W.; Chai, L.; Qiu, G. Bioleaching of heavy metals from contaminated alkaline sediment by auto- and heterotrophic bacteria in stirred tank reactor. *Trans. Nonferrous Met. Soc. China* **2014**, *24*, 2969–2975. [CrossRef]
17. Hao, X.; Zhu, P.; Zhang, H.; Liang, Y.; Yin, H.; Liu, X.; Bai, L.; Liu, H.; Jiang, H. Mixotrophic acidophiles increase cadmium soluble fraction and phytoextraction efficiency from cadmium contaminated soils. *Sci. Total Environ.* **2019**, *655*, 347–355. [CrossRef] [PubMed]
18. González-Toledo, S.; Domínguez-Domínguez, J.; García-Almendárez, B.; Prado-Barragán, L.; Regalado-González, C. Optimization of nisin production by *Lactococcus lactis* UQ2 using supplemented whey as alternative culture medium. *J. Food Sci.* **2010**, *75*, 347–353. [CrossRef] [PubMed]
19. Brennan, L.; Owende, P. Biofuels from microalgae-A review of technologies for production, processing, and extractions of biofuels and co-products. *Renew. Sustain. Energy Rev.* **2010**, *14*, 557–577. [CrossRef]
20. Ren, J.; Lin, W.; Shen, Y.; Wang, J.; Luo, X.; Xie, M. Optimization of fermentation media for nitrite oxidizing bacteria using sequential statistical design. *Bioresour. Technol.* **2008**, *99*, 7923–7927. [CrossRef]
21. Modiri, S.; Kermanshahi, R.; Soudi, M.; Dad, N.; Ebadi, M.; Zahiri, H.; Noghabi, K. Growth optimization of *Lactobacillus acidophilus* for production of antimicrobial peptide acidocin 4356: Scale up from flask to lab-scale fermenter. *Iran. J. Biotechnol.* **2021**, *19*, e2686.
22. Ren, H.; Zentek, J.; Vahjen, W. Optimization of production parameters for probiotic *Lactobacillus* strains as feed additive. *Molecules* **2012**, *24*, 3286. [CrossRef]
23. Zhang, B.; Zhang, Y.; Teng, Y.; Fan, M. Sulfate radical and its application in decontamination technologies. *Crit. Rev. Environ. Sci. Technol.* **2015**, *45*, 1756–1800. [CrossRef]
24. Yuan, B.; Huang, L.; Liu, X.; Bai, L.; Liu, H.; Jiang, H.; Zhu, P.; Xiao, Y.; Geng, J.; Liu, Q.; et al. Application of mixotrophic acidophiles for the bioremediation of cadmium-contaminated soils elevates cadmium removal, soil nutrient availability, and rice growth. *Ecotoxicol. Environ. Saf.* **2022**, *236*, 113499. [CrossRef]
25. Harris, J.; Kelley, S.; Pace, N. New perspective on uncultured bacterial phylogenetic division OP11. *Appl. Environ. Microbiol.* **2004**, *70*, 845–849. [CrossRef]
26. Cheung, M.; Au, C.; Chu, K.; Kwan, H.; Wong, C. Composition and genetic diversity of picoeukaryotes in subtropical coastal waters as revealed by 454 pyrosequencing. *ISME J.* **2010**, *4*, 1053–1059. [CrossRef] [PubMed]
27. Hao, X.; Bai, L.; Liu, X.; Zhu, P.; Liu, H.; Xiao, Y.; Geng, J.; Liu, Q.; Huang, L.; Jiang, H. Cadmium speciation distribution responses to soil properties and soil microbes of plow layer and plow pan soils in cadmium-contaminated paddy fields. *Front. Microbiol.* **2021**, *12*, 774301. [CrossRef] [PubMed]
28. Magoč, T.; Salzberg, S.L. FLASH: Fast length adjustment of short reads to improve genome assemblies. *Bioinformatics* **2011**, *27*, 2957–2963. [CrossRef] [PubMed]
29. Caporaso, J.G.; Kuczynski, J.; Stombaugh, J.; Bittinger, K.; Bushman, F.D.; Costello, E.K.; Fierer, N.; Peña, A.G.; Goodrich, J.K.; Gordon, J.I.; et al. QIIME allows analysis of high-throughput community sequencing data. *Nat. Methods* **2010**, *7*, 335–336. [CrossRef]
30. Edgar, R.C. UPARSE: Highly accurate OTU sequences from microbial amplicon reads. *Nat. Methods* **2013**, *10*, 996. [CrossRef]
31. Wang, Q. Naive Bayesian classifier for rapid assignment of rRNA sequences into the new bacterial taxonomy. *Appl. Environ. Microbiol.* **2007**, *73*, 5261–5267. [CrossRef] [PubMed]
32. Quast, C.; Pruesse, E.; Yilmaz, P.; Gerken, J.; Glckner, F.O. The SILVA ribosomal RNA gene database project: Improved data processing and web-based tools. *Nucleic Acids Res.* **2012**, *41*, 590–596. [CrossRef]
33. Manzoor, A.; Qazi, J.; Haq, I.; Mukhtar, H.; Rasool, A. Significantly enhanced biomass production of a novel bio-therapeutic strain *Lactobacillus plantarum* (AS-14) by developing low cost media cultivation strategy. *J. Biol. Eng.* **2017**, *11*, 17. [CrossRef] [PubMed]
34. Nguyen, T.H.; Won, S.; Ha, M.; Duc Nguyen, D.; Kang, H.Y. Bioleaching for environmental remediation of toxic metals and metalloids: A review on soils, sediments, and mine tailings. *Chemosphere* **2021**, *282*, 131108. [CrossRef] [PubMed]
35. Deng, X.; Yang, Z.; Chen, R. Study of characteristics on metabolism of *Penicillium chrysogenum* F1 during bioleaching of heavy metals from contaminated soil. *Can. J. Microbiol.* **2019**, *65*, 629–641. [CrossRef] [PubMed]
36. Vera, M.; Schippers, A.; Sand, W. Progress in bioleaching: Fundamentals and mechanisms of bacterial metal sulfide oxidation-part A. *Appl. Microbiol. Biotechnol.* **2013**, *97*, 7529–7541. [CrossRef] [PubMed]

37. Brejová, B.; Lichancová, H.; Brázdovič, F.; Cillingová, A.; Neboháčová, M.; Tomáška, Ľ.; Vinař, T.; Nosek, J. Draft genome sequence of an obligate psychrophilic yeast, *Candida psychrophila* NRRL Y-17665T. *Genome Announc.* **2017**, *5*, e817–e851. [CrossRef] [PubMed]
38. Ko, M.; Park, H.; Kim, K.; Lee, J. The role of Acidithiobacillus ferrooxidans and Acidithiobacillus thiooxidans in arsenic bioleaching from soil. *Environ. Geochem. Health* **2013**, *35*, 727–733. [CrossRef] [PubMed]
39. Guezennec, A.; Joulian, C.; Jacob, J.; Bodenan, F.; D'Hugues, P.; Hedrich, S. Influence of CO_2 supplementation on the bioleaching of a copper concentrate from Kupferschiefer Ore. *Solid State Phenom.* **2017**, *262*, 242–245. [CrossRef]
40. Witne, J.Y.; Phillips, C.V. Bioleaching of Ok Tedi copper concentrate in oxygen- and carbon dioxide-enriched air. *Miner. Eng.* **2001**, *14*, 25–48. [CrossRef]
41. Jerez, C. Bioleaching and biomining for the industrial recovery of metals. *Compr. Biotechnol.* **2017**, *3*, 758–771.
42. Peng, T.; Zhou, D.; Liu, Y.; Yu, R.; Qiu, G.; Zeng, W. Effects of pH value on the expression of key iron/sulfur oxidation genes during bioleaching of chalcopyrite on thermophilic condition. *Ann. Microbiol.* **2019**, *69*, 627–635. [CrossRef]
43. Yang, S.; Li, Y.; Si, S.; Liu, G.; Yun, H.; Tu, C.; Li, L.; Luo, Y. Feasibility of a combined solubilization and eluent drainage system to remove Cd and Cu from agricultural soil. *Sci. Total Environ.* **2022**, *807*, 150733. [CrossRef] [PubMed]
44. Liu, C.; Lin, Y. Reclamation of copper-contaminated soil using EDTA or citric acid coupled with dissolved organic matter solution extracted from distillery sludge. *Environ. Pollut.* **2013**, *178*, 97–101. [CrossRef] [PubMed]
45. Guo, Z.; Zhang, L.; Cheng, Y.; Xiao, X.; Pan, F.; Jiang, K. Effects of pH, pulp density and particle size on solubilization of metals from a Pb/Zn smelting slag using indigenous moderate thermophilic bacteria. *Hydrometallurgy* **2010**, *104*, 25–31. [CrossRef]

toxics

MDPI

Article

Oxidative Dissolution Process of Sphalerite in $Fe_2(SO_4)_3$-O_3 System: Implications for Heavy Metals Removal and Recovery

Mingtong Zhang [1], Hongbo Zhao [1,*], Yisheng Zhang [1], Xin Lv [1], Luyuan Zhang [1], Li Shen [1], Liang Hu [1], Jiankang Wen [2,*], Louyan Shen [3] and Xianping Luo [4]

1 School of Minerals Processing & Bioengineering, Central South University, Changsha 410083, China; zmtincsu@csu.edu.cn (M.Z.); zys666@csu.edu.cn (Y.Z.); bqt2100301002@student.cumtb.edu.cn (X.L.); zhangluyuan@csu.edu.cn (L.Z.); lishen@csu.edu.cn (L.S.); huliang2018@csu.edu.cn (L.H.)

2 National Engineering Research Center for Environment-Friendly Metallurgy in Producing Premium Non-Ferrous Metals, GRINM Group Co., Ltd., Beijing 100088, China

3 China Nerin Engineering Co., Ltd., Nanchang 330103, China; shenlouyan@nerin.com

4 College of Resources and Environment, Jiangxi University of Science and Technology, Ganzhou 341000, China; luoxianping9491@163.com

* Correspondence: zhbalexander@csu.edu.cn (H.Z.); wjk@grinm.com (J.W.)

Abstract: Metal sulfides in waste rocks and tailings are susceptible to serious soil and water contamination due to the generation of acid mine drainage (AMD) during stockpiling. The hydrometallurgical process is one of the most essential heavy metal remediation technologies through harmless disposal and resource utilization of the waste sulfides. However, atmospheric hydrometallurgy of sulfides still faces great challenges due to low leaching efficiency and high cost. In this work, we proposed a cooperative leaching system ($Fe_2(SO_4)_3$-O_3) and investigated the oxidative dissolution process of sphalerite (ZnS). Under the optimal conditions, the extracted zinc reached 97.8%. Reactive oxygen species (ROS) ($\cdot OH$, $^1O_2$ and $\cdot O_2^-$) were identified in the radical quenching experiments. The dissolution of sphalerite did not show passivation due to the ozone's capability to oxidize the sulfur in sphalerite to sulfate. In addition, stirring rate, O_3 inlet concentration, and $Fe_2(SO_4)_3$ concentration had a significant effect on the dissolution of sphalerite. Meanwhile, the apparent activation energy was 24.11 kJ/mol based on kinetic fitting, which indicated that the controlling step of the reaction was mainly a diffusion process. This work demonstrated the cooperative effect of sphalerite leaching in the O_3-$Fe_2(SO_4)_3$ system and provided a theoretical reference for efficient and atmospheric dissolution of sphalerite.

Keywords: heavy metals; leaching; dissolution process; metal sulfides; sphalerite

1. Introduction

Currently, froth flotation is the most widely used method of separating sulfide ores to obtain a sulfide concentrate for smelting [1]. The waste ore after flotation is called flotation tailings, containing large amounts of heavy metal ions and sulfides. As a result, in recent decades, this method has produced billions of tons of flotation tailings along with the operation of mining plants around the world [2]. Moreover, the mining process also produces large quantities of sulphureous mining waste every year. These sulfide metals mainly include pyrite, sphalerite, chalcopyrite, and galena [3]. As sulphureous waste ore is oxidized during stockpiling, the acid mine drainage (AMD) produced could lead to serious soil and water pollution through the ground or surface water cycle.

Tailing dam and surface paste methods have been developed for more economical and safe storage of tailing [4]. However, with the increasing number of tailings storage depots, tailing management has become more difficult and the risk of harm to the environment has increased. In addition, there are still a considerable number of valuable metals in these tailings that are worth extracting and utilizing. In recent years, a number of new methods are being investigated to recover valuable metal components and to reduce the hazards of sulphureous waste ore [5,6]. Hydrometallurgy is often used to dissolve low-grade minerals of complex composition. Meanwhile, the process is one of the most essential heavy metal remediation technologies through harmless disposal (heavy metals removal) and resource utilization (heavy metals recovery) of the waste sulfides.

The bioleaching process has been used in experimental and industrial-scale studies [7,8] due to low cost, large production scale and low mineral grade requirements. *Acidithiobacillus ferrooxidans* is commonly investigated and used microorganisms for leaching sulfide ores [9]. In the presence of this bacteria, pyrite (FeS_2), the prime component of the sulfide tailings, oxidized to trivalent iron and sulphate [4]. Moreover, many studies have shown that zinc, iron, and lead sulfide undergo bioleaching to produce the relevant sulfates and elemental sulfur [3]. However, elemental sulfur can lead to the results of slow reaction rate and inadequate reaction [10–12], because of the formation of the passivation layer. The sulfur layer is most likely formed when the sulfur in sulfide ores is oxidized by trivalent iron. Santos, Rivera-Santillán, Talavera-Ortega, and Bautista [13] studied sphalerite leaching at a pH of 1.0 and 0.5 M Fe^{3+} at 70 °C, using 44–53 μm particle sizes, and a Zn extraction of 70% after 7 h was obtained, and scanning electron microscopy images showed the presence of a passive layer on the surface of the undissolved sphalerite. Likewise, Nikkhou, Xia, and Deditius [12] dissolved sphalerite particles < 22 μm with 2.06 M $Fe_2(SO_4)_3$ in a 1 M citric acid solution between 35–130 °C, and achieved Zn extraction < 80%. Furthermore, the sulfur layer and hydrated ferrous sulfate were confirmed by surface analysis of the samples.

Powerful oxidation can be used to remove the passivation of the sulfur layer. For instance, Santos, Rivera-Santillán, Talavera-Ortega, and Bautista [13] dissolved 106–150 μm sphalerite grains using 2.06 M $Fe(NO_3)_3$ at 90 °C, and achieved complete extraction within 5 h. Nitrate plays a role in converting elemental sulfur to sulfate. Likewise, [13] leached pyrite by bioleaching (*Leptospirillum ferrooxidans* and *Acidithiobacillus thiooxidans*) and found that the passivation layer generated caused inhibition of dissolution. After ozone activation, the sulfur layer on the surface was destroyed and the total iron leaching rate increased from about 34% to about 50%. Compared to other oxidizers, ozone has significant advantages such as high efficiency, environmental friendliness, and ease of generation. At the same time, these characteristics are also very conducive to the realization of the resource utilization of sulfide tailings.

Ozone is a powerful and environmentally friendly oxidant, with a redox potential of 2.07 V (SHE) in acidic solutions, which is widely used in the treatment of wastewater as an advanced oxidant [13,14]. Not only does it eliminate the sulfur layer, but it also does not introduce impurities into the leaching solution. In addition, the leachate can be continuously used for the sulfide tailings dissolution, which reduces the volume of wastewater. In recent years, many researchers have studied the utility of ozone in hydrometallurgy and the treatment of mine wastewater. These include gold ore [15,16], silver-bearing pyrite [17], pyrite [13], stibnite [18], sphalerite [19], wastewater containing cyanide [20], and thiosalts [21], etc.

However, the reaction mechanism of sulfide tailings with ozone is very complex because tailings contain multiple metal sulfide ores and the reaction process involves gas–liquid and liquid–solid mass transfer. In addition, there are two reaction pathways for ozone: direct reaction via ozone molecules and indirect reaction by ·OH (E = 2.80 V), which is generated by its decomposition [22]. Therefore, the generation of sulfur layers can be avoided by whichever pathway predominates. Although in the majority of cases, hydroxyl radicals are generated under alkaline conditions [22], they can be generated under acidic conditions in the presence of some catalysts [23]. Example include copper sulfide [24], galena [25], ferrous ions [14,26], and trivalent ferric ions [23].

If ozone and sulfide tailings are chosen as reactants, it is difficult to obtain a clear reaction mechanism. Thus, one sulphide ore can be selected to react with ozone to analyze the mechanism step by step. In this study, sphalerite was selected for research. In addition, the effect of Fe^{3+} should be taken into account in the leaching process, since the oxidation of pyrite by ozone produces Fe^{3+} with oxidizing properties [13]. Furthermore, Fe^{3+} may affect the pathway of ozone oxidation of sulfide ores due to more difficult gas–solid mass transfer. The original reaction path for the sphalerite oxygen pressure leaching process is Equation (1), and with the addition of Fe^{3+} it is Equations (2) and (3). A significant increase in the leaching rate with the addition of Fe^{3+} can be attributed to the increase in the rate of mass transfer.

$$ZnS + H_2SO_4 + 0.5O_2 \rightarrow ZnSO_4 + H_2O + \frac{1}{8}S_8 \quad (1)$$

$$ZnS + Fe_2(SO_4)_3 \rightarrow ZnSO_4 + 2FeSO_4 + \frac{1}{8}S_8 \quad (2)$$

$$2FeSO_4 + H_2SO_4 + 0.5O_2 \rightarrow Fe_2(SO_4)_3 + H_2O \quad (3)$$

Overall, the dissolution of sphalerite was systematically investigated in this study. The impacts of different dissolving conditions on the extracted zinc were evaluated. The reactive oxygen species in the O_3-$Fe_2(SO_4)_3$ system were identified by quenching experiments. The reuse of leachate in O_3-$Fe_2(SO_4)_3$ system was also evaluated by cyclic tests. In addition, the properties of the residue surface and electrochemical properties were systematically studied. The reaction mechanism and dissolution kinetics in the sphalerite dissolution process were explored. This study provides a valuable reference for the leaching of other sulfide minerals from tailings.

2. Materials and Methods

2.1. Materials

The sphalerite samples were obtained from Changsha, Hunan Province. These samples were then crushed, ground, sieved, and used for leaching experiments with particle sizes less than 38 μm. Moreover, the XRF (X-Ray Fluorescence) analysis showed that these sphalerite samples contained 63.76 wt% Zn, 26.01 wt% S, and 1.53 wt% Fe. The XRD (X-Ray Diffraction) results were consistent with this result, indicating that these samples are almost pure ZnS.

Ozone gas was generated from oxygen in the air using an ozone generator (Xianglu Environmental Co., Ltd., Changsha, China; XLK-G20). $Fe_2(SO_4)_3$, H_2SO_4, KOH, benzoquinone (p-BQ), L-histidine, and tert-butyl alcohol (TBA) were analytical grade and purchased from Sinopharm Chemical Reagent Beijing Co., Ltd., Beijing, China. Distilled water was used to prepare the leach solution and the relevant properties of the distilled water samples are shown in Table 1.

Table 1. The water quality parameters of distilled water.

Aqueous Solution	pH	Total Dissolved Solids (mg/L)	Electrical Conductivity (μS/cm)	Total Hardness
distilled water-1	7.15	0.44	0.072	NA
distilled water-2	7.07	0.29	0.065	NA

2.2. Dissolution Experiments

For less evaporation, all leaching experiments were conducted in a 500 mL beaker covered with a watch glass. The beaker was placed in a water bath with magnetic stirring to maintain the temperature and stirring rate. Ozone was injected into the solution through a venting stone to increase the gas–liquid interface area. The leachates were taken regularly for inductively coupled plasma optical emission spectroscopy (ICP–OES) (ICAP 7400, Thermo Fisher Scientific Co., Waltham, MA, USA). Deionized water was used to compensate for the evaporated loss. The leach residues were filtered for surface analysis and electrochemical studies.

The aim was to evaluate whether efficient and environmentally friendly leaching of sphalerite can be achieved by introducing ozone into acidic ferric sulfate solutions. Leaching experiments were carried out to determine the effect of the zinc dissolution rate by varying the stirring speed (from 300 to 750 r/min), sulfuric acid addition (from 10 to 40 g/L), temperature (from 30 to 70 °C), ozone inlet concentration (from 0 mg/L to 95 mg/L), and ferric ion concentration (from 0.2 to 0.8 mol/L). All leaching experiments were maintained with a leaching solution of 200 mL, a ventilation rate of 1.2 L/min, a slurry solid–liquid ratio of 2%, and a leaching time of 6 h. Moreover, the concentration of ferrous ions was measured by enzyme standardization using o-phenanthroline as the indicator, and that of ferric ions was calculated.

In order to explore the reaction mechanism and kinetics of sphalerite dissolved in O_3 + $Fe_2(SO_4)_3$ system, the residues oxidized by different oxidation factors were used for surface analysis electrochemical tests. In addition, comparative experiments were conducted concerning ozone and ferric ions, without ozone and without ferric ions under otherwise identical leaching conditions. Then, their leachate and leachate residue were analyzed accordingly. Moreover, to determine whether reactive oxygen species were generated, scavenger experiments with benzoquinone, L-histidine and tert-butyl alcohol as the scavenger were conducted. All experiments were run in duplicate.

2.3. Analysis of the Residue Surface

The residue samples were filtered, dried, and used for surface analysis, which were dissolved separately for two hours in $Fe_2(SO_4)_3$, O_3, and O_3 + $Fe_2(SO_4)_3$ system, respectively.

The surface images of the residues were analyzed by scanning electron microscopy [27]. The S of different valence states on the leaching residue surface was detected by X-ray photoelectron spectroscopy (XPS) (ESCALAB 250Xi). The tests were conducted using an Al K Alpha X-ray source in a standard lens model. Besides, pass energies of 100 eV and 0.1 eV/step were adopted in the constant analyzer. Thermo Avantage 5.957 was used to fit the XPS spectra. Before fitting, the spectra were calibrated with C 1 s at a binding energy of 284.8 eV, and background subtraction was performed using the Smart method. Then, the spectra were peak fitted using the Gauss-Lorentz line (SGL) function according to the binding energy of the specific sulfur form.

The mineral composition of the residues was detected by X-ray diffraction (XRD, Advance D8/Bruker, Billerica, MA, USA). The microstructure and material composition of the residue samples surface were investigated by scanning electron microscopy (JSM-6490LV/JEOL, Tokyo, Japan) and energy dispersive spectroscopy (EDS) (Nepture XM 4/EDAX, Pleasanton, CA, USA). The possible reaction products on the surface of the residues were characterized by Raman spectroscopy (Horiba Scientific LabRAM HR Evolution, Piscataway, NJ, USA). An exciting source at 632 nm (He Ne laser source) was used and energy level was 5 mW.

2.4. Electrochemical Tests

Carbon paste electrodes made from sphalerite ore powder were processed for 2 h in $Fe_2(SO_4)_3$, O_3, and O_3 + $Fe_2(SO_4)_3$ system, respectively.

Carbon paste electrodes are made using mineral powder, including residues and untreated sphalerite, homogeneously mixed and pressed with graphite powder and paraffin wax in a ratio of 7:2:1. A three-electrode system was built, and a carbon paste electrode, Ag/AgCl, and carbon rod played the role of the working electrode, reference electrode, and counter electrode, respectively. Furthermore, the reference electrode was connected to the glass cell through a Luggin capillary with saturated KCl solution. Moreover, the electrolyte was a 10 g/L sulfuric acid solution in all tests. An electrochemical workstation (CHI700E, CH Instruments, Inc., Bee Cave, TX, USA) was used to complete these tests including open-circuit potential-time (OCPT), electrochemical impedance spectroscopy (EIS), and potentiodynamic polarization (Tafel plot).

3. Results and Discussion

3.1. Comparision of Sphalerite Leaching in Various Systems

3.1.1. Dissolution of Sphalerite

As shown in Figure 1, the extracted zinc in the $Fe_2(SO_4)_3$, O_3, and O_3 + $Fe_2(SO_4)_3$ systems was approximately 16.3%, 84.4%, and 97.8%, respectively. The distinction between these systems is whether or not they contain one or both of $Fe_2(SO_4)_3$ and O_3. Their concentrations and other experimental conditions were identical. The pH values of the leachate in the $Fe_2(SO_4)_3$, O_3, and O_3 + $Fe_2(SO_4)_3$ systems were 1.06, 0.94, and 1.06, respectively. H_2SO_4 addition was 10 g/L for all systems, but pH was higher in the presence of $Fe_2(SO_4)_3$. This may be due to the complexation of SO_4^{2-} with H^+ reducing the free H^+ concentration. Comparatively, the extracted Zn of O_3 system was significantly higher than that of system $Fe_2(SO_4)_3$. The combination of O_3 and $Fe_2(SO_4)_3$ had a substantial promoting effect on the dissolution of sphalerite. Especially in the first three hours, the extracted Zn of the O_3 + $Fe_2(SO_4)_3$ system is greater than the sum of that of O_3 and $Fe_2(SO_4)_3$ systems. On the one hand, we verified that the powerful oxidation effect of O_3 can be used for sphalerite dissolution. On the other hand, we can speculate that there is a cooperative effect between O_3 and $Fe_2(SO_4)_3$. Likewise, O_3 was used to enhance the dissolution of pyrite by oxidizing the passivation layer in a bioleaching system [13]. Furthermore, several studies indicated that Fe (II) and Fe (III) can promote the decolorization of Reactive Red 2 with O_3 through catalysis [26,28]. It was proposed that more reactive substances are generated during the catalytic process, such as ·OH. Hence, the catalysis of O_3 by $Fe_2(SO_4)_3$ may be the reason why sphalerite dissolved faster in the O_3 + $Fe_2(SO_4)_3$ system. In addition, $Fe_2(SO_4)_3$ plays a crucial role in enhancing oxygen mass transfer in pressure leaching of sphalerite [29]. Thus, it may also enhance O_3 mass transfer in the dissolution of sphalerite in the O_3 + $Fe_2(SO_4)_3$ system.

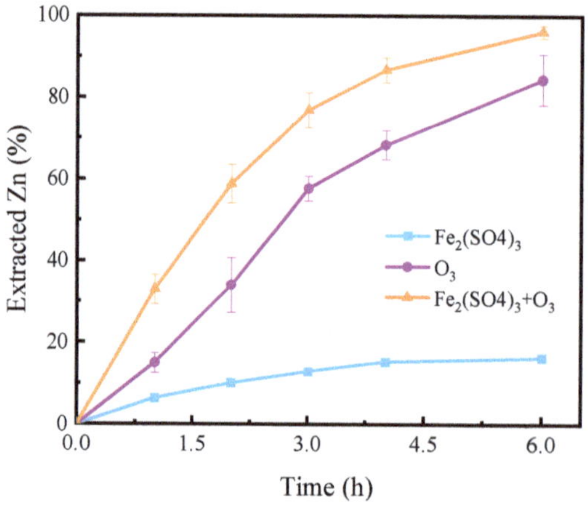

Figure 1. Dissolution of sphalerite in different systems. Experimental conditions: [$Fe_2(SO_4)_3$] = 0.2 M, O_3 inlet = 95 mg/L and 1.2 L/min, H_2SO_4 addition = 10 g/L, T = 40 °C, stirring rate = 650 rpm.

3.1.2. Role of Reactive Oxygen Species

Reactive oxygen species (ROS) (including ·OH, $^1O_2$ and $\cdot O_2^-$) that are produced by the decomposition of ozone may play an essential role in dissolution of sphalerite in the O_3 + $Fe_2(SO_4)_3$ system [30]. In order to obtain insights into the mechanism of sphalerite dissolution, these typical ROS generated in the O_3 + $Fe_2(SO_4)_3$ system were identified by a radical quenching experiment. Normally, tert-butanol (TBA) is considered a sensitive scavenger for ·OH ($k_{(\cdot OH, TBA)} = 6.0 \times 10^8\ M^{-1}\ S^{-1}$), and L-histidine is an excellent quencher of ·OH and $^1O_2$ ($k_{(\cdot OH, L\text{-}histidine)} = 5.0 \times 10^9\ M^{-1}\ S^{-1}$, $k_{(1O2, L\text{-}histidine)} = 1.5 \times 10^8\ M^{-1}\ S^{-1}$) [31]. In addition, para-benzoquinone (p-BQ) is typically selected as ·OH and $\cdot O_2^-$ scavenger

($k_{(\cdot OH,\ p\text{-}BQ)} = 4.5 \times 10^9\ M^{-1}\ S^{-1}$, $k_{(\cdot O2-,\ p\text{-}BQ)} = (0.9 - 1.0) \times 10^9\ M^{-1}\ S^{-1}$) [30]. Hence, TBA, L-histidine, and p-BQ were applied to identify $\cdot OH$, $\cdot OH + {}^1O_2$, and $\cdot OH + \cdot O_2^-$.

As illustrated in Figure 2a, the addition of 0.1 M TBA relatively inhibited sphalerite dissolution, especially within 1–3 h, which indicated that $\cdot OH$ played a relatively vital role in sphalerite dissolution in the $O_3 + Fe_2(SO_4)_3$ system. Although in the majority of cases, $\cdot OH$ is generated under alkaline conditions [22], it can be generated under acidic conditions in the presence of some catalysts (such as Fe (II) and Fe (III)) [23,26,28]. Furthermore, the addition of 0.1 M L-histidine resulted in a further inhibition of sphalerite dissolution, suggesting the presence of ${}^1O_2$. The addition of 0.1 M p-BQ severely suppressed sphalerite dissolution, illustrating that $\cdot O_2^-$ is the dominant reactive oxygen species in the $O_3 + Fe_2(SO_4)_3$ system (Figure 2a). It was assumed that the quantity of $\cdot OH$ generated and scavenged was the same in each set of experiments. Thus, it can be considered that the extracted Zn of "No scavenger" is contributed by $O_3 + Fe_2(SO_4)_3$ (direct effect), $\cdot OH$, ${}^1O_2$ and $\cdot O_2^-$, "L-histidine" is contributed by $O_3 + Fe_2(SO_4)_3$ (direct effect), ${}^1O_2$ and $\cdot O_2^-$, "TBA" is contributed by $O_3 + Fe_2(SO_4)_3$ (direct effect), and $\cdot O_2^-$, and "p-BQ" is contributed by $O_3 + Fe_2(SO_4)_3$ (direct effect). Calculations show that the contributions of $O_3 + Fe_2(SO_4)_3$ (direct effect), $\cdot OH$, ${}^1O_2$, and $\cdot O_2^-$ are about 33.2%, 6.3%, 6.5%, and 50.1% in the $O_3 + Fe_2(SO_4)_3$ system, respectively (Figure 2b). However, it is noticeable that ROS scavenging experiments can be interfered with by many factors. Hence, the results are recommended for qualitative instead of quantitative evaluation.

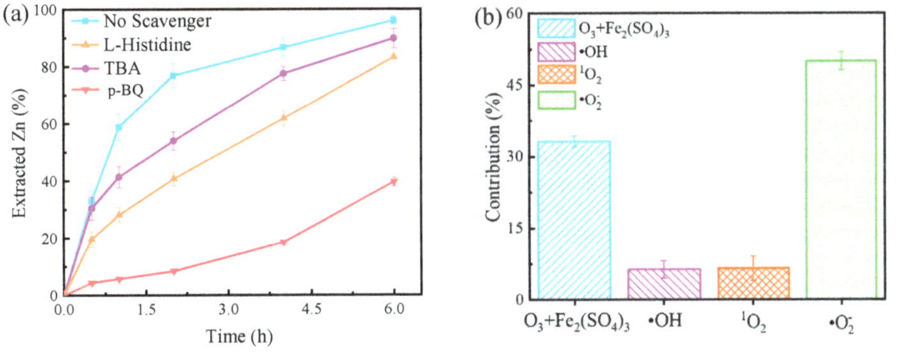

Figure 2. (**a**) Dissolution of sphalerite in the $O_3 + Fe_2(SO_4)_3$ system with different scavengers. (**b**) The contributions of $O_3 + Fe_2(SO_4)_3$ and ROS ($\cdot OH$, ${}^1O_2$, $\cdot O_2^-$) to the extracted Zn. Experimental conditions: $[Fe_2(SO_4)_3] = 0.2$ M, O_3 inlet = 95 mg/L and 1.2 L/min, H_2SO_4 addition = 10 g/L, T = 40 °C, stirring rate = 650 rpm; [L-histidine] = 0.1 M, [TBA] = 0.1 M, [p-BQ] = 0.1 M.

3.1.3. The Reuse of Leachate

To verify that the leachate can be reused in the $O_3 + Fe_2(SO_4)_3$ system, a sphalerite continuous dissolution experiment was conducted. As exhibited in Figure 3a, the Zn^{2+} concentration reached 11.9 g/L after 6 h dissolution, followed by the addition of minerals and continued dissolving for 6 h to reach 22.7 g/L. This result indicated that the solution after dissolving sphalerite in the $O_3 + Fe_2(SO_4)_3$ system had the potential to be reused several times. Furthermore, for understanding the mechanism, the concentration of Fe^{3+} was monitored during sphalerite dissolution in the $O_3 + Fe_2(SO_4)_3$ and $Fe_2(SO_4)_3$ systems. As displayed in Figure 3b, the $[Fe^{3+}]$ was always maintained around the initial concentration (0.4 M) in the $O_3 + Fe_2(SO_4)_3$ system, while in the $Fe_2(SO_4)_3$ system the $[Fe^{3+}]$ decreased continuously from the initial concentration. This equilibrium may be explained by the fact that ozone constantly produced Fe^{3+} from Fe^{2+}. Hence, this system can be easily maintained in a state where sphalerite can be dissolved efficiently. Additionally, in order to further purify valuable metals, separation and wastewater treatment are usually required after the leaching process. The process that allows continuous leaching of minerals not only improves the efficiency of dissolution, but also reduces the cost of separation and wastewater treatment. Therefore, dissolving sphalerite or other sulfide ores in the O_3 + $Fe_2(SO_4)_3$ system is significant for the entire process of metal extraction.

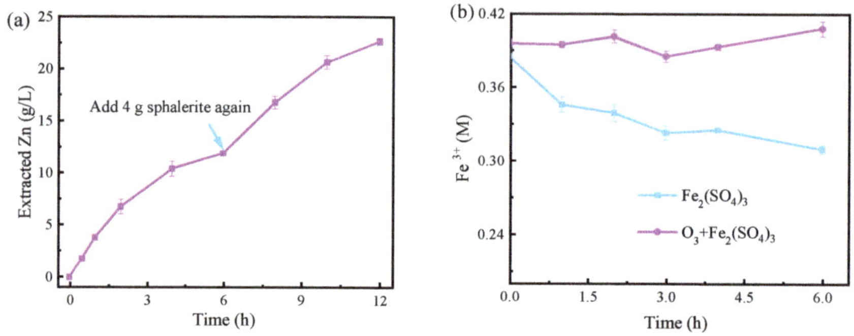

Figure 3. (**a**) Variation of the extracted Zn by reusing leachate to dissolve sphalerite in the $O_3 + Fe_2(SO_4)_3$ system. (**b**) Variation of Fe^{3+} concentration during sphalerite leaching in the $O_3 + Fe_2(SO_4)_3$ and $Fe_2(SO_4)_3$ systems. Experimental conditions: $[Fe_2(SO_4)_3]$ = 0.2 M, O_3 inlet = 95 mg/L and 1.2 L/min, H_2SO_4 addition = 10 g/L, T = 40 °C, stirring rate = 650 rpm.

3.2. The Effect of Different Dissolving Conditions on the Extracted Zinc

The effect of stirring rate on the extracted zinc was evaluated, as illustrated in Figure 4a. The extraction of zinc increases with increasing stirring rate between 300 and 650 rpm. This trend can be attributed to the enhancement of O_3 mass transfer in the leach solution [15], since the renewal rate of the gas–liquid contact surface increases with the stirring rate. In addition, the extracted zinc at 750 rpm was higher than that at 650 rpm during the first two hours, and then reduced to a value comparable to that at 450 rpm. The reduction may be because the higher stirring rate cause minerals to accumulate on the vessel walls. Thus, a rotation rate of 650 rpm was selected as one of the optimal conditions for further research.

The effect of the sulfuric acid (H_2SO_4) addition on the extracted zinc was investigated, as shown in Figure 4b. The addition of H_2SO_4 slightly improved sphalerite dissolution in the O_3 + $Fe_2(SO_4)_3$ system from 10 g/L to 40 g/L. The observed facilitation can be attributed to two reasons. First, the addition of H_2SO_4 reduced the hydrolysis of $Fe_2(SO_4)_3$, and may have decreased the formation of insoluble precipitates. Second, H_2SO_4 is capable of dissolving sphalerite directly. However, the addition of 0.5 M H_2SO_4 in the presence of 0.3 M $Fe_2(SO_4)_3$ had little effect on the extracted Zn [32]. In addition, the study showed that as the concentration of H_2SO_4 increases, the solubility of O_3 decreases [19]. Consequently, this promotion is not linear, in the order of 20 > 10 > 40 > 0 g/L H_2SO_4 addition. Overall, the addition of H_2SO_4 had a weak promoting effect on sphalerite in the O_3 + $Fe_2(SO_4)_3$ system. Hence, subsequent experiments set the H_2SO_4 addition rate at 10 g/L.

As exhibited in Figure 4c, the effect of temperature (20–45 °C) on sphalerite dissolution was evaluated in the O_3 + $Fe_2(SO_4)_3$ system. The values of extracted zinc between 30 and 45 °C were in the order of 40 > 35 ≈ 45 > 30 > 25 ≈ 20 °C. The observed results can be attributed to two overlapping reasons. On the one hand, the reaction constant of sphalerite in the O_3 + $Fe_2(SO_4)_3$ system increases with increasing temperature, and on the other hand, the solubility of O_3 decreases. Likewise, Vinals, Juan, Ruiz, Ferrando, Cruells, Roca, and Casadao [16] leached gold and palladium with aqueous ozone and obtained the fastest leaching rate at approximately 40 °C. In general, temperature had a slightly promoting effect on sphalerite in the O_3 + $Fe_2(SO_4)_3$ system within 20 to 45 °C. Subsequent experiments were conducted with the temperature set at 40 °C.

As depicted in Figure 4d, the effect of O_3 inlet concentration (0–95 mg/L) on sphalerite dissolution was evaluated in the O_3 + $Fe_2(SO_4)_3$ system. It is clear that O_3 can significantly increase the rate and degree of dissolution of sphalerite with $Fe_2(SO_4)_3$. Indeed, the extracted zinc increased from 17.6% to 74.5% by increasing the O_3 inlet concentration from 0 to 30 mg/L. Furthermore, the extracted zinc significantly increases with increasing O_3 inlet concentration, indicating that it plays a crucial role in the dissolution of sphalerite in the O_3 + $Fe_2(SO_4)_3$ system. Subsequent experiments were conducted with the O_3 inlet concentration set at 95 mg/L.

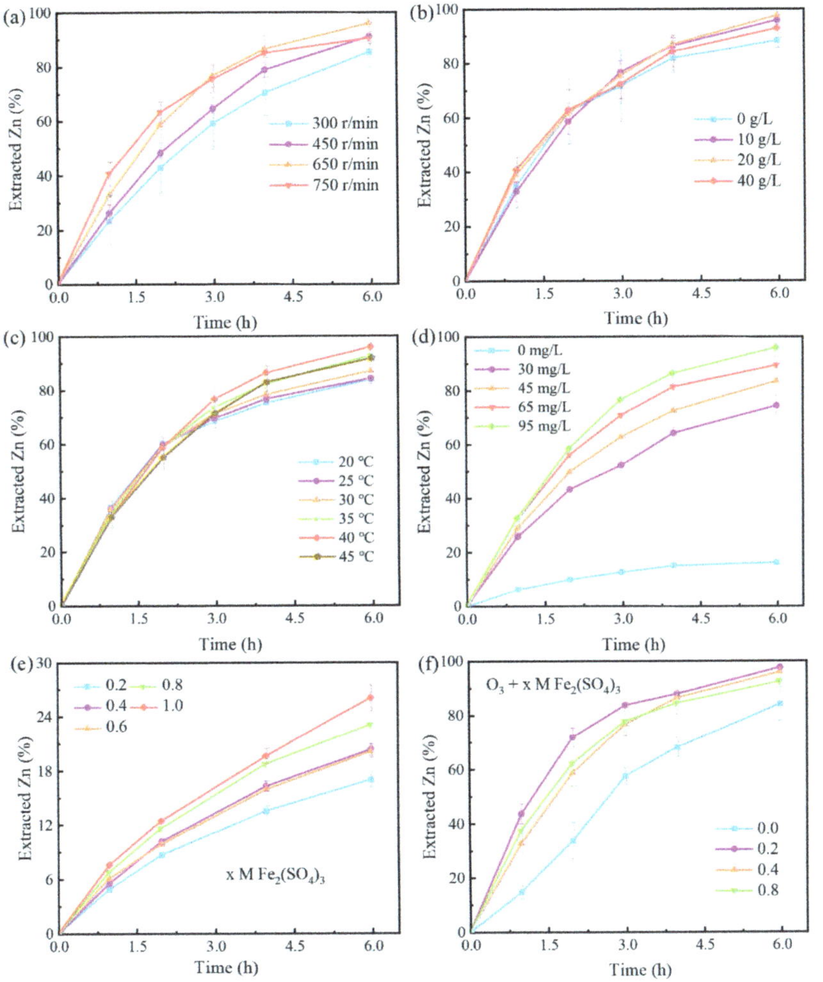

Figure 4. The effect of different factors on the extracted zinc: (**a**) stirring rate (rpm); (**b**) sulfuric acid addition; (**c**) temperature; (**d**) ozone inlet concentration; (**e**) $Fe_2(SO_4)_3$ concentration, no ozone inlet; (**f**) $Fe_2(SO_4)_3$ with 95 mg/L and 1.2 L/min O_3 inlet.

As shown in Figure 4e,f, the effect of $Fe_2(SO_4)_3$ concentration on sphalerite dissolution was evaluated in the $Fe_2(SO_4)_3$ and O_3 + $Fe_2(SO_4)_3$ systems. The extracted zinc increased linearly with increasing $Fe_2(SO_4)_3$ concentration in the $Fe_2(SO_4)_3$ system. The results can be attributed to the fact that the extracted zinc grew as the dose of oxidizer increased [32]. In the O_3 + $Fe_2(SO_4)_3$ system, dissolution of sphalerite was significantly facilitated by the addition of $Fe_2(SO_4)_3$ (0.2–0.8 M). However, the trend of the extracted Zn with increasing $Fe_2(SO_4)_3$ concentration was opposite to that of the $Fe_2(SO_4)_3$ system. The results can be attributed to several reasons. The first reason could be that higher concentrations of ferric ions will complex with sulfate to form insoluble sulfates which prevent the reaction from continuing [12,33]. Second, there were multiple reaction pathways in the O_3 + $Fe_2(SO_4)_3$ system, and an increase in the $Fe_2(SO_4)_3$ concentration promoted the slower reaction pathways. The Fe^{3+} in the system acts as both a catalyst and direct reactor, and its direct dissolution of sphalerite is the slower reaction pathway. Moreover, there is extra ozone depletion due to S_8 from the direct reaction of sphalerite and $Fe_2(SO_4)_3$. Finally, excessive Fe^{2+} may consume hydroxyl radicals with high redox potential, as Zhang, Dong, and Yang [27] mentioned in Equation (4). Thus, 0.2 M $Fe_2(SO_4)_3$ was chosen as one of the optimal conditions during sphalerite dissolution in the O_3 + $Fe_2(SO_4)_3$ system.

$$Fe^{2+} + \cdot OH \rightarrow Fe^{3+} + OH^- \quad (4)$$

3.3. Properties of the Residue Surface

3.3.1. SEM Images and EDS

The surface morphology of the residues after 2 h processed in different systems were observed by SEM (Figure 5a,d). It is easy to obtain the following information. First, Figure 5c,d illustrate that different surface characteristics will form when sphalerite is dissolved by ozone and ferric sulfate. The former are pits with well-defined edges, and the latter are covered by layer products. The unique corrosion surface in Figure 5c may be related to the adsorption site of ozone. Second, the similar surface morphologies of Figure 5b,d indicated that the direct reaction with sphalerite may be mainly $Fe_2(SO_4)_3$, when ozone is used in combination with $Fe_2(SO_4)_3$. In addition, many granular products were produced when sphalerite was dissolved in the O_3 + $Fe_2(SO_4)_3$ system. However, in the $Fe_2(SO_4)_3$ system, the product is layered and adsorbed on the sphalerite surface. With the proceeding reaction, the product layer will impede mass transfer, thus causing incomplete leaching of sphalerite. For example, Nikkhou, Xia, and Deditius [12] leached 106–150 μm sphalerite particles using 1.03 M $Fe_2(SO_4)_3$ in a 1 M citrate solution in the temperature range of 35–130 °C and obtained maximum Zn extractions of 46.1%.

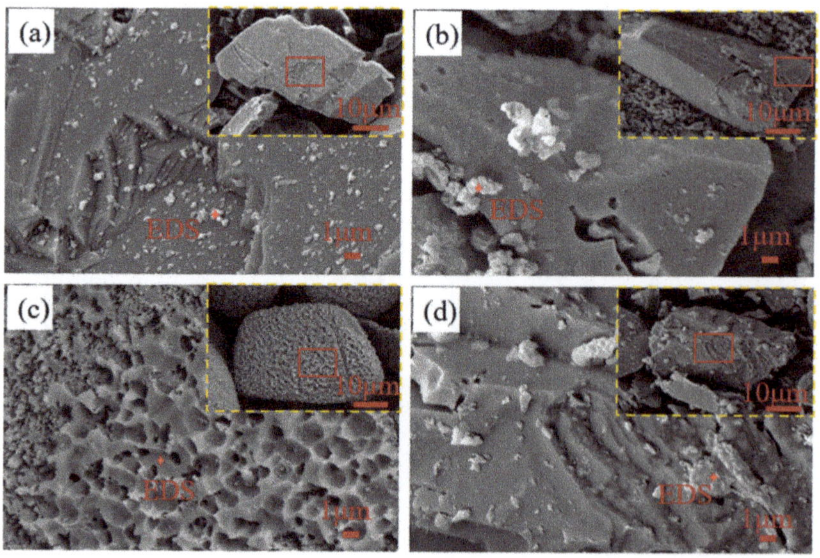

Figure 5. SEM images of sphalerite and dissolved residues from different systems: (**a**) unprocessed; (**b**) O_3 + $Fe_2(SO_4)_3$ system; (**c**) O_3 system; (**d**) $Fe_2(SO_4)_3$ system; dissolved conditions: $[Fe_2(SO_4)_3]$ = 0.2 M, O_3 inlet = 95 mg/L and 1.2 L/min, H_2SO_4 addition = 10 g/L, T = 40 °C, stirring rate = 650 rpm, reaction time = 2 h.

EDS testing further explored the main elemental composition of the residue surface (Table 2). To increase the electrical conductivity of the sample, a layer of platinum was sprayed on the surface. The surface elemental composition of the residues leached by ozone is close to that of unprocessed sphalerite. This may be due to the direct oxidation of sulfur in sphalerite to soluble sulfate by ozone. However, the mass ratio of elemental sulfur increased from 28.28% to 41.53% and 38.22% after leaching in the O_3 + $Fe_2(SO_4)_3$ and $Fe_2(SO_4)_3$ systems, respectively. This result suggests that the product may be elemental sulfur.

Table 2. Elemental composition of the residue surface.

Parameters	Zn (wt%)	S (wt%)
a	66.34	28.28
b	50.60	41.53
c	67.52	27.89
d	58.61	38.22

3.3.2. Spectral Properties

The main composition of the surface products of the residues were confirmed by XRD and Raman spectra tests (Figure 6a,b). The primary crystals in the residues are sphalerite, and a small amount of elemental sulfur is produced when $Fe_2(SO_4)_3$ is present (Figure 6a). In addition, the amount of elemental sulfur produced is significantly more in the O_3 + $Fe_2(SO_4)_3$ system than the $Fe_2(SO_4)_3$ system. The results are mainly because the introduction of ozone plays an essential role in maintaining the concentration of $Fe_2(SO_4)_3$. The continuous reaction of $Fe_2(SO_4)_3$ with sphalerite produced large amounts of S_8. However, as can be seen from the SEM image, this elemental sulfur does not cover the sphalerite surface.

Figure 6b presents the Raman bands of particles on the surface of residues after 2 h of leaching under different oxidation factors. The Raman bands of sphalerite are at 298 (dominant), 309, 340, and 350 cm^{-1}, where the band of 350 cm^{-1} is the Zn-S band and the rest are Fe-S bands [34]. However, the band at 350 cm^{-1} in this experiment is the dominant band. This phenomenon can be attributed to the fact that the sphalerite used in this experiment contains only 1.53 wt% Fe. In addition, the study of White [35] indicated that S_8 had dominant bands at 153, 419, and 472 cm^{-1} and minor bands in 187, 246, and 437 cm^{-1}. The positions of these bands are consistent with the last two Raman spectra in Figure 6b. Therefore, it is more certain that the elemental sulfur is generated both in the $Fe_2(SO_4)_3$ and the O_3 + $Fe_2(SO_4)_3$ systems.

The chemical state of elemental sulfur is essential to explore the mechanism of sphalerite dissolution. Hence, an XPS test of elemental sulfur was performed. Figure 6c–f illustrate XPS (S 2p) spectra of the surface of residues after 2 h of leaching under different oxidation factors, and well-fitted results. The 2p orbital usually shows split peaks ($2p_{1/2}$ and $2p_{3/2}$). Firstly, S^{2-} counts of the samples were significantly reduced in the $Fe_2(SO_4)_3$ and O_3 + $Fe_2(SO_4)_3$ systems. This decrease indicates that a large number of other sulfur products had been produced. Secondly, polysulfides (S_n^{2-}, $n \geq 2$) were produced on the surface of all residues. This can be attributed to the occurrence of surface relaxation. Because of the unbalanced chemical bond forces of the atoms on new surfaces, S atoms move outward and Zn atoms move inward, forming S-enriched surfaces [35]. Thirdly, while both Raman and XRD spectra showed the presence of S_8 in the residue surfaces leached in the $Fe_2(SO_4)_3$ and O_3 + $Fe_2(SO_4)_3$ systems. However, S_8 was only present in $Fe_2(SO_4)_3$ system in the XPS spectra. This difference may be attributed to the fact that elemental sulfur evaporates in a vacuum for T > 200 K [36]. The S_8 tested in this XPS may be the residual part after evaporation. Thus, the S_8 can be identified as one of the reaction products. Finally, the XPS peak for SO_4^{2-} appears independently because a significant chemical shift occurs in the electron orbitals of S in the presence of four oxygen atoms. The amount of SO_4^{2-} in the different leaching conditions is in the order of $Fe_2(SO_4)_3$ system > O_3 + $Fe_2(SO_4)_3$ system > O_3 system. This phenomenon may be attributed to the fact that the amount of S_8 covering the residue surfaces is in the order of $Fe_2(SO_4)_3$ system > O_3 + $Fe_2(SO_4)_3$ system > O_3 system = 0. Furthermore, SO_4^{2-} was adsorbed in the porous elemental sulfur. In addition, no S_8 was present in the solution after complete leaching in the O_3 and O_3 + $Fe_2(SO_4)_3$ systems. This is probably because S_8 can be oxidized to SO_4^{2-} by O_3 [13].

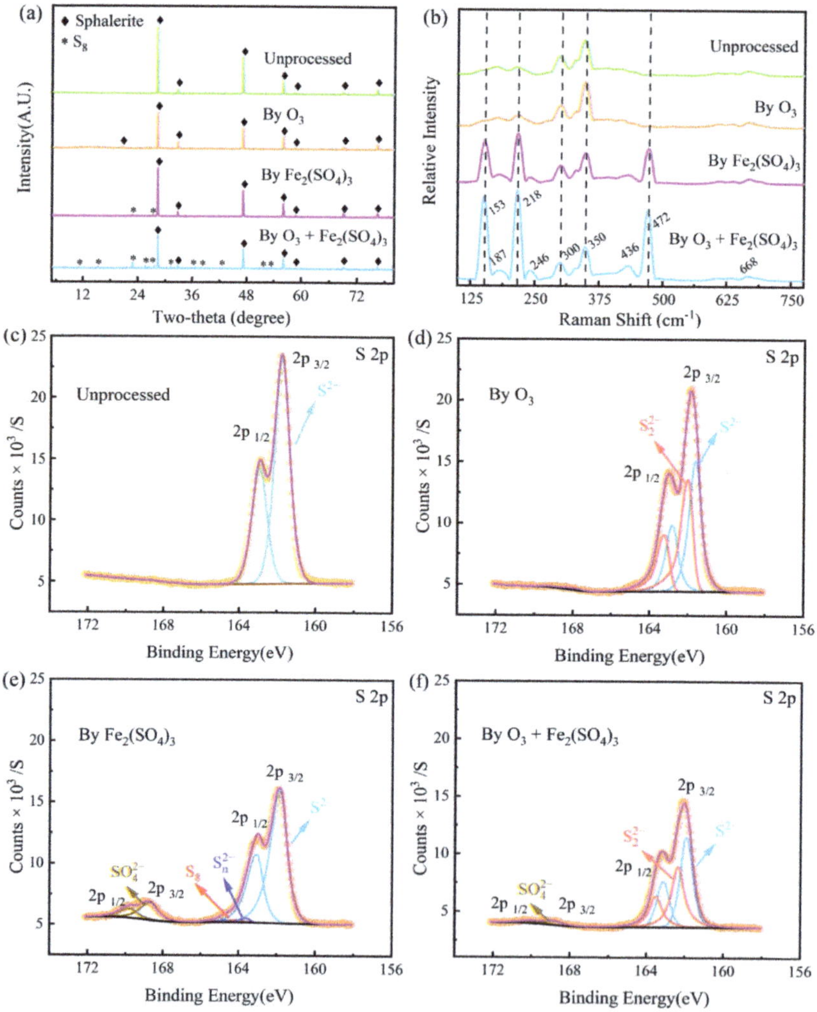

Figure 6. Spectra properties of sphalerite and dissolved residues from different systems: (**a**) XRD spectra; (**b**) Raman spectra; (**c**–**f**) XPS (S 2p). Dissolved conditions: $[Fe_2(SO_4)_3]$ = 0.2 M, O_3 inlet = 95 mg/L and 1.2 L/min, H_2SO_4 addition = 10 g/L, T = 40 °C, stirring rate = 650 rpm, reaction time = 2 h.

3.4. Electrochemical Characterization

Loss of electrons is less likely to occur at the electrode with a larger open circuit potential (OCP), i.e., the tendency to be oxidized is smaller [13,37]. Therefore, sphalerite carbon paste electrodes with higher OCPs are more difficult to dissolve by oxidation. Figure 7a illustrates that the most easily oxidized is the unprocessed sphalerite carbon paste electrode. Next is the electrode processed in the O_3 and O_3 + $Fe_2(SO_4)_3$ systems. The OCP of the latter is slightly larger, but much smaller compared to the OCP of the electrode processed in the $Fe_2(SO_4)_3$ system. This difference in OCP may be due to whether a sulfur layer is produced on the electrode surface.

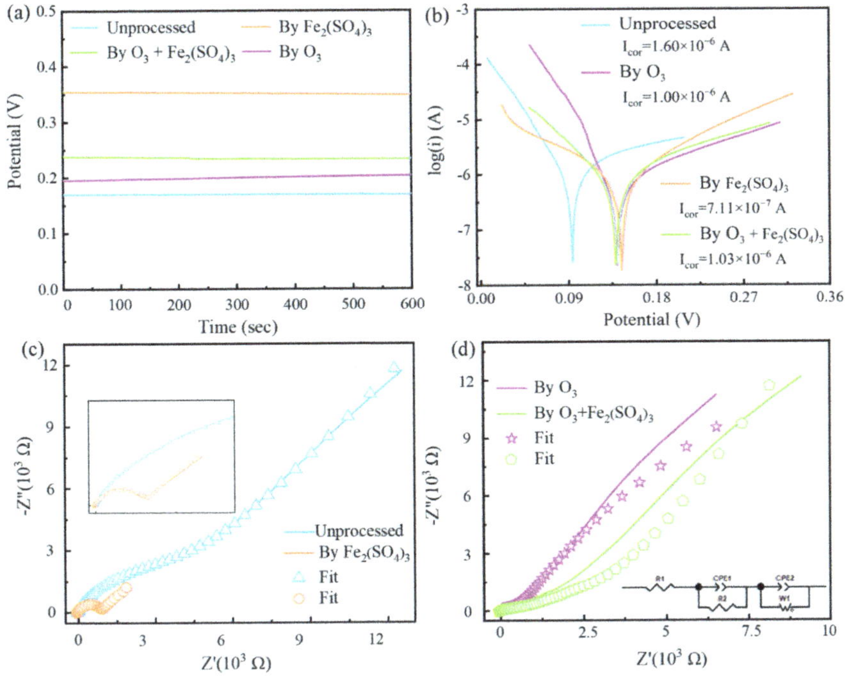

Figure 7. Electrochemical properties of sphalerite carbon paste electrode processed in different systems: (**a**) open circuit potential; (**b**) Tafel curves; (**c**,**d**) Nyquist impedance spectra and equivalent circuit. Processed conditions: $[Fe_2(SO_4)_3]$ = 0.2 M, O_3 inlet = 95 mg/L and 1.2 L/min, H_2SO_4 addition = 10 g/L, T = 40 °C, stirring rate = 650 rpm, reaction time = 2 h.

The corrosion kinetics of minerals can be analyzed by the Tafel test (Figure 7b). The corrosion kinetics are proportional to the corrosion current (I_{corr}) [13]. These I_{corr} of electrodes are from larger to smaller unprocessed and processed in the O_3 + $Fe_2(SO_4)_3$, O_3, and $Fe_2(SO_4)_3$ systems. These results indicate that the corrosion kinetics of the electrode oxidized in the O_3 + $Fe_2(SO_4)_3$ system are faster than those in the $Fe_2(SO_4)_3$ system (Table 3). In addition, the deviation of the corrosion potential (E_{corr}) from the OCP may be attributed to the effect of polarization. However, the order of these E_{corr} values in magnitude is consistent with that of OCP. Hence, these corrosion currents are valid for comparing their corrosion kinetics.

Table 3. Parameter values of the Tafel test curves.

Parameters	Cat Slp (1/v)	Ano Slp (1/v)	Lin Pol R (ohms)	Corr I (A)
Unprocessed	25.0	4.0	534,38	1.600×10^{-6}
By O_3	25.0	6.5	146,393	1.000×10^{-6}
By $Fe_2(SO_4)_3$	10.5	10.0	194,038	7.112×10^{-7}
By O_3 + $Fe_2(SO_4)_3$	15.5	6.0	127,008	1.033×10^{-6}

The transfer model of charge can be analyzed by electrochemical impedance spectroscopy (EIS). Moreover, Nyquist impedance spectra (Figure 7c,d) and Bode plots were well fitted based on the equivalent circuit (in Figure 7d), and the fitted parameters are shown in Table 4. In the equivalent circuit, R_1, R_2, R_3, and W_1 represent the solution resistor, surface electric double layer resistance, passivation layer resistance and Warburg impedance, respectively. The resistance and other parameters of the electrodes were relatively similar for different oxidation conditions, except for the electrodes oxidized in the $Fe_2(SO_4)_3$ system. In particular, the passivation resistance (R_3) is much higher than that of other electrodes under oxidation conditions. These results indicate that a significant passivation layer was generated at the electrode oxidized in the $Fe_2(SO_4)_3$ system. And it severely impeded the charge transfer. In contrast, the passivation resistance (R_3) is less than 10^{-5} at the electrode oxidized in the O_3 + $Fe_2(SO_4)_3$ and O_3 systems. The results indicated that ozone prevents the generation of passivation layers. In addition, the negligible R_3 of the unprocessed sphalerite electrode is probably due to polishing. Therefore, good charge transfer efficiency can be provided when sphalerite is oxidized with ozone and iron(III) sulfate.

Table 4. Impedance parameter values of the equivalent circuit.

Parameters	R_1	CPE-T_1	CPE-P_1	R_2	CPE-T_2	CPE-P_2	R_3	W_1-R	W_1-T	W_1-P
Unprocessed	7.24	6.84×10^{-7}	1.118	2.59	4.93×10^{-5}	0.778	0.002	8432	5.94	0.297
By $Fe_2(SO_4)_3$	6.70	3.98×10^{-6}	1.211	63.58	1.62×10^{-5}	1.042	733	8996	456.21	0.496
By O_3	6.06	1.03×10^{-6}	1.516	8.60	3.72×10^{-7}	1.464	$<10^{-5}$	5418	5.88	0.477
By O_3 + $Fe_2(SO_4)_3$	6.09	2.04×10^{-7}	1.476	26.55	3.93×10^{-7}	1.357	$<10^{-5}$	7600	6.33	0.390

3.5. Dissolution Kinetics

The dissolution of sphalerite in the $Fe_2(SO_4)_3$-O_3 system is intrinsically a liquid–solid reaction. Furthermore, the reaction rate was significantly affected by O_3 concentration and stirring rate, and slightly by temperature. Thus, the rate-controlling step of the reaction may be the diffusion process of O_3. To verify this speculation, the Jander (cylindrical diffusion) equation (Equation (5)) was used to fit the leaching data at different temperatures and different O_3 concentrations [38]. As seen in Figure 8a,c, the good fit of this equation to the leaching data suggests that it can be used to explain the kinetics of sphalerite dissolution in the $Fe_2(SO_4)_3$-O_3 system. The apparent activation energy of the reaction process was calculated to be 24.11 kJ/mol based on the Arrhenius equation, since ln(k) versus 1/T is excellently linear in the range of 20–40 °C, as seen in Figure 8b. The values of activation energy indicate that the dissolution reaction of sphalerite in the $Fe_2(SO_4)_3$-O_3 system is mainly controlled by diffusion.

$$\left(1 - (1-\alpha)^{1/2}\right)^2 = kt \tag{5}$$

To further understand the effect of O_3 concentration on sphalerite dissolution in the $Fe_2(SO_4)_3$-O_3 system, the plot of ln(k) versus ln[O_3] was fitted. As exhibited in Figure 8c an excellent linear relationship was observed between them, confirming the linear effect of O_3 concentration on the extracted zinc. In addition, the slope value (0.83) represents the empirical order of sphalerite dissolution regarding the O_3 inlet concentration in the $Fe_2(SO_4)_3$-O_3 system (Figure 8d). Furthermore, it can be seen from the SEM images that there is no product layer in the presence of ozone, indicating that it is not product layer diffusion. Therefore, it is more certain that the dissolution rate of sphalerite in the $Fe_2(SO_4)_3$-O_3 system is mainly controlled by ozone diffusion processes.

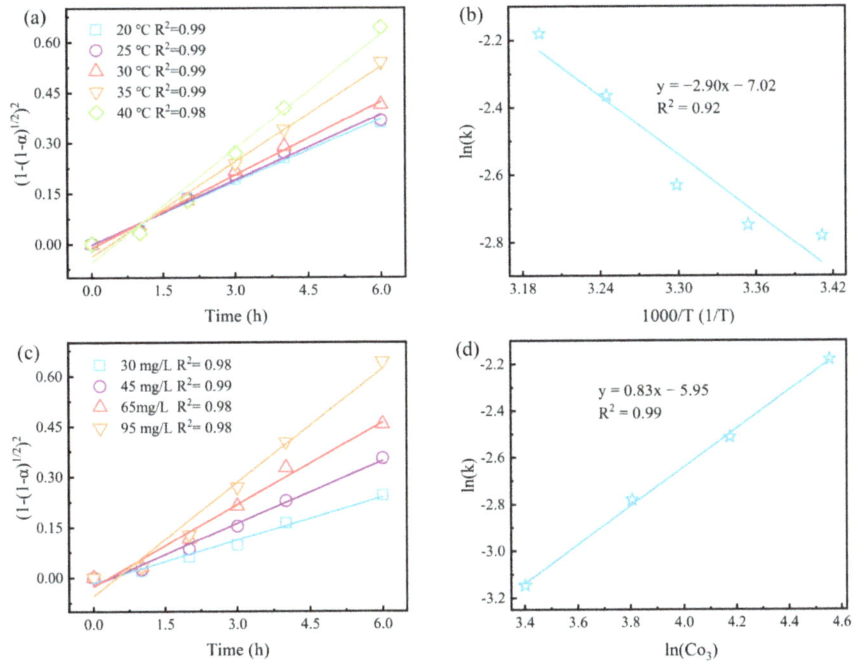

Figure 8. Kinetic fitting of sphalerite dissolving in $Fe_2(SO_4)_3$-O_3 system: (**a**,**b**) kinetic fit and graph of ln(k)–1/T in 20–40 °C; (**c**,**d**) kinetic fit and plot of ln(k)–ln[O_3] in 30–95 mg/L O_3 inlet. Experimental conditions: [$Fe_2(SO_4)_3$] = 0.4 M, O_3 inlet = 95 mg/L and 1.2 L/min, H_2SO_4 addition = 10 g/L, T = 40 °C, stirring rate = 650 rpm, reaction time = 2 h.

3.6. Possible Reaction Mechanisms

According to the above surface properties and experimental results, a possible mechanism for sphalerite dissolution in the $Fe_2(SO_4)_3$-O_3 system has been proposed, as shown in Figure 9. (1) Direct oxidation of O_3 and $Fe_2(SO_4)_3$ with sphalerite. $Fe_2(SO_4)_3$ can react directly with sphalerite and the characteristic product is S_8. The reaction of O_3 with sphalerite produces little or no elemental sulfur, and the products are likely to be soluble sulfur oxides ($SO_4{}^{2-}$). Furthermore, O_3 oxidizes the Fe^{2+} produced by sphalerite reduction to Fe^{3+} and S_8 to soluble sulfur oxides. (2) ROS ($\cdot OH$, ${}^1O_2$ and $\cdot O_2{}^-$) generated by Fe(II)- and Fe(III)-catalyzed O_3 oxidizing sphalerite. These ROS act similarly to O_3, but have a more powerful oxidizing capacity than it. (3) The dissolution reaction of sphalerite in the $Fe_2(SO_4)_3$-O_3 system is mainly controlled by diffusion, and the rate of sphalerite dissolution depends mainly on the concentration of dissolved O_3.

The specific process of ROS generation is shown in Equations (6)–(12) [26,30]. The generation of $\cdot OH$ from Fe^{2+} and Fe^{3+} with O_3 is essentially an acid-consuming process, and thus they can be used as catalysts under acidic conditions. Moreover, there may be two main reasons why $\cdot O_2{}^-$ among the ROS played a dominant role in sphalerite dissolution in the $Fe_2(SO_4)_3$-O_3 system. First, Fe^{2+} and Fe^{3+} readily form surface hydroxyl groups with water, which react with ozone to produce large quantities of $\cdot O_2{}^-$ and $\cdot OH_2$. Second, $\cdot OH_2$ is capable of producing $\cdot O_2{}^-$ and $\cdot OH$ in different pathway. More $\cdot OH_2$ may be converted to $\cdot O_2{}^-$ during sphalerite dissolution in the $Fe_2(SO_4)_3$-O_3 system.

$$Fe^{2+} + O_3 \rightarrow Fe^{3+} + \cdot O_3^- \tag{6}$$

$$O_3^- + H^+ \rightarrow O_2 + \cdot OH \tag{7}$$

$$Fe^{3+} + O_3 + H_2O \rightarrow FeO^{2+} + H^+ + \cdot OH + O_2 \quad (8)$$

$$O_3 + OH^- \rightarrow \cdot O_2^- + \cdot HO_2 \quad (9)$$

$$HO_2 + O_3 \rightarrow 2O_2 + \cdot OH \quad (10)$$

$$HO_2 \rightarrow \cdot O_2^- + H^+ \quad (11)$$

$$OH + \cdot O_2^- \rightarrow ^1O_2 + OH^- \quad (12)$$

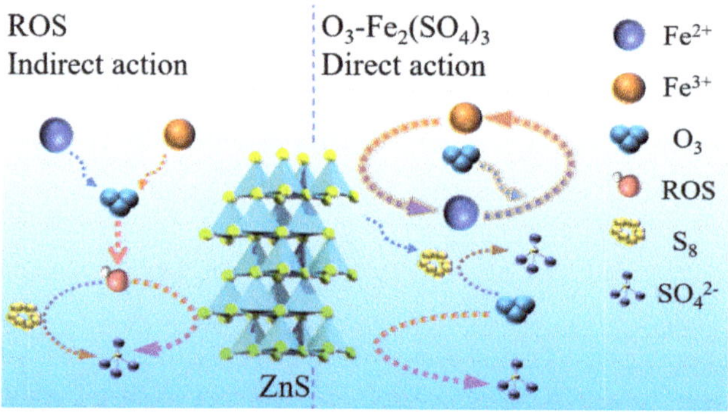

Figure 9. Mechanism model for dissolution of sphalerite in the $Fe_2(SO_4)_3$-O_3 system.

4. Conclusions

The proposed O_3-$Fe_2(SO_4)_3$ leaching process is an excellent alternative for the efficient and eco-friendly recovery of valuable metals from sphalerite at atmospheric pressure. Under the optimum conditions ($[Fe_2(SO_4)_3]$ = 0.2 M, O_3 inlet = 95 mg/L and 1.2 L/min, H_2SO_4 addition = 10 g/L, T = 40 °C, stirring rate = 650 rpm), the extracted zinc achieved 97.8% and was significantly higher than systems O_3 (82.1%) and $Fe_2(SO_4)_3$ (16.4%). At the same time, the electrochemical properties of ozonated sphalerite are similar to those of sphalerite, which indicates that ozone prevents the generation of passivation layers.

ROS ($\cdot OH$, $^1O_2$ and $\cdot O_2^-$) from O_3 decomposition played a significant role in the dissolution of sphalerite, especially $\cdot O_2^-$. Furthermore, the sulfur products of sphalerite in the $Fe_2(SO_4)_3$ and O_3 systems are S_8 and sulfates, respectively. S_8 was also produced in the O_3-$Fe_2(SO_4)_3$ system, but without inhibiting the dissolution of sphalerite. Moreover, the apparent activation energy was 24.11 kJ/mol based on kinetic fitting, which indicated that the controlling step of the reaction was mainly a diffusion process. In addition, the empirical order (0.83) of sphalerite dissolution regarding the O_3 inlet concentration was obtained. These kinetic results revealed that the O_3 concentration in the solution is the critical factor in determining the rate of sphalerite dissolution in the O_3-$Fe_2(SO_4)_3$ system.

Author Contributions: Conceptualization, H.Z.; formal analysis, M.Z.; funding acquisition, H.Z. and L.S. (Li Shen); investigation, M.Z.; resources, H.Z.; supervision, H.Z.; visualization, M.Z.; writing—original draft, M.Z.; writing—review & editing, M.Z., H.Z., Y.Z., L.Z., X.L. (Xin Lv), L.H., J.W., L.S. (Louyan Shen) and X.L. (Xianping Luo). All authors have read and agreed to the published version of the manuscript.

Funding: This work was supported by the National Natural Science Foundation of China (project No. 52174266), Open Foundation of State Key Laboratory of Complex Nonferrous Metal Resources Clean Utilization (CNMRCUKF2109).

Institutional Review Board Statement: Not applicable.

Informed Consent Statement: Not applicable.

Data Availability Statement: Some or all data that support the findings of this study are available from the corresponding author upon reasonable request.

Acknowledgments: The authors would like to thank Shiyanjia Lab (www.shiyanjia.com) for the support of the XRD test.

Conflicts of Interest: Author Jiankang Wen was employed by the company GRINM Group Co., Ltd., and author Louyan Shen was employed by the company China Nerin Engineering Co., Ltd. The remaining authors declare that they have no known competing financial interests or personal relationships that could have appeared to influence the work reported in this paper.

References

1. Peleka, E.N.; Gallios, G.P.; Matis, K.A. A perspective on flotation: A review. *J. Chem. Technol. Biotechnol.* **2018**, *93*, 615–623. [CrossRef]
2. Xie, X.-D.; Min, X.-B.; Chai, L.-Y.; Tang, C.-J.; Liang, Y.-J.; Li, M.; Ke, Y.; Chen, J.; Wang, Y. Quantitative evaluation of environmental risks of flotation tailings from hydrothermal sulfidation–flotation process. *Environ. Sci. Pollut. Res.* **2013**, *20*, 6050–6058. [CrossRef] [PubMed]
3. Mahmoud, A.; Cézac, P.; Hoadley, A.F.A.; Contamine, F.; D'Hugues, P. A review of sulfide minerals microbially assisted leaching in stirred tank reactors. *Int. Biodeterior. Biodegrad.* **2017**, *119*, 118–146. [CrossRef]
4. Romero-García, A.; Iglesias-González, N.; Romero, R.; Lorenzo-Tallafigo, J.; Mazuelos, A.; Carranza, F. Valorisation of a flotation tailing by bioleaching and brine leaching, fostering environmental protection and sustainable development. *J. Clean. Prod.* **2019**, *233*, 573–581. [CrossRef]
5. Zhang, R.Y.; Hedrich, S.; Römer, F.; Goldmann, D.; Schippers, A. Bioleaching of cobalt from Cu/Co-rich sulfidic mine tailings from the polymetallic Rammelsberg mine, Germany. *Hydrometallurgy* **2020**, *197*, 105443. [CrossRef]
6. Hubau, A.; Guezennec, A.G.; Joulian, C.; Falagán, C.; Dew, D.; Hudson-Edwards, K.A. Bioleaching to reprocess sulfidic polymetallic primary mining residues: Determination of metal leaching mechanisms. *Hydrometallurgy* **2020**, *197*, 105484. [CrossRef]
7. Mäkinen, J.; Salo, M.; Khoshkhoo, M.; Sundkvist, J.E.; Kinnunen, P. Bioleaching of cobalt from sulfide mining tailings; a mini-pilot study. *Hydrometallurgy* **2020**, *196*, 105418. [CrossRef]
8. Lorenzo-Tallafigo, J.; Iglesias-González, N.; Romero-García, A.; Mazuelos, A.; del Amo, P.R.; Romero, R.; Carranza, F. The reprocessing of hydrometallurgical sulphidic tailings by bioleaching: The extraction of metals and the use of biogenic liquors. *Miner. Eng.* **2022**, *176*, 107343. [CrossRef]
9. Valdés, J.; Pedroso, I.; Quatrini, R.; Dodson, R.J.; Tettelin, H.; Blake, R.; Eisen, J.A.; Holmes, D.S. metabolism: From genome sequence to industrial applications. *BMC Genom.* **2008**, *9*, 597. [CrossRef]
10. Markus, H.; Fugleberg, S.; Valtakari, D.; Salmi, T.; Murzin, D.Y.; Lahtinen, M. Reduction of ferric to ferrous with sphalerite concentrate, kinetic modelling. *Hydrometallurgy* **2004**, *73*, 269–282. [CrossRef]
11. Aydogan, S.; Aras, A.; Canbazoglu, M. Dissolution kinetics of sphalerite in acidic ferric chloride leaching. *Chem. Eng. J.* **2005**, *114*, 67–72. [CrossRef]
12. Nikkhou, F.; Xia, F.; Deditius, A.P. Variable surface passivation during direct leaching of sphalerite by ferric sulfate, ferric chloride, and ferric nitrate in a citrate medium. *Hydrometallurgy* **2019**, *188*, 201–215. [CrossRef]
13. Lv, X.; Zhao, H.; Zhang, Y.; Yan, Z.; Zhao, Y.; Zheng, H.; Liu, W.; Xie, J.; Qiu, G. Active destruction of pyrite passivation by ozone oxidation of a biotic leaching system. *Chemosphere* **2021**, *277*, 130335. [CrossRef] [PubMed]
14. Zeng, Z.; Zou, H.; Li, X.; Sun, B.; Chen, J.; Shao, L. Ozonation of Phenol with O_3/Fe(II) in Acidic Environment in a Rotating Packed Bed. *Ind. Eng. Chem. Res.* **2012**, *51*, 10509–10516. [CrossRef]
15. Vinals, J.; Juan, E.; Ruiz, M.; Ferrando, E.; Cruells, M.; Roca, A.; Casadao, J. Leaching of gold and palladium with aqueous ozone in dilute chloride media. *Hydrometallurgy* **2006**, *81*, 142–151. [CrossRef]
16. Li, Q.; Li, D.; Qian, F. Pre-oxidation of high-sulfur and high-arsenic refractory gold concentrate by ozone and ferric ions in acidic media. *Hydrometallurgy* **2009**, *97*, 61–66. [CrossRef]
17. Rodriguez-Rodriguez, C.; Nava-Alonso, F.; Uribe-Salas, A. Silver leaching from pyrargyrite oxidation by ozone in acid media. *Hydrometallurgy* **2014**, *149*, 168–176. [CrossRef]
18. Tian, Q.; Wang, H.; Xin, Y.; Yang, Y.; Li, D.; Guo, X. Effect of selected parameters on stibnite concentrates leaching by ozone. *Hydrometallurgy* **2016**, *165*, 295–299. [CrossRef]

19. Mubarok, M.Z.; Sukamto, K.; Ichlas, Z.T.; Sugiarto, A.T. Direct sulfuric acid leaching of zinc sulfide concentrate using ozone as oxidant under atmospheric pressure. *Miner. Metall. Process.* **2018**, *35*, 133–140. [CrossRef]
20. Carrillo-Pedroza, F.R.; Nava-Alonso, F.; Uribe-Salas, A. Cyanide oxidation by ozone in cyanidation tailings: Reaction kinetics. *Miner. Eng.* **2000**, *13*, 541–548. [CrossRef]
21. Gervais, M.; Dubuc, J.; Paquin, M.; Gonzalez-Merchan, C.; Genty, T.; Neculita, C.M. Comparative efficiency of three advanced oxidation processes for thiosalts oxidation in mine-impacted water. *Miner. Eng.* **2020**, *152*, 106349. [CrossRef]
22. Wang, J.; Chen, H. Catalytic ozonation for water and wastewater treatment: Recent advances and perspective. *Sci. Total Environ.* **2020**, *704*, 135249. [CrossRef] [PubMed]
23. Beltran, F.J.; Rivas, F.J.; Montero-de-Espinosa, R. Iron type catalysts for the ozonation of oxalic acid in water. *Water Res.* **2005**, *39*, 3553–3564. [CrossRef] [PubMed]
24. Pirgalıoğlu, S.; Özbelge, T.A. Comparison of non-catalytic and catalytic ozonation processes of three different aqueous single dye solutions with respect to powder copper sulfide catalyst. *Appl. Catal. A Gen.* **2009**, *363*, 157–163. [CrossRef]
25. Fu, P.F.; Wang, L.H.; Lin, X.F.; Ma, Y.H.; Hou, Z.W. Ozonation of recalcitrant O-isopropyl-N-ethylthionocarbamate catalyzed by galena in flotation effluents and its dissolution behaviors. *Miner. Eng.* **2021**, *165*, 106859. [CrossRef]
26. Zhang, X.B.; Dong, W.Y.; Yang, W. Decolorization efficiency and kinetics of typical reactive azo dye RR2 in the homogeneous Fe(II) catalyzed ozonation process. *Chem. Eng. J.* **2013**, *233*, 14–23. [CrossRef]
27. Saveliev, A.A.; Galeeva, E.V.; Semanov, D.A.; Galeev, R.R.; Aryslanov, I.R.; Falaleeva, T.S.; Davletshin, R.R. Adaptive noise model based iteratively reweighted penalized least squares for fluorescence background subtraction from Raman spectra. *J. Raman Spectrosc.* **2022**, *53*, 247–255. [CrossRef]
28. Wu, C.H.; Kuo, C.Y.; Chang, C.L. Homogeneous catalytic ozonation of C.I. Reactive Red 2 by metallic ions in a bubble column reactor. *J. Hazard. Mater.* **2008**, *154*, 748–755. [CrossRef] [PubMed]
29. Berezowsky, R.M.G.S.; Collins, M.J.; Kerfoot, D.G.E.; Torres, M.S. The commercial status of pressure leaching technology. *JOM* **1991**, *43*, 9–15. [CrossRef]
30. Zhang, D.P.; Liu, Y.D.; Song, Y.Y.; Sun, X.B.; Liu, W.; Duan, J.; Cai, Z.Q. Synergistic effect of Fe and Ce on Fe doped CeO for catalytic ozonation of amoxicillin: Efficiency evaluation and mechanism study. *Sep. Purif. Technol.* **2023**, *313*, 123430. [CrossRef]
31. Wang, Y.S.; Li, N.; Fu, Q.L.; Cheng, Z.J.; Song, Y.J.; Yan, B.B.; Chen, G.Y.; Hou, L.A.; Wang, S.B. Conversion and impact of dissolved organic matters in a heterogeneous catalytic peroxymonosulfate system for pollutant degradation. *Water Res.* **2023**, *241*, 120166. [CrossRef] [PubMed]
32. Dutrizac, J.E. The dissolution of sphalerite in ferric sulfate media. *Met. Mater. Trans. B* **2006**, *37*, 161–171. [CrossRef]
33. Zhao, H.; Zhang, Y.; Sun, M.; Ou, P.; Zhang, Y.; Liao, R.; Qiu, G. Catalytic mechanism of silver in the oxidative dissolution process of chalcopyrite: Experiment and DFT calculation. *Hydrometallurgy* **2019**, *187*, 18–29. [CrossRef]
34. White, S.N. Laser Raman spectroscopy as a technique for identification of seafloor hydrothermal and cold seep minerals. *Chem. Geol.* **2009**, *259*, 240–252. [CrossRef]
35. Liu, J.; Wen, S.-M.; Xian, Y.-J.; Bai, S.-J.; Chen, X.-M. First-principle study on the surface atomic relaxation properties of sphalerite. *Int. J. Miner. Metall. Mater.* **2012**, *19*, 775–781. [CrossRef]
36. Harmer, S.L.; Goncharova, L.V.; Kolarova, R.; Lennard, W.N.; Muñoz-Márquez, M.A.; Mitchell, I.V.; Nesbitt, H.W. Surface structure of sphalerite studied by medium energy ion scattering and XPS. *Surf. Sci.* **2007**, *601*, 352–361. [CrossRef]
37. Zhang, Y.S.; Zhao, H.B.; Zhang, Y.J.; Liu, H.W.; Yin, H.Q.; Deng, J.S.; Qiu, G.Z. Interaction mechanism between marmatite and chalcocite in acidic (microbial) environments. *Hydrometallurgy* **2020**, *191*, 105217. [CrossRef]
38. Dickinson, C.F.; Heal, G.R. Solid–liquid diffusion controlled rate equations. *Thermochim. Acta* **1999**, *340–341*, 89–103. [CrossRef]

toxics

MDPI

Article

Removal of Pb from Contaminated Kaolin by Pulsed Electrochemical Treatment Coupled with a Permeable Reactive Barrier: Tuning Removal Efficiency and Energy Consumption

Yinyin Zhang, Libin Zang *, Yuyan Zhao, Qiaoqiao Wei and Jiangtao Han

College of GeoExploration Science and Technology, Jilin University, Changchun 130026, China; zyyaaaaaa@163.com (Y.Z.); hanjt@jlu.edu.cn (J.H.)

* Correspondence: zanglb13@jlu.edu.cn

Abstract: Lead contamination in soil has emerged as a significant environmental concern. Recently, pulse electrochemical treatment (PECT) has garnered substantial attention as an effective method for mitigating lead ions in low-permeability soils. However, the impact of varying pulse time gradients, ranging from seconds to hours, under the same pulse duty cycle on lead removal efficiency (LRE) and energy consumption in PECT has not been thoroughly investigated. In this study, a novel, modified PECT method is proposed, which couples PECT with a permeable reaction barrier (PRB) and adds acetic acid to the catholyte. A comprehensive analysis of LRE and energy consumption is conducted by transforming pulse time. The results show that the LREs achieved in these experiments were as follows: PCb-3 s (89.5%), PCb-1 m (91%), PCb-30 m (92.9%), and PCb-6 h (91.9%). Importantly, these experiments resulted in significant reductions in energy consumption, with decreases of 68.5%, 64.9%, 51.8%, and 47.4% compared to constant voltage treatments, respectively. It was observed that LRE improved with an increase in both pulse duration and voltage gradient, albeit with a corresponding rise in energy consumption. The results also revealed that corn straw biochar as a PRB could enhance LRE by 6.1% while adsorbing migrating lead ions. Taken together, the present data highlights the potential of modified PECT technology for remediation of lead-contaminated soil, which provides an optimal approach to achieve high LRE while minimizing energy consumption.

Keywords: lead contamination; pulse electrochemical treatment; permeability reaction barrier; removal efficiency; energy consumption

1. Introduction

The accumulation of lead (Pb) contaminants on both the surface and within the soil matrix represents a significant threat to human health and the natural soil ecosystem [1]. Extensive research efforts have been devoted to addressing the remediation of soil contaminated with lead using physical, chemical, and biological approaches [2–5]. Nevertheless, these conventional techniques often entail prolonged operation, high energy consumption and significant labor. Consequently, there has been a growing exploration of low-energy, eco-friendly, and sustainable electrochemical treatment (ECT) technologies for remediating heavy metal–contaminated soils, especially those characterized by low-permeability properties [6]. Recently, ECT technology has achieved successful implementation at both laboratory and pilot scales for the removal of heavy metals from soil [7]. ECT technology is founded on three fundamental processes: electroosmosis, electromigration, and electrophoresis [4]. Upon the insertion of electrodes into the soil, an electric field is generated, which imparts energy to charged metal ions, compelling them to migrate in the opposite direction of their inherent charge, a phenomenon referred to as electromigration [8]. Electroosmosis pertains to the movement of pore water within the soil matrix's porosity, induced by the presence of an electric double layer when an electric field is applied [9]. Electrophoresis, on the other hand, involves the motion of charged particles within an electric field [10]. The electrolysis reactions occurring at the anode and cathode lead to the generation of hydrogen ions and hydroxide ions [11]. During this electrolysis process, hydrogen ions are released and exchanged with positively charged metal ions located on the surface of soil particles, thereby facilitating the desorption and dissolution of metal ions within the vicinity of the anode [11]. Meanwhile, hydroxide ions produced at the cathode tend to accumulate and create complexes with heavy metal ions within the soil. This accumulation triggers precipitation and crystallization, thereby influencing the migration of metal ions [12].

In recent decades, ECT technology has emerged as a promising technique for remediating soil pollution [5,8,11,13–16]. Many recent advancements have been made in this field [17–19]. However, Ryu et al. confirmed that achieving a 75.7% removal efficiency for lead necessitates a treatment duration of 720 h and incurs an energy consumption of 1205 kWh/ton in the traditional ECT [14]. High removal efficiency and low energy consumption are major challenges for the field application of traditional ECT [20]. In order to balance the removal efficiency and associated costs, pulse electrochemical treatment (PECT) was modified from the traditional ECT [21,22]. Yuan et al. indicated that PECT technology involves intermittently controlling the flow of electric current at timed intervals, thereby periodically applying an electric field in the soil to drive the migration of heavy metal ions, achieving the removal of the latter [21,22]. Unlike traditional constant voltage treatment, PECT demonstrated clear advantages in reducing energy consumption and enhancing removal efficiency [23]. Ryu pointed out that pulse voltage could decrease energy consumption by adjusting power-off intervals while also reducing polarization effects and focusing phenomena, ultimately leading to improved removal efficiency [23]. Sun et al. achieved a significant reduction in energy consumption by applying a pulsed current of 0.2 mA/cm^2 for the treatment of heavy metal-contaminated soil [24]. In a specific soil environment, the removal of heavy metals primarily relies on two factors: (1) the ability of electrolytic hydrogen ions to replace metal ions adsorbed on the soil surface and (2) the electromigration capability of metal ions in soil pores [25]. With regard to acid enhancers, numerous studies have substantiated that the inclusion of acidic enhancers could substantially enhance the removal efficiency of lead from soil [21,26,27]. Zhang et al. confirmed that removal efficiency increased by 20% in the ECT of Pb-contaminated kaolin by adding acetic acid [28]. Moreover, in recent years, various eco-friendly adsorbent materials have been employed in remediating heavy metal-contaminated soil by acting as permeable reactive barriers (PRBs) [29]. These materials exhibit strong adsorption capabilities, possess large specific surface areas, and contain abundant functional groups [19]. Biochar, as a superior PRB filling material, can be derived from a variety of organic and inorganic sources, including agricultural residues, forest residues, algal biomass, waste tires, and heavy crude oil [30]. He et al. utilized a new sheet PRB material, resulting in a significant reduction of residual heavy metals [26]. The comparison of lead removal efficiency and energy consumption under different electrochemical treatment conditions is displayed in Table S1 for ECT taking Pb and other heavy metals as an example.

Single ECT methods that can achieve an LRE of over 90% are rare [22], and there is even less research on PECT that simultaneously considers removal efficiency and electrical energy consumption for lead ion removal from soil [21,22]. Moreover, the concurrent application of various composite technologies often tends to prolong treatment time and increase voltage gradient, inevitably leading to an increase in time and electric energy costs [21,22]. Therefore, it is meaningful to explore a treatment technology that efficiently removes Pb from the soil while minimizing energy consumption. Yuan et al. employed a 12 h ON and 12 h OFF pulse treatment cycle, achieving an impressive removal efficiency of 93.5%. However, limited attention was given to the time gradient, and the 705 h treatment duration proved to be relatively lengthy, increasing time-related costs [22]. Based on previous research on the removal of Pb from the soil, a novel, modified PECT technology by coupling with PRB and adding acetic acid to catholyte was applied in this study. The varying pulse time gradients, ranging from seconds to hours, under the same pulse duty cycle were investigated. Zhou et al. showed that the pulse interval of 30 min ON/30 min OFF can achieve the highest removal efficiency for fluoride removal in soil [31]. Consequently, the effects of different pulse time gradients on heavy metal removal efficiency and energy consumption were compared by using a 1:2 pulse duty cycle. The pulse time gradient included seconds, minutes, and hours and consisted of four pulse interval periods: 3 s ON/3 s OFF, 1 min ON/1 min OFF, 30 min ON/30 min OFF, and 6 h ON/6 h OFF. In this study, corn straw biochar was utilized as a PRB and positioned adjacent to the catholyte compartment within the reactor. It exhibited a porous structure and a substantial specific surface area, enabling it to adsorb metal ions that were transported through pore water from the soil through electrostatic interactions, van der Waals forces, and capillary action [32]. Moreover, biochar contained numerous soluble ions within its structure, and its inherent carbonaceous nature served as a conductor for ions. Therefore, PRB not only adsorbed lead ions leached from the soil but also increased the electric current, playing a significant role in enhancing the removal efficiency [18,32].

This work focused on assessing the efficiency of lead removal from contaminated kaolin by using pulse voltage, with corn straw biochar serving as a PRB and acetic acid incorporated into the catholyte. The initial concentration of lead in the soil sample was 1153 mg kg^{-1}. The primary objective was to conduct a comprehensive analysis of LRE and energy consumption by varying pulse time gradients in the PECT system coupled with the PRB. Furthermore, the study evaluated the potential of biochar in preventing secondary pollution in the catholyte and provided innovative design concepts for future remediation of lead-contaminated soil.

2. Materials and Methods

2.1. Chemicals and Materials

The kaolin sample used in this study was obtained from Lingshou County Dehang Mineral Products Inc. (Shijiazhuang, China); the calcination temperature of kaolin was between 700 °C and 1300 °C. The kaolin was calcined to a particle size of 200 mesh and artificially contaminated with Pb $(CH_3COO)_2{\cdot}3H_2O$ solution at a concentration of 1 g L^{-1} of metal, with the soil moisture content maintained at 50%. The mixture of kaolin and solution was thoroughly stirred multiple times to ensure a homogeneous distribution of contaminants, and this process was continued for one month. The soil sample was then naturally dried and sequentially filtered using an 80-mesh sieve in a cool and dry laboratory environment. The initial concentration of Pb in the soil sample was determined to be 1153 mg kg^{-1}. The soil pH was 5.77, and the electrical conductivity of the soil was 86.45 µs/cm, after mixing with lead acetate. The biochar used in this study was prepared from corn straw, which was purchased from Henan Lize Environmental Protection Technology Inc. (Shangqiu, China). The direct current (DC) power supplies were obtained from Chengde County Yuantao Trading Inc. (Chengde, China) and provided controllable DC voltage for the experiments. The relays were obtained from Leqing Lingyu E-commerce Inc. (Wenzhou, China). The electrochemical experimental reactor used in the study was manufactured by Hangzhou Zun-quan Acrylic Inc. (Hangzhou, China). The chemical properties of the initial samples and biochar are presented in Tables 1 and 2.

Table 1. Initial chemical properties of kaolin.

The Properties of Kaolin	Value
pH	5.77 ± 0.21
EC (µs/cm)	86.45 ± 25.74
SiO_2 (%)	54.42 ± 0.11
Al_2O_3 (%)	42.68 ± 0.09
TiO_2 (%)	1.77 ± 0.01
Fe_2O_3 (%)	0.58 ± 0.009
CaO (%)	0.23 ± 0.007
K_2O (%)	0.13 ± 0.005

Table 1. *Cont.*

The Properties of Kaolin	Value
P_2O_5 (%)	0.08 ± 0.002
MgO (%)	0.06 ± 0.001
SrO (%)	0.03 ± 0.001
Cr_2O_3 (%)	0.02 ± 0.001

Table 2. Initial chemical properties of biochar.

The Properties of Biochar	Value
Organic carbon content (%)	42.21 ± 0.21
Total nitrogen content (%)	8.34 ± 0.05
Total phosphorus content (%)	2.31 ± 0.01
Total potassium content (%)	16.12 ± 0.09
Ash content (%)	7.23 ± 0.03
Others (%)	23.79 ± 0.11
pH	9.46 ± 0.05

2.2. Experimental Setup

A schematic of the electrochemical experimental reactor is shown in Figure 1. The reactor was constructed using plexiglass and had dimensions of 24 cm (length) × 4 cm (width) × 6 cm (depth). It consisted of four compartments: the soil compartment, two electrolytic compartments, and the biochar compartment. The corn straw biochar was utilized in the biochar compartment as a PRB and positioned adjacent to the catholyte compartment within the reactor. It can adsorb lead ions to prevent excessive heavy metals from entering the catholyte compartment [33]. The two electrolytic compartments received fresh electrolytes for the electrochemical progress and discharged the waste liquid at both ends of the electrolytic compartments. The biochar compartment was positioned adjacent to the soil compartments and the cathodic electrolytic compartment. The soil compartment refers to the specific area where the soil sample was placed and was divided into five equal sections (S1–S5) from the anode toward the cathode. A fiber filter paper was positioned in the partition layer of the compartment to facilitate filtration. To ensure a homogeneous distribution of the electric field, sheet-shaped graphite electrodes measuring 0.2 cm × 7 cm × 4 cm were inserted into the two electrolytic compartments. A wire connected in parallel to the longitudinal section of the compartment served to link the DC power source with the multimeter, allowing for the measurement of the electrical parameters. The relays were utilized to control the periodic ON/OFF function of the DC power. By regulating the ON/OFF time of the power, the relay facilitated the formation of a pulse-type ECT system in the soil. A multimeter was installed by wire between the DC power and the soil reaction device for monitoring the current. The study evaluated a total of nine experimental groups and compared the performance of different pulse voltage supply types, the presence of biochar-based PRB, voltage gradients, treatment durations, and interval periods of pulse voltage in ECT; the details of each experimental group are presented in Table 3.

Table 3. Experimental parameters of the ECT programs.

No.	PRB	Power Type	Power On/Off Interval Periods	Voltage Gradient (V/cm)	Treatment Time (h)
DC0-S	no PRB	DC power	Constant voltage	4	72
DC0	no PRB	DC power	Constant voltage	4	204
DCb	biochar	DC power	Constant voltage	4	204
DC2b	biochar	DC power	Constant voltage	2	204
PCb-3 s	biochar	PC power	3 s/3 s	4	204
PCb-1 m	biochar	PC power	1 min /1 min	4	204
PCb-30 m	biochar	PC power	30 min/30 min	4	204
PCb-6 h	biochar	PC power	6 h/6 h	4	204
PC2b-6 h	biochar	PC power	6 h/6 h	2	204

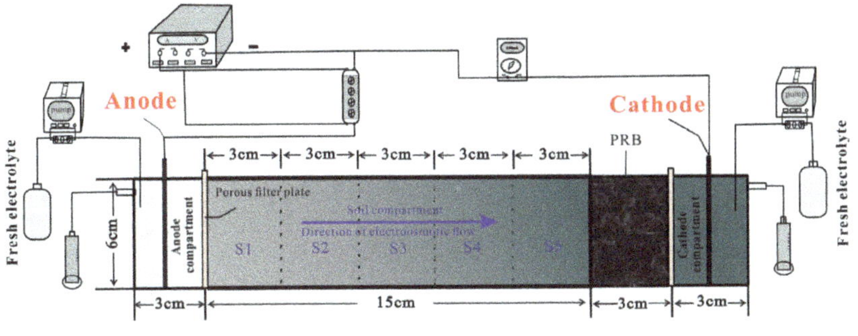

Figure 1. Experimental setup for pulse voltage power system.

2.3. Experimental Procedure

In this study, the soil samples were loaded into an experimental reactor designed for electrochemical experiments and were compacted layer by layer. Each layer of soil was compressed uniformly to minimize the presence of empty spaces between soil particles, ensuring consistent soil properties in the direction of the applied voltage. The compaction process continued until the soil surface was level with the drainage hole. To incorporate biochar, it was wrapped in filter paper and placed in the designated biochar compartment within the reactor. The anodic compartment was filled with deionized water while the cathodic compartment was filled with a 0.3 mol/L acetic acid solution. These electrolytes were intended to induce the desired electrochemical reactions and facilitate the electrochemical process within the experimental system. Graphite electrodes were used in the experimental setup and positioned in the respective electrolytic compartments [34,35]. To prevent direct contact between the electrodes and the soil samples, porous plexiglass plates and filter paper were employed as separation barriers. These materials allowed for the passage of electric current while maintaining a physical separation between the electrodes and the soil samples. This setup ensured the effective and safe application of the ECT to the soil samples. Before treatment, the soil samples were saturated with deionized water for 12 h. This step was conducted to ensure that the soil samples reached their maximum moisture capacity. By fully saturating the soil, the water content in the soil pores and voids was optimized.

All experiments were conducted at room temperature. The duration of the experiment was set to 72 h and 204 h, respectively. The soil compartment had a total length of 15 cm, with the biochar compartment and electrolytic compartments having a length of 5 cm. The experiment was designed to apply voltage gradients of 2 V/cm and 4 V/cm, corresponding to voltages of 42 V and 84 V, respectively. The experimental group employed a pulse voltage power supply system controlled by relays. Zhou et al. showed that the pulse interval of 30 min ON/30 min OFF achieved the highest removal efficiency for fluoride removal in soil [31], and Mu'azu et al. showed that heavy metal removal efficiency increased with the pulse duty cycle and voltage gradient increases [36]. Based on previous studies on the pulse duty cycle in the PECT system, the effects of different pulse time gradients on heavy metal removal efficiency and energy consumption were compared using a 1:2 pulse duty cycle. The pulse time gradient included seconds, minutes, and hours. It consisted of four interval periods of pulse voltage: 3 s ON/3 s OFF, 1 min ON/1 min OFF, 30 min ON/30 min OFF, and 6 h ON/6 h OFF. A peristaltic pump was used to introduce fresh electrolytes into the bottom of the electrolytic compartments. The electrolyte was regularly replenished to compensate for electrolytic and electroosmotic losses, which helped neutralize the accumulated anions in the electrolyte and regulate the pH of the electrolyte. Throughout the experiment, the pH of the anolyte and catholyte was measured every three hours. Current data from the multimeter were recorded, and the electroosmotic flow was measured every 12 h. The cathodic and anodic waste liquid was collected into storage bottles daily, and the concentration of Pb in the liquid was measured using an atomic absorbance spectrometer. After ECT, soil samples labeled S1, S2, S3, S4, and S5 were extracted. These soil samples were naturally dried for subsequent testing. The removal efficiency of lead in the soil was determined using Formula (1) [37]:

$$R_{ave} = \frac{\sum_{k=1}^{n} (C_{ini} - C_{sk})/C_{ini}}{n} * 100\% \tag{1}$$

where the concentration of the initial soil sample is C_{ini}; the concentrations of S1, S2, S3, S4, and S5 region are denoted as C_{sk}, with k = 1, 2, 3, 4 and 5; n is the number of sample points at the same distance to the anode and is 5 in this study.

2.4. Instrumental Analysis

During the ECT process, all ECT experiments were carried out using a DC power (DP310, MAISITAIKE, Dongguan, China) supply to maintain a constant voltage gradient. The current intensity during the experiments was measured using a digital multimeter (DEM12, DELIXI, Wuhu, China). A multi-channel peristaltic pump (BT100-1L, LANGE, Shanghai, China) was used to supply fresh electrolytes to the electrolytic compartments at a steady flow rate of 505.4 μL/min. The evaluation of electroosmotic flow was conducted by connecting graduated cylinders to the cathodic compartment. The pH measurements of the soil samples and electrolyte were carried out at a temperature of 25 °C using a pH meter (PB-21, SARTORIUS, Gottingen, Germany). Before measurement, calibration of the pH meter with standard solutions was required. Following the standard for pH, 10 g of dry soil was blended with 25 milliliters of distilled water and agitated for 5 min. Subsequently, the suspension was allowed to settle for 30 min, and the pH of the supernatant solution was measured [38,39]. The total content of Pb in the soil samples was extracted using a microwave-assisted digestion system with a mixture of HNO_3, HF, and $HClO_4$, following the method described by Carignan and Tessier in 1988 [34]. The concentration of Pb within soil samples was measured by the X-ray Fluorescence Spectrometer (Soil handheld XRF Analyzer Explorer 9000, TIANRUI, Kunshan, China). In the measurement process, the soil samples need to undergo high-intensity pressing for 15 s using a bead machine. During measurement, it was crucial to ensure the stable placement of the instrument and verify that the probe had not been contaminated. The concentration of Pb ions in the liquid was measured by an atomic absorbance spectrometer (A3-AFG-12, PUXI, Beijing, China) [40]. When measuring the concentration of Pb in the solution, it was essential to preheat the instrument for 20 min. Afterward, the fume hood was opened, and the acetylene gas ignited. A visual inspection of the flame color was necessary to confirm its normalcy. Before measurement, calibration of the instrument was mandatory, and blank samples were set up. The preparation of standard samples should be accurate, ensuring reasonable concentration levels. Moisture content was determined through thermogravimetric analysis using an electric vacuum drying oven (GRX-12, SHANGHAIJINGHONG, Shanghai, China) for 24 h. The electrical conductivity of the soil samples was measured using a soil-to-water ratio of 1:2.5 and a calibrated conductivity meter (DDS-307A, REX, Shanghai, China) [40].

3. Results

3.1. Electrical Current

Electrical current serves as an important indicator for assessing the movement of ions within soil pores [4]. All ECT experiments displayed a characteristic pattern of initially increasing and then decreasing electrical current. Figure 2a reveals that constant voltage coupled with biochar elicited a high current response. In DCb and DC2b, the peak value of electrical current appeared relatively early and was higher than in DC0 and PC2b-6 h. This suggested that the presence of multiple movable ions in biochar increased the overall number of ions capable of movement in the circuit, while the continuous and stable electric field generated by direct current accelerated ion movement in the soil. Especially for the DCb group, the current quickly reached a peak value in a short period, and the peak value was much higher than other groups. The DCb group used a voltage gradient of

4 V/cm with biochar. A high voltage gradient significantly increased the current density in soil pore water in a short period, and the conductive ions inherent in the biochar also increased the initial current density. A majority of ions in the soil and biochar quickly migrated to the cathodic region under the influence of a high field strength. This resulted in a reduction in the quantity of mobile ions in the soil pore water, leading to a significant downward trend in the current of the DCb group after 12 h [41]. The current stabilizes when the count of mobile ions remains constant. Figure 2b demonstrates that, initially, the current peak value exhibited the following order: PCb-6 h > PCb-30 m > PCb-1 m > PCb-3 s. A prolonged pulse time gradient enabled the continuous energy output for mobile ions over a specific duration, thereby activating the mobile ions adsorbed on the soil surface and biochar. The increase in the count of mobile ions results in an augmentation of current [2,25,42,43]. Lastly, the digital multimeter detected extremely small currents in the PCb-6 h group and PCb-30 m group during the power outage period. This phenomenon bore a similarity to the induced polarization method commonly employed in mineral exploration. The soil, characterized by its moisture content and the presence of heavy metals, can be considered a polarizable material. The entire soil matrix, in conjunction with the power source, effectively forms a closed electrical circuit. Notably, when a pulsed electric field was applied and subsequently turned off, a small electric current continued to flow within a certain range. This observation signified that when an electric current passed through, energy was stored within the soil medium. Typically, after the applied electric field was deactivated, this stored energy was released by maintaining the flow of electric current [44].

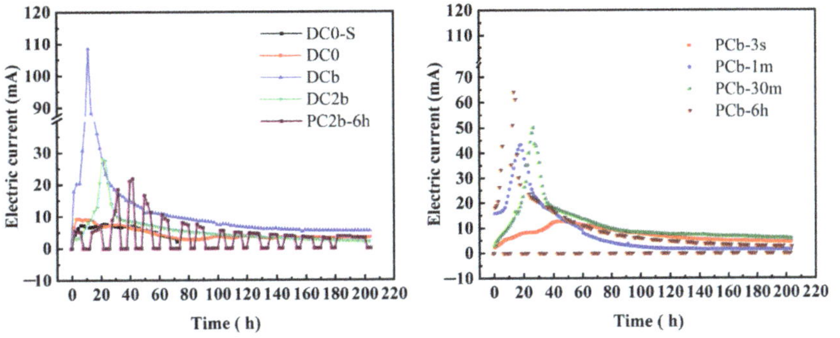

(a) (b)

Figure 2. Electric current during electrochemical treatment (**a**,**b**).

3.2. Electroosmotic Flow

In many studies, it has been observed that the electroosmotic flow (EOF) generally occurs from the anode to the cathode during treatment when the porous matrix has a negative surface charge and a low-intensity direct electric current is applied [45]. The principle of EOF was described by the Helmholtz–Smoluchowski theory. The electroosmotic flow rate q_{eo} [m^3/s] is calculated using Equation (2) [7,46]:

$$q_{eo} = k_{eo} I \sigma \tag{2}$$

where k_{eo} ((m^2/s)/V) is the electroosmotic permeability coefficient, I is the electric current, and σ is the conductivity (Siemens/cm).

As shown in Figure 3a, the electroosmotic flow in DO0 exhibited a sharp increase, reaching a significantly high value of 2499 mL compared to other groups. The observed sequence of electroosmotic flow was DO0 > DCb > DC2b > PC2b-6 h, indicating that high electroosmotic flow could be generated under two conditions: a voltage gradient of 4 V/cm and a constant voltage power supply. In Figure 3b, during the initial stages, there was a slower rise in electroosmotic flow in the PCb-3 s group and PCb-30 m group compared to the other groups. This phenomenon can be inferred as a result of the rapid increase in current during the early stage, causing an acceleration in the electroosmotic flow. From 36 h to 65 h, the electroosmotic flow in the PCb-3 s group exhibited a small fluctuation, initially increasing and then decreasing. This was because the electroosmotic flow showed a reversal during the ECT process [47]. Based on the current trends, it could be inferred that the soil surface potential changed from negative to positive, which further contributed to the reversal of the electroosmotic flow under higher applied voltage [4,48]. Additionally, the intermittent interruption of the electric current during this period led to the cessation of electroosmotic flow. After 65 h, the electroosmotic flow gradually increased, and the electroosmotic flow rate followed the order of PCb-1 m > PCb-3 s > PCb-30 m > PCb-6 h. The underlying cause of this phenomenon can be explained by the trend in current distribution. When the current intensity was high, it led to reverse electroosmotic flow, which means that electroosmotic flow occurred from the cathode to the anode. However, the overall current trend consistently exhibited a prevailing anode-to-cathode flow tendency, which was more pronounced than the counterflow of electroosmotic fluid. Consequently, this caused a higher electroosmotic flow in the PCb-1 m group compared to the PCb-30 m and PCb-6 h groups. In contrast, the overall current intensity in the PCb-3 s group was weaker than that in the PCb-1 m group, resulting in a lower electroosmotic flow in the PCb-3 s group compared to the higher electroosmotic flow in the PCb-1 m group.

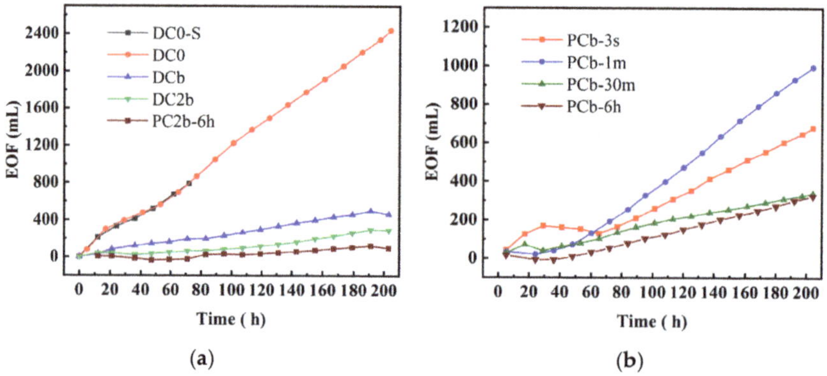

Figure 3. Electroosmotic flow during electrochemical treatment (**a**,**b**).

3.3. Moisture Content of Treated Soil

The moisture content of treated soil is a crucial factor in facilitating electromigration and electroosmosis, which are important mechanisms for the migration of heavy metals during the ECT process [49]. A previous survey indicated that the migration of heavy metals could occur when the soil moisture content in the treatment device exceeds 15% [48]. Changes in the moisture content of the treated soil in different regions of the experiment are shown in Figure 4a–e. The initial soil moisture content was 27.4%. In the presence of constant voltage power (DC0, DCb, DC2b) at 204 h, the variation in moisture content of treated soil was influenced by the processing time, voltage gradient, and the presence of biochar. Biochar has a water-buffering capacity and absorbs some water from adjacent soil areas. It also acts as a barrier to water molecules, limiting their penetration into the soil area. As a result, the soil moisture content of the DCb with biochar was lower than DC0. Furthermore, the generation of high electroosmotic flow, which was driven by a high voltage gradient, led to the transfer of moisture from the soil to the catholyte compartment. This was one of the reasons why the soil moisture content of DC2b was higher than that of DCb. The soil moisture content of DC0-S at 72 h of treatment time exhibited inhomogeneous distribution characteristics compared to DC0 due to the short processing time and incomplete physical and chemical reactions. Additionally, from the distribution of moisture content in the treated soil for DC2b and PC2b-6 h, it could be observed that the influence of the power supply type on the moisture content of treated soil was not significant under a voltage gradient of 2 V/cm. This phenomenon might be attributed to the comparatively lower voltage gradients in the DC2b and PC2b-6 h groups in comparison to the other groups. When the applied voltage was low, the electroosmotic flow rate within the soil pores tended to be more gradual, resulting in fewer concentration and polarization effects. Consequently, this led to a more consistent distribution of soil moisture content after treatment, with no significant variations.

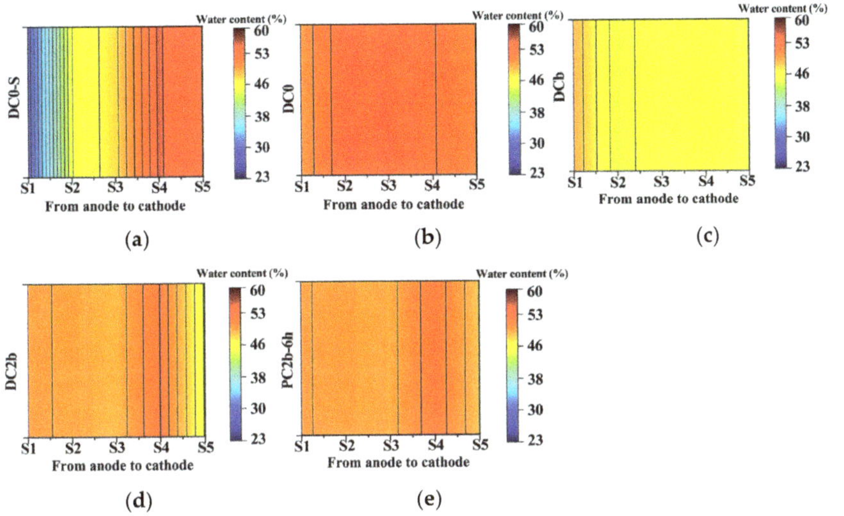

Figure 4. Moisture content of treated soil (**a**–**e**).

3.4. Electrolyte pH

Figure 5a–d illustrates the pH variation of the anolyte and catholyte. Numerous studies have investigated the impact of electrolyte pH during the ECT for treating contaminated soil [50]. Generally, in the absence of acidic additives for pH control, the electrolytic reaction in the ECT device leads to different levels of pH at each end of the soil compartment. The catholyte exhibits an alkaline pH, while the anolyte shows an acidic pH [51]. This difference in pH is caused by hydrolysis, which involves the generation of hydroxide ions at the cathode ($2H_2O + 4e^- \rightarrow 2H_2(g) \uparrow + 4OH^-$) and hydrogen ions at the anode ($2H_2O - 4e^- \rightarrow O_2(g) \uparrow + 4H^+$) [37]. Figure 5a,c presents the pH distribution of the anolyte over time for all electrochemical experiments. Initially, the pH of the anolyte decreased and then increased, ultimately stabilizing within a narrow range from 1.83 to 4.08. During the initial period of power on, the anolyte pH of the DCb, DC2b, PCb-1 m, PCb-30 m, and PCb-6 h experienced a significant decrease, which corresponded to the increased production of hydrogen ions when the current intensified. In Figure 5b,d, it could be observed that the pH of the catholyte of the DCb group and PCb-6 h group reached the highest value of pH at 12 h and 7 h, respectively, aligning with the enhanced production of hydroxide ions when the current substantially increased.

3.5. Soil pH and Soil Conductivity

The soil pH variation after ECT is depicted in Figure 6a,b. Initially, the soil pH of the Pb-contaminated soil was 5.77. After ECT progress, the soil samples were divided into five sections, labeled S1 to S5, based on their distance from the cathode. In Figure 6a, it can be observed that the pH of treated soil remained stable throughout the entire zone, ranging from 4.71 to 5.35. This stability was attributed to the adjustment of catholyte pH using a 0.3 mol/L acetic acid solution. The hydrogen ions provided by the acetic acid neutralize the hydroxide ions produced in the catholyte during electrolysis. Furthermore, the hydroxide ions generated in the middle of the soil combine with the hydrogen ions produced at the anode, resulting in a decrease in soil pH. This, in turn, led to a more uniform distribution of pH throughout the entire soil region. Figure 6b demonstrates the distribution of soil pH from S1 to S5 across the four experimental groups of pulse voltage. In the PCb-1 m and PCb-6 h treatment, the soil pH range fell between 5.04 and 5.97. In the PCb-3 s and PCb-30 m treatments, the soil pH range was between 4.51 and 5.11.

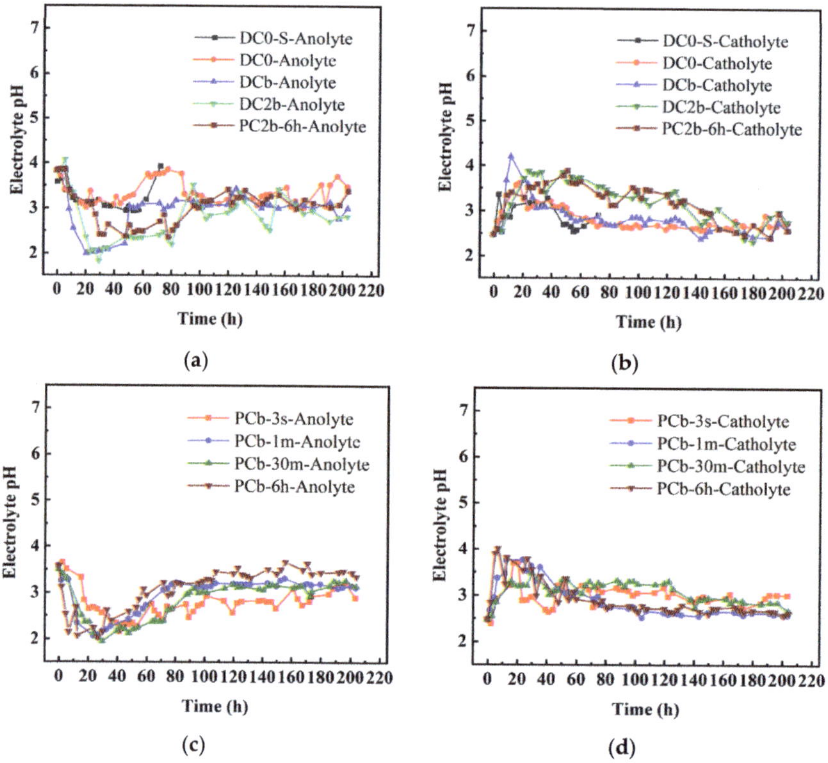

Figure 5. Electrolyte pH during electrochemical treatment (**a**–**d**).

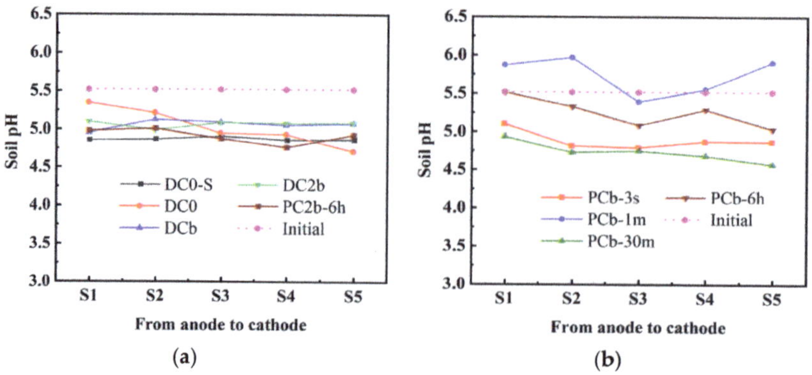

Figure 6. Soil pH of treated soil and initial soil pH (**a**,**b**).

In Figure 7a,b, the conductivity of soil after the ECT is illustrated. The initial conductivity of the soil was 86.45 μs/cm. Except for the PCb-6 h treatment, which had a high value in the S5 region, the distribution of electrical conductivity in the treated soil was homogeneous. In the experiments of pulse voltage at 204 h, the electrical conductivity ranged from 8.6 μs/cm to 25.75 μs/cm. The conductivity of the soil was influenced by the presence of conductive ions. The addition of acetic acid as the catholyte neutralizes hydroxide ions in the soil through diffusion, promoting the desorption of soluble ions that were adsorbed on the soil surface [4,40,52]. These ions then migrate to different soil areas under the influence of the electric field, resulting in a decrease in conductive ions in the soil. However, there were significant differences in the distribution of electrical conductivity in the treated soil of the DC0-S. Based on these observations, it could be concluded that the short electric treatment time affected the chemical and physical reactions of various substances in the soil. When the treatment time was short, these reactions could be incomplete, leading to inconsistencies in the distribution pattern of electrical conductivity across different soil regions.

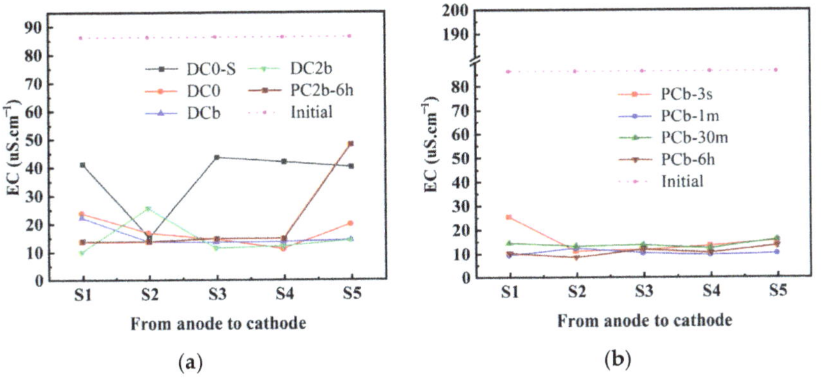

Figure 7. Conductivity of treated soil after electrochemical treatment (**a**,**b**).

3.6. Residual Content of Pb in Collected Catholyte

Since Pb ions primarily migrate to the cathode through electromigration and electroosmotic flow, the monitoring of Pb ions was carried out in the catholyte, while the content of Pb ions in the waste liquid discharged from the anolyte compartment was not detected. Figure 8a illustrated the trends in the content of Pb ions in the catholyte during the treatment of constant voltage. Initially, a substantial number of Pb ions were discharged from the soil on the first day, followed by a slow migration in the DC0-S and DC0 treatment. On the other hand, the voltage gradient had a slightly positive effect, with the content of Pb ions in the catholyte being higher under a voltage gradient of 4 V/cm compared to 2 V/cm during the early stages of the experiment. In the pulse voltage systems, the trends in the content of Pb ions in the catholyte discharge liquid differed from those in the constant voltage systems. The initial content of Pb ions in the catholyte was low, and only trace amounts were discharged after the third day in the PC2b-6 h group. Figure 8b demonstrates that the PCb-6 h group reached the peak of Pb ion release two days earlier than the PCb-30 m group. However, the overall trend was to increase first and then decrease. The results revealed that the PRB had a notable and negative effect on the cumulative content of Pb ions in the catholyte. The biochar material utilized in the PRB effectively adsorbed Pb ions, impeding their migration from the PRB compartment to the catholyte [53,54]. These findings highlighted the significant impact of the PRB on the migration of Pb ions from the PRB compartment to the catholyte compartment. The presence of the PRB impeded the Pb ions movement, causing them to accumulate in the PRB and be gradually released over time. Furthermore, the voltage gradient also played a role in influencing the content of Pb ions in the catholyte. The specific mechanisms behind these observations require further analysis and investigation.

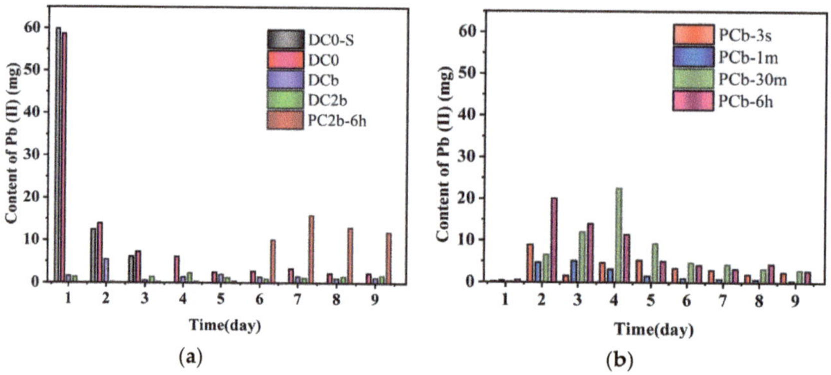

Figure 8. Residual Pb ion content (mg) in collected catholyte over time (day) (**a**,**b**).

3.7. Removal Efficiency of Pb

The removal efficiency of Pb is affected by various experimental conditions [7,47,55,56]. The experiment investigated five variable conditions, including the type of power supply (constant voltage or pulse voltage), the presence of a PRB made of biochar, different voltage gradients, different treatment times, and different interval periods of pulse voltage. The experimental results in Figure 9a present the average removal efficiency of Pb from soil. Firstly, it was observed that the average removal efficiency of Pb in the continuously energized group with constant voltage (DC2b) and the group subjected to pulse voltage (PC2b-6 h) with a 6 h interval period was essentially the same. The mechanism of Pb migration facilitated by biochar as a PRB was analyzed by comparing the features observed in DC0 and DCb. In general, biochar not only adsorbed migrated Pb ions but also enhanced the current intensity in the soil circuits. This increased current intensity promoted the migrated rate of Pb, resulting in a higher removal efficiency for the DCb group compared to the DC0 group. Furthermore, when biochar was combined with a voltage gradient of 4 V/cm, the DCb group exhibited higher removal efficiency for Pb compared to the DC2b group, which operated under a voltage gradient of 2 V/cm. Previous research has shown that long-term treatment is beneficial for improving the removal efficiency of heavy metals [57]. In this study, the removal efficiency of DC0 was 17% higher than that of the corresponding group with short treatment times (DC0-S). Furthermore, the removal efficiency of Pb was remarkably high at a higher voltage gradient of 4 V/cm and a treatment time of 204 h, using biochar as the PRB in the DCb, PCb-3 s, PCb-1 m, PCb-30 m, and PCb-6 h groups. The corresponding removal efficiency for these groups was 94.1%, 89.5%, 91%, 92.9%, and 91.9%, respectively. There were differences in the average removal efficiency among the PCb-3 s, PCb-1 m, PCb-30 m, and PCb-6 h groups in the pulse voltage system. Specifically, the average removal efficiency of Pb in the PCb-1 m, PCb-30 m, and PCb-6 h groups was 1.5%, 3.4%, and 2.4% higher, respectively, than that of the PCb-3 s group.

The removal efficiency of Pb in each soil region from S1 to S5 is presented in Figure 9b. In the S1 region of soil, the removal efficiency of Pb exceeded 97% in the groups at a treatment time of 204 h. The removal efficiencies of the nine experimental groups from S1 to S5, were high, surpassing 80%. Notably, the removal efficiency of the S5 region was a crucial factor that influenced the average removal efficiency. In the S5 region of the DC0-S groups, there was a noticeable inflection point, attributed to the difference in ECT duration. Most experimental groups underwent a 204 h ECT, whereas the DC0-S group had a shorter treatment duration of 72 h. Unlike the other experimental groups, applying a high voltage gradient over a shorter treatment period caused mobile charged ions within the soil to move in the opposite direction of the electrical current within the soil pores. Specifically, positively charged lead ions migrated toward the cathode direction, leading to their gradual accumulation in the S5 region. However, the shorter treatment duration was not conducive to the lead ions in the S5 region migrating into the catholyte compartment, resulting in a significant accumulation of lead ion within the S5 region. This observation suggests that the system of pulse voltage was effective in desorbing heavy metal ions from the soil surface; the brief power-off period in the system of pulse voltage reduced the polarization effect in the soil [4,38,58,59]. Consequently, it allowed the acid ions from the catholyte to gradually diffuse toward the middle of the soil within the fluid present in the soil pores. This diffusion process was beneficial for decreasing the alkalinity in the portion of the soil middle and facilitated the desorption and migration of lead, which was adsorbed on the soil surface, towards the catholyte during the subsequent cycle of the electric field.

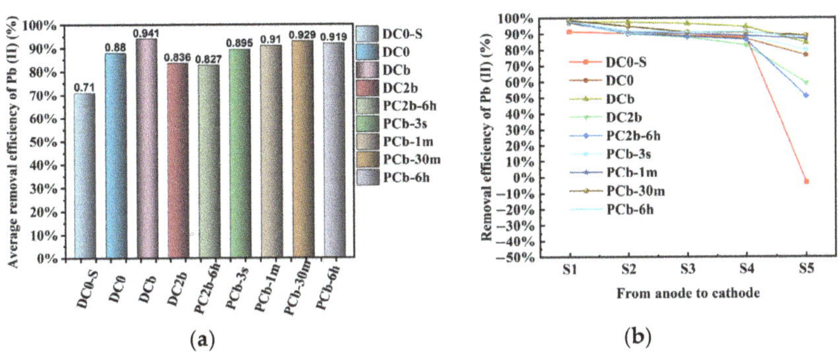

Figure 9. The average removal efficiency (%) of Pb in soil and the removal efficiency (%) of Pb in each soil region (**a**,**b**).

3.8. Electrical Energy Consumption

The electrical energy consumption can be calculated based on the applied current, voltage, and cross-sectional area of the soil in the vertical current direction. The calculation formula is as follows [60]:

$$E = \frac{1}{V_s}\int UI\,dt \tag{3}$$

where E is the energy consumption per unit volume of soil (kWh/m^3), V_s represents the soil volume (m^3), U represents the voltage (V), I represents the current (A), and t represents the treatment time (h).

The energy consumption and lead removal efficiency in all experiment groups are presented in Table 4. The electrical energy consumption during ECT using corn straw biochar as a PRB system is shown in Figure 10a,b. In Figure 10a, the energy consumption of the DC0-S, DC0, DCb, DC2b, and PC2b-6 h groups was 99.6 kWh/ m^3, 213.9 kWh/m^3, 563.2 kWh/m^3, 133.5 kWh/m^3, and 68.4 kWh/m^3, respectively. The DCb group was significantly higher than the other experiments. This was likely due to the high electric current observed in the group with a high voltage gradient and biochar used as PRB under constant voltage. This indicated that a high voltage gradient and treatment time were important factors, contributing to high energy consumption. It is noteworthy that in Figure 10a, the PC2b-6 h group exhibited lower energy consumption compared to the DC2b group. The result suggests that the interval of power outage in the experiment reduced the effective treatment time, which was beneficial for reducing energy consumption. As shown in Figure 10b, the energy consumption of PCb-6 h, PCb-30 m, PCb-1 m, and PCb-3 s was 296 kWh/m^3, 271.6 kWh/m^3, 197.9 kWh/m^3, and 177.6 kWh/m^3. The order of energy consumption in the experimental group was as follows: PCb-6 h > PCb-30 m > PCb-1 m > PCb-3 s. Compared to the traditional experiment (DCb) using a constant voltage coupled with PRB for treatment of Pb-contaminated soil, the groups using pulse voltage power, including PCb-3 s, PCb-1 m, PCb-30 m, and PCb-6 h, reduced energy consumption by 68.5%, 64.9%, 51.8%, and 47.4%, respectively.

Table 4. The removal efficiency of Pb (%) and energy consumption in all experiment groups.

Exp.	S1 (%)	S2 (%)	S3 (%)	S4 (%)	S5 (%)	Average Removal (%)	Energy Consumption (kWh/m^3)
DC0-S	91.5	90.2	88.9	87.9	−3.4	71	99.6
DC0	97	90.9	88.6	86.9	76.6	88	213.9
DCb	97.6	97.4	96.5	94.6	84.4	94.1	563.2
DC2b	98.1	89.9	87.8	82.8	59.2	83.6	133.5
PCb-3 s	97.6	90.6	90.4	88.8	80.2	89.5	177.6
PCb-1 m	97.4	91.6	90.1	88.8	86.9	91	197.9
PCb-30 m	98.4	94.9	91.3	91.1	88.9	92.9	271.6
PCb-6 h	97.9	91.7	91.3	91.4	87.5	91.9	296
PC2b-6 h	97.4	90.2	88.3	86.4	51.1	82.7	68.4

The soil sample region was delineated by the distance from the anode toward the cathode, designated as S1, S2, S3, S4, and S5.

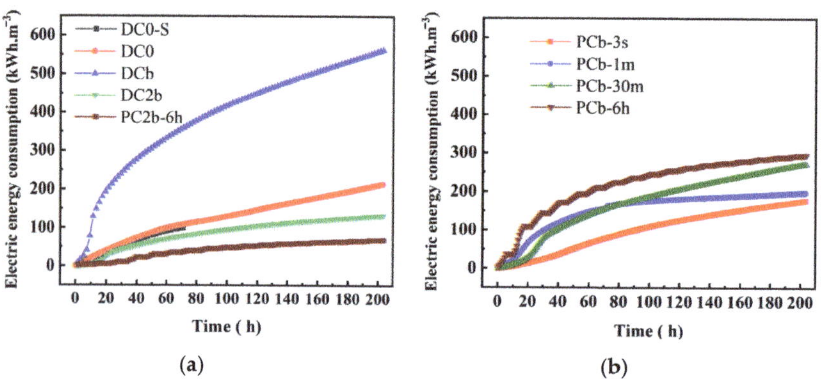

Figure 10. Electric energy consumption during electrochemical treatment (**a**,**b**).

The utilization of the pulse voltage involved integrating different electrodynamic enhancement technologies [61–64], such as the addition of acetic acid in the catholyte, the establishment of a PRB, and the application of a voltage gradient of 4 V/cm. These approaches collectively aimed to reduce the polarization effects that typically occur in the traditional ECT [13,40,43,59]. Through this process, they enhanced the outcomes in Pb removal by stimulating the desorption of Pb from the soil surface. Moreover, these methods aided in reducing the electrical conductivity of soil.

4. Conclusions

This study confirmed the high efficiency of the PECT system coupled with the PRB in lead removal from soil. The LREs were found to be 89.5% (PCb-3 s), 91% (PCb-1 m), 92.9% (PCb-30 m), and 91.9% (PCb-6 h), with particularly remarkable results in the S1 region, where removal efficiency reached 97%. Furthermore, under the same pulse duty cycle conditions, shorter pulse time gradient in seconds and minutes (3 s ON/3 s OFF and 1 min ON/1 min OFF) demonstrated lower current requirements during the ECT process, resulting in reducing energy consumption. Specifically, energy consumption in the experiments for PCb-3 s, PCb-1 m, PCb-30 m, and PCb-6 h was 177.6 kWh/m^3, 197.9 kWh/m^3, 271.6 kWh/m^3, and 296 kWh/m^3, which represented a significant reduction when compared to the constant voltage coupled with PRB experiment (DCb), where energy consumption reached as high as 563.2 kWh/m^3 under some voltage gradients.

Consequently, the energy consumption was reduced, with percentage decreases of 68.5%, 64.9%, 51.8%, and 47.4%, respectively.

In contrast to previous PECT methods for treating lead-contaminated soil, the present research revealed that the PECT system coupled with PRB, designed with shorter pulse time gradients, effectively achieved high LRE while significantly reducing energy consumption and shortening the treatment duration. Additionally, concerning the treatment of secondary pollution introduced by the catholyte, the study highlighted the effective adsorption capabilities of corn straw biochar. The temporal evolution of lead content in the catholyte indicated that the optimal time to replace biochar falls within the first days of pulse treatment.

In summary, the present data provided an innovative ECT method for lead-contaminated soil, offering high removal efficiency, lower time requirements, and energy savings. However, further research is necessary to investigate the optimal strategies for remediating different soil types and using modified biochar to achieve maximal adsorption of lead ions from the catholyte.

Supplementary Materials: The following supporting information can be downloaded at: https://www.mdpi.com/article/10.3390/toxics11120961/s1, Table S1: The comparison of lead removal efficiency and energy consumption under different electrochemical treatment conditions.

Author Contributions: Conceptualization, L.Z., J.H. and Y.Z. (Yinyin Zhang); methodology, L.Z. and Y.Z. (Yinyin Zhang); software, Y.Z. (Yinyin Zhang); validation, Y.Z. (Yinyin Zhang); formal analysis, Y.Z. (Yinyin Zhang); investigation, Y.Z. (Yinyin Zhang); resources, L.Z., Y.Z. (Yuyan Zhao) and Q.W.; data curation Y.Z. (Yinyin Zhang); writing—original draft preparation, Y.Z. (Yinyin Zhang); writing—review and editing, L.Z. and Y.Z. (Yinyin Zhang); visualization, Y.Z. (Yinyin Zhang); supervision, L.Z., J.H. and Y.Z. (Yuyan Zhao). All authors have read and agreed to the published version of the manuscript.

Funding: This research was funded by Lixin Excellent Young Teacher Training Program in Jilin University (419080520490).

Institutional Review Board Statement: Not applicable.

Informed Consent Statement: Not applicable.

Data Availability Statement: Some or all data, models, or codes that support the findings of this study are available from the corresponding author upon reasonable request.

Conflicts of Interest: The authors declare no conflict of interest.

References

1. Amrate, S.; Akretche, D.E.; Innocent, C.; Seta, P. Removal of Pb from a calcareous soil during EDTA-enhanced electrokinetic extraction. *Sci. Total Environ.* **2005**, *349*, 56–66. [CrossRef] [PubMed]
2. Lin, H.; Wang, Z.; Liu, C.; Dong, Y. Technologies for removing heavy metal from contaminated soils on farmland: A review. *Chemosphere* **2022**, *305*, 135457. [CrossRef] [PubMed]
3. Hu, W.; Cheng, W.; Wen, S.; Kang, N. Revealing underlying mechanisms affecting electrokinetic remediation of an artificially Cu- and Pb-contaminated loess using the external regulatory system with adsorbent. *Front. Mater.* **2022**, *9*, 967871. [CrossRef]
4. Taneja, S.; Karaca, O.; Haritash, A.K. Treatment of Pb-contaminated soil by electrokinetics: Enhancements by varying voltage, chelant, and electrode material. *J. Geochem. Explor.* **2023**, *250*, 107240. [CrossRef]
5. Shareef, Z.N.; Yatiya, S.; Sulaiman, M.M. Enhanced of electro-kinetic technology for removal of lead ions from silty clay contaminated soil using EDTA and multi anode electrode. *S. Afr. J. Chem. Eng.* **2023**, *45*, 120–126. [CrossRef]
6. Wu, Y.; Wang, S.; Cheng, F.; Guo, P.; Guo, S. Enhancement of electrokinetic-bioremediation by ryegrass: Sustainability of electrokinetic effect and improvement of n-hexadecane degradation. *Environ. Res.* **2020**, *188*, 109717. [CrossRef] [PubMed]
7. Kim, D.-H.; Jo, S.-U.; Yoo, J.-C.; Baek, K. Ex situ pilot scale electrokinetic restoration of saline soil using pulsed current. *Sep. Purif. Technol.* **2013**, *120*, 282–288. [CrossRef]
8. Acar, Y.B.; Gale, R.J.; Alshawabkeh, A.N.; Marks, R.E.; Puppala, S.; Bricka, M.; Parker, R. Electrokinetic remediation: Basics and technology status. *J. Hazard. Mater.* **1995**, *40*, 117–137. [CrossRef]
9. Bessaim, M.M.; Missoum, H.; Bendani, K.; Laredj, N.; Bekkouche, M.S. Sodic-saline soil remediation by electrochemical treatment under uncontrolled pH conditions. *Arab. J. Geosci.* **2020**, *13*, 199. [CrossRef]

10. Kim, D.-H.; Jo, S.-U.; Choi, J.-H.; Yang, J.-S.; Baek, K. Hexagonal two dimensional electrokinetic systems for restoration of saline agricultural lands: A pilot study. *Chem. Eng. J.* **2012**, *198–199*, 110–121. [CrossRef]
11. Altin, A.; Degirmenci, M. Lead (II) removal from natural soils by enhanced electrokinetic remediation. *Sci. Total. Environ.* **2005**, *337*, 1–10. [CrossRef] [PubMed]
12. Xu, L.; Yu, C.; Mao, Y.; Zong, Y.; Zhang, B.; Chu, H.; Wu, D. Can flow-electrode capacitive deionization become a new in-situ soil remediation technology for heavy metal removal? *J. Hazard. Mater.* **2021**, *402*, 123568. [CrossRef] [PubMed]
13. Niinae, M.; Nishigaki, K.; Aoki, K. Removal of Lead from Contaminated Soils with Chelating Agents. *Mater. Trans.* **2008**, *49*, 2377–2382. [CrossRef]
14. Ryu, B.G.; Park, G.-Y.; Yang, J.-W.; Baek, K. Electrolyte conditioning for electrokinetic remediation of As, Cu, and Pb-contaminated soil. *Sep. Purif. Technol.* **2011**, *79*, 170–176. [CrossRef]
15. Benamar, A.; Ammami, M.; Song, Y.; Portet-Koltalo, F. Scale-up of electrokinetic process for dredged sediments remediation. *Electrochim. Acta* **2020**, *352*, 136488. [CrossRef]
16. Chen, R.; Zhou, L.; Wang, W.; Cui, D.; Hao, D.; Guo, J. Enhanced Electrokinetic Remediation of Copper-Contaminated Soil by Combining Steel Slag and a Permeable Reactive Barrier. *Appl. Sci.* **2022**, *12*, 7981. [CrossRef]
17. Wang, J.; Gao, W.; Zhu, J.; Yang, Y.; Niu, Y. Remediation of Cd and Cu Contaminated Agricultural Soils near Oilfields by Biochar Combined with Sodium Humate-Wood Vinegar. *Agronomy* **2023**, *13*, 1009. [CrossRef]
18. Zhu, Z.; Kong, Y.; Yang, H.; Tian, Y.; Zhou, X.; Zhu, Y.; Fang, Z.; Zhang, L.; Tang, S.; Fan, Y. Effects of Pretreatment and Polarization Shielding on EK-PRB of Fe/Mn/C-LDH for Remediation of Arsenic Contaminated Soils. *Nanomaterials* **2023**, *13*, 325. [CrossRef]
19. Silva, K.N.; Henrique, J.M.; Vilar, V.J.; A Martínez-Huitle, C.; dos Santos, E.V. Cork-based permeable reactive barriers coupled to electrokinetic processes for interrupting pollutants reaching groundwater: A case study of lead-contaminated soil. *J. Chem. Technol. Biotechnol.* **2022**, *97*, 2861–2870. [CrossRef]
20. Sun, Z.; Zhao, M.; Chen, L.; Gong, Z.; Hu, J.; Ma, D. Electrokinetic remediation for the removal of heavy metals in soil: Limitations, solutions and prospection. *Sci. Total. Environ.* **2023**, *903*, 165970. [CrossRef]
21. Yuan, L.; Xu, X.; Li, H.; Wang, Q.; Wang, N.; Yu, H. The influence of macroelements on energy consumption during periodic power electrokinetic remediation of heavy metals contaminated black soil. *Electrochim. Acta* **2017**, *235*, 604–612. [CrossRef]
22. Usman, A.K.; Mu'azu, N.D.; Lukman, S.; Essa, M.; Bukhari, A.A.; Al-Malack, M.H. Removal of Lead and Copper from Contaminated Mixed Clay Soils Using Pulsed Electrokinetics. *Soil Sediment Contam. Int. J.* **2020**, *29*, 465–480. [CrossRef]
23. Ryu, B.-G.; Yang, J.-S.; Kim, D.-H.; Baek, K. Pulsed electrokinetic removal of Cd and Zn from fine-grained soil. *J. Appl. Electrchem.* **2020**, *40*, 1039–1047. [CrossRef]
24. Sun, T.R.; Ottosen, L.M. Effects of pulse current on energy consumption and removal of heavy metals during electrodialytic soil remediation. *Electrochim. Acta* **2012**, *86*, 28–35. [CrossRef]
25. Song, P.; Xu, D.; Yue, J.; Ma, Y.; Dong, S.; Feng, J. Recent advances in soil remediation technology for heavy metal contaminated sites: A critical review. *Sci. Total. Environ.* **2022**, *838*, 156417. [CrossRef] [PubMed]
26. He, C.; Hu, A.; Wang, F.; Zhang, P.; Zhao, Z.; Zhao, Y.; Liu, X. Effective remediation of cadmium and zinc co-contaminated soil by electrokinetic-permeable reactive barrier with a pretreatment of complexing agent and microorganism. *Chem. Eng. J.* **2021**, *407*, 126923. [CrossRef]
27. Vizcaíno, R.L.; Yustres, A.; Asensio, L.; Saez, C.; Cañizares, P.; Rodrigo, M.; Navarro, V. Enhanced electrokinetic remediation of polluted soils by anolyte pH conditioning. *Chemosphere* **2018**, *199*, 477–485. [CrossRef]
28. Zhang, T.; Zou, H.; Ji, M.; Li, X.; Li, L.; Tang, T. Enhanced electrokinetic remediation of lead-contaminated soil by complexing agents and approaching anodes. *Environ. Sci. Pollut. Res.* **2013**, *21*, 3126–3133. [CrossRef]
29. Pei, L.; Zhang, X.; Yuan, Z. Remediation of Cu Contaminated Soil by Fe78Si9B13AP Permeability Reaction Barrier Combined with Electrokinetic Method. *J. Renew. Mater.* **2023**, *11*, 2969–2983. [CrossRef]
30. Wang, J.; Wang, S. Preparation, modification and environmental application of biochar: A review. *J. Clean. Prod.* **2019**, *227*, 1002–1022. [CrossRef]
31. Zhou, M.; Zhu, S.; Liu, F.; Zhou, D. Pulse-enhanced electrokinetic remediation of fluorine-contaminated soil. *Korean J. Chem. Eng.* **2014**, *31*, 2008–2013. [CrossRef]
32. Wang, H.X.; Teng, H.W.; Wang, X.Y.; Xu, J.L.; Sheng, L.X. Physicochemical modification of corn straw biochar to improve performance and its application of constructed wetland substrate to treat city tail water. *J. Environ. Manag.* **2022**, *310*, 114758. [CrossRef] [PubMed]
33. Zhou, H.; Xu, J.; Lv, S.; Liu, Z.; Liu, W. Removal of cadmium in contaminated kaolin by new-style electrokinetic remediation using array electrodes coupled with permeable reactive barrier. *Sep. Purif. Technol.* **2020**, *239*, 116544. [CrossRef]
34. Wang, Q.; Shaheen, S.M.; Jiang, Y.; Li, R.; Slaný, M.; Abdelrahman, H.; Kwon, E.; Bolan, N.; Rinklebe, J.; Zhang, Z. Fe/Mn- and P-modified drinking water treatment residuals reduced Cu and Pb phytoavailability and uptake in a mining soil. *J. Hazard. Mater.* **2021**, *403*, 123628. [CrossRef]
35. Wang, Q.; Slaný, M.; Gu, X.; Miao, Z.; Du, W.; Zhang, J.; Gang, C. Lubricity and Rheological Properties of Highly Dispersed Graphite in Clay-Water-Based Drilling Fluids. *Materials* **2022**, *15*, 1083. [CrossRef]
36. Mu'azu, N.D.; Usman, A.; Jarrah, N.; Alagha, O. Pulsed Electrokinetic Removal of Chromium, Mercury and Cadmium from Contaminated Mixed Clay Soils. *Soil Sediment Contam. Int. J.* **2016**, *25*, 757–775. [CrossRef]

37. Zhou, H.; Liu, Z.; Li, X.; Xu, J. Remediation of lead (II)-contaminated soil using electrokinetics assisted by permeable reactive barrier with different filling materials. *J. Hazard. Mater.* **2021**, *408*, 124885. [CrossRef]
38. Liu, Y.; Zhuang, Y.-F.; Xiao, F.; Liu, Z. Mechanism for reverse electroosmotic flow and its impact on electrokinetic remediation of lead-contaminated kaolin. *Acta Geotech.* **2023**, *18*, 124885. [CrossRef]
39. Beyrami, H. Effect of different treatments on electrokinetic remediation of Zn, Pb and Cd from a contaminated calcareous soil. *Chin. J. Chem. Eng.* **2021**, *38*, 255–265. [CrossRef]
40. Liu, X.-Y.; Xu, L.-H.; Zhuang, Y.-F. Effect of electrolyte, potential gradient and treatment time on remediation of hexavalent chromium contaminated soil by electrokinetic remediation and adsorption. *Environ. Earth Sci.* **2023**, *82*, 40. [CrossRef]
41. Xie, N.; Chen, Z.; Wang, H.; You, C. Activated carbon coupled with citric acid in enhancing the remediation of Pb-Contaminated soil by electrokinetic method. *J. Clean. Prod.* **2021**, *308*, 127433. [CrossRef]
42. Li, Y.; Zhao, H.-P.; Zhu, L. Remediation of soil contaminated with organic compounds by nanoscale zero-valent iron: A review. *Sci. Total Environ.* **2021**, *760*, 143413. [CrossRef] [PubMed]
43. Moghadam, M.J.; Moayedi, H.; Sadeghi, M.M.; Hajiannia, A. A review of combinations of electrokinetic applications. *Environ. Geochem. Health* **2016**, *38*, 1217–1227. [CrossRef] [PubMed]
44. Marshall, D.J.; Madden, T.R. Induced polarization, a study of its causes. *Geophysics* **1959**, *24*, 790–816. [CrossRef]
45. Cameselle, C.; Reddy, K.R. Development and enhancement of electro-osmotic flow for the removal of contaminants from soils(Conference Paper). *Electrochim. Acta* **2012**, *86*, 10–22. [CrossRef]
46. Li, Z.; Yu, J.E.; Neretnieks, I. Removal of CU(II) and CR(III) from naturally contaminated loam by electromigration. *J. Environ. Sci. Health-Toxic/Hazard. Subst. Environ. Eng.* **1997**, *32*, 1293–1308. [CrossRef]
47. Qiao, R.; Aluru, N.R. Charge Inversion and Flow Reversal in a Nanochannel Electro-osmotic Flow. *Phys. Rev. Lett.* **2004**, *92*, 198301. [CrossRef] [PubMed]
48. Liu, L.; Li, W.; Song, W.; Guo, M. Remediation techniques for heavy metal-contaminated soils: Principles and applicability. *Sci. Total. Environ.* **2018**, *633*, 206–219. [CrossRef]
49. Li, X.; Yang, Z.; He, X.; Liu, Y. Optimization analysis and mechanism exploration on the removal of cadmium from contaminated soil by electrokinetic remediation. *Sep. Purif. Technol.* **2020**, *250*, 117180. [CrossRef]
50. Gao, J.; Luo, Q.; Zhang, C.; Li, B.; Meng, L. Enhanced electrokinetic removal of cadmium from sludge using a coupled catholyte circulation system with multilayer of anion exchange resin. *Chem. Eng. J.* **2013**, *234*, 1–8. [CrossRef]
51. Li, G.; Li, Y.; Zhang, J.; Liu, J. Study on Electrokinetic Remediation of Pb-contaminated Saturated Sand. *Int. J. Electrochem. Sci.* **2020**, *15*, 1486–1497. [CrossRef]
52. Zhou, D.-M.; Deng, C.-F.; Cang, L. Electrokinetic remediation of a Cu contaminated red soil by conditioning catholyte pH with different enhancing chemical reagents. *Chemosphere* **2004**, *56*, 265–273. [CrossRef]
53. Xiao, J.; Pang, Z.; Zhou, S.; Chu, L.; Rong, L.; Liu, Y.; Li, J.; Tian, L. The mechanism of acid-washed zero-valent iron/activated carbon as permeable reactive barrier enhanced electrokinetic remediation of uranium-contaminated soil. *Sep. Purif. Technol.* **2020**, *244*, 116667. [CrossRef]
54. Suzuki, T.; Kawai, K.; Moribe, M.; Niinae, M. Recovery of Cr as Cr(III) from Cr(VI)-contaminated kaolinite clay by electrokineics coupled with a permeable reactive barrier. *J. Hazard. Mater.* **2014**, *278*, 297–303. [CrossRef] [PubMed]
55. Li, G.; Guo, S.; Li, S.; Zhang, L.; Wang, S. Comparison of approaching and fixed anodes for avoiding the 'focusing' effect during electrokinetic remediation of chromium-contaminated soil. *Chem. Eng. J.* **2012**, *203*, 231–238. [CrossRef]
56. Rojo, A.; Hansen, H.K.; Monárdez, O. Electrokinetic remediation of mine tailings by applying a pulsed variable electric field. *Miner. Eng.* **2014**, *55*, 52–56. [CrossRef]
57. Mena, E.; Villaseñor, J.; Rodrigo, M.A.; Cañizares, P. Electrokinetic remediation of soil polluted with insoluble organics using biological permeable reactive barriers: Effect of periodic polarity reversal and voltage gradient. *Chem. Eng. J.* **2016**, *299*, 30–36. [CrossRef]
58. Rezaee, M.; Asadollahfardi, G.; Gomez-Lahoz, C.; Villen-Guzman, M.; Paz-Garcia, J.M. Modeling of electrokinetic remediation of Cd- and Pb-contaminated kaolinite. *J. Hazard. Mater.* **2018**, *366*, 630–635. [CrossRef] [PubMed]
59. Qiu, C.; He, G.; Shi, W.; Zou, M.; Liu, C. The polarization characteristics of lithium-ion batteries under cyclic charge and discharge. *J. Solid State Electrochem.* **2019**, *23*, 1887–1902. [CrossRef]
60. Shen, Z.; Chen, X.; Jia, J.; Qu, L.; Wang, W. Comparison of electrokinetic soil remediation methods using one fixed anode and approaching anodes. *Environ. Pollut.* **2007**, *150*, 193–199. [CrossRef]
61. Xu, J.; Liu, C.; Hsu, P.-C.; Zhao, J.; Wu, T.; Tang, J.; Liu, K.; Cui, Y. Remediation of heavy metal contaminated soil by asymmetrical alternating current electrochemistry. *Nat. Commun.* **2019**, *10*, 2440. [CrossRef] [PubMed]
62. Zhang, P.; Jin, C.; Sun, Z.; Huang, G.; She, Z. Assessment of Acid Enhancement Schemes for Electrokinetic Remediation of Cd/Pb Contaminated Soil. *Water Air Soil Poll.* **2016**, *227*, 217. [CrossRef]
63. Kim, S.; Moon, S.-H.; Kim, K.-W. Removal of Heavy Metals from Soils using Enhanced Electrokinetic Soil Processing. *Water Air Soil Poll.* **2001**, *125*, 259–272. [CrossRef]
64. Liu, Q.; Zhang, Q.; Jiang, S.; Du, Z.; Zhang, X.; Chen, H.; Cao, W.; Nghiem, L.-D.; Ngo, H.-H. Enhancement of lead removal from soil by in-situ release of dissolved organic matters from biochar in electrokinetic remediation. *J. Clean. Prod.* **2022**, *361*, 132294. [CrossRef]

toxics

MDPI

Article

Leaching Behavior of As and Pb in Lead–Zinc Mining Waste Rock under Mine Drainage and Rainwater

Ziwen Guo, Jiejie Yang, Kewei Li, Jiaxin Shi, Yulong Peng, Emmanuel Konadu Sarkodie, Bo Miao, Hongwei Liu, Xueduan Liu and Luhua Jiang *

Key Laboratory of Biometallurgy, School of Minerals Processing and Bioengineering, Ministry of Education, Central South University, Changsha 410083, China; kidgzw@hotmail.com (Z.G.); jiejieyang@csu.edu.cn (J.Y.); 15116475510@163.com (K.L.); 215611008@csu.edu.cn (J.S.); 225611009@csu.edu.cn (Y.P.); emmasark@csu.edu.cn (E.K.S.); miaobo@csu.edu.cn (B.M.); hongweiliu@csu.edu.cn (H.L.); xueduanliu@csu.edu.cn (X.L.)
* Correspondence: jiangluhua@csu.edu.cn

Abstract: At present, the pollution of arsenic (As) and lead (Pb) is becoming increasingly serious. The pollution caused by the release of As and Pb from lead–zinc mines has seriously affected the water and soil environment and threatened human health. It is necessary to reveal the release characteristics of As and Pb. The actual scene of mine drainage (MD) and rainwater (RW) leaching waste rocks is the one of the main reasons for the release of As and Pb. However, the leaching behavior of As and Pb in these waste rocks under MD and RW suffered from a lack of in-depth research. In this study, we investigated the occurrence of As and Pb in waste rocks (S1–S6) by using X-ray diffraction (XRD) and time-of-flight secondary ion mass spectrometry (TOF-SIMS), and then, the changes in As and Pb concentration and the hydrochemical parameter in leaching solution were systematically studied. Furthermore, the correlation between the release of As and Pb and mineral composition was also evaluated. Results showed that these waste rocks were mainly composed of carbonate and sulfide minerals. As and Pb were mainly bounded or associated with sulfide minerals such as arsenopyrite, pyrite, chalcopyrite, and galena in these waste rocks, and small parts of As and Pb were absorbed or encased by clay minerals such as kaolinite and chlorite. Under MD and RW leaching, the pH, redox potential (Eh), and electric conductivity (EC) of each waste rock tended to be consistent due to their buffering ability; the leachate pH of waste rocks with more carbonate minerals was higher than that of sulfide minerals. Both As and Pb were released most under MD leaching in comparison to RW, reaching 6.57 and 60.32 mg/kg, respectively, due to MD's low pH and high Eh value. However, As in waste rock released more under alkaline conditions because part of the arsenic was in the form of arsenate. As and Pb release were mainly positively correlated with the proportions of sulfide minerals in these waste rocks. MD leaching significantly promoted the release of As and Pb from waste rocks, which would cause a great threat to the surrounding environment, and control measures were imperative. This paper not only reveals the As and Pb pollution mechanism around the lead–zinc mining area but also provides a theoretical basis for the prevention and control of As and Pb pollution in the future.

Keywords: waste rock; lead–zinc mine; toxic metal(loid)s; occurrence characteristics; leaching behavior

1. Introduction

Non-ferrous metal minerals resources are indispensable and play an important role in economic development [1]. In the process of the exploitation of mineral resources, a large amount of mining waste rock has been produced [2]. According to the report on National Mineral Resources *Conservation and Comprehensive Utilization (2020)* issued by the Ministry of Natural Resources of the People's Republic of China, the stripping ratio of non-ferrous metal minerals ranged from 2.14 to 10.86 t/t in 2018 [3]; this means that a large number of mining wastes including tailings and waste rocks will be generated, although part of them can be used as resources. In the absence of strict environmental control, the weathered oxidation of these wastes will produce acid mine drainage (MD). Acid MD will lead to toxic metal(loid) pollution, posing a threat to the surrounding environment and human health [4].

Lead–zinc minerals are important non-ferrous metal mineral resources, being mostly distributed in parts of southern China such as the Yunnan, Hunan, and Fujian provinces [5]. The toxic metal(loid) pollution of lead–zinc mine areas usually contains Pb, As, Cd, Cr, Cu, Zn, and Ni [6–8]. Among these toxic metal(loid) elements, the most serious pollutants are As and Pb [9,10]. For example, Chenzhou is a typical lead–zinc mining area in Hunan Province, and the total concentrations of As and Pb in the soil of an abandoned lead–zinc mine in this area could reached up to 84.85 and 802.58 mg/kg, respectively, which were far beyond the national risk control values for development land soil [11]. Inner Mongolia is an important non-ferrous metal-producing area in northern China; both As and Pb in the soil around a typical lead–zinc mine in Chifeng, Inner Mongolia, China are seriously polluted, and the concentrations of As and Pb were found to be 82 and 600 mg/kg, respectively [12,13]. In Nigeria, Pb pollution in agricultural soil around the Ishiagu lead–zinc mining mine was serious, and the total concentration of Pb could reach up to 5305.5 mg/kg [14]. The pollution of As and Pb in the soil surrounding the lead–zinc mining area in northern Turkey was also serious and their concentrations surprisingly reached 1171 and 3725 mg/kg, respectively [15]. Therefore, it is urgent to study the release characteristics of As and Pb in the mining wastes at lead–zinc mines, which is of great significance for its environmental impact assessment and pollution control.

Usually, mining wastes are common toxic metal(loid)s bearing matrices that occur at lead–zinc mining sites [16]. The leaching behaviors of toxic metal(loid)s in different kinds of mining wastes are different due to the differences in the processing process [17]. For example, finer tailings have a larger contact area than the coarser characteristics of waste rocks, which makes it easier for them to oxidize and release toxic metal(loid)s [18,19]. Due to the tailings having a greater tendency to release toxic metal(loid)s than waste rocks, many studies have investigated their environmental risk, mobility, and safe disposal at present [20–22]. However, they found that tailings in most mining sites were often deposited in ponds to prevent seepage, and waste rocks were usually neglected and stored in open sites [3,23]. The pollution of As and Pb in waste rocks could also not be ignored. For example, the pollution of As and Pb in the soil of a lead–zinc mine in northern Guangxi, China mainly came from the mining waste rocks [24]. The waste rocks from underground polymetallic mines in Balya, Turkey caused a high concentration of As and Pb pollution in the Kocacay River during the wet season [25]. Due to the different net acid generation levels, redox reactions, and interfacial activity of toxic metal(loid)s in waste rocks, the environmental hazards caused by toxic metal(loid)s were also different. At present, only a few studies have investigated the release of toxic metal(loid)s from waste rocks. Therefore, the release behavior of toxic metal(loid)s in waste rocks at lead–zinc mines is worth exploring.

In past studies, laboratory static leaching tests and dynamic tests were commonly used to evaluate the release characteristics of toxic metal(loid)s in mining wastes [26–28]. Leaching solution pH, waste particle size, and solid–liquid ratio in the reaction system were the most studied factors influencing the release behavior of toxic metal(loid)s in mining wastes [29–31]. MD and rainwater (RW) were the most common media solutions that waste rocks were exposed to in mining areas. To reveal the oxidative dissolution behavior of waste rocks under MD and RW leaching, most studies have simulated RW and MD by using deionized water or acid solutions (nitric acid, acetic acid, sulfuric acid, hydrochloric acid) to leach waste rocks. The results have shown that all of them promoted the release of toxic metal(loid)s [32–34]. Using synthetic RW and acid MD as a leaching solution could partly reflect the oxidative dissolution of wastes in mining sites. However, the hydrochemistry of actual MD and RW is rather different from that of a simulation, leading to influencing the release of As and Pb in waste rocks. It is necessary to select MD and RW that exist in an actual scenario to explore the leaching behavior of As and Pb. Thus, a representative lead–zinc mine in Shanggao County, Jiangxi Province, China was investigated in this study and six waste rock samples were collected from this mine. The objective of this study was to analyze the release characteristics of As and Pb from waste rocks under MD and RW leaching using an inductively coupled plasma optical emission spectrometer (ICP-OES), time-of-flight secondary ion mass spectrometry (TOF-SIMS), and X-ray diffraction (XRD) and investigate the correlation between the release of As and Pb and the mineral components in the waste rocks through redundancy (RDA) analysis.

2. Materials and Methods

2.1. Chemicals

NaOH used in preparing the alkaline leaching solution was analytically pure and purchased from Sinopharm Chemical Reagent Co., Ltd, Shanghai, China. pH standard buffers (4.00, 6.86, and 9.18) used for pH meter calibration were obtained from Shanghai Lei Ci Co., Ltd, Shanghai, China. Electric conductivity (EC) standard solution (1413 μS/cm) used for EC meter calibration was provided by Mettler Toledo, Co., Ltd, Shanghai, China. The water for solution preparation was high-grade purified water (18.25 Ω/cm) prepared by Elga Purelab Chorus2, Lane End, High Wycombe, the UK.

2.2. Waste Rock Sample Collection and Pretreatment

The waste rocks used in this study were obtained from a lead–zinc mining area in Shanggao County, Jiangxi Province, China (114°52′30″–114°55′00″ E, 28°09′00″–28°10′50″ N). According to the stacking time and waste rock type, six kinds of waste rock named S1–S6 were obtained from this mine. The overview sampling information is shown in Figure 1. After transporting samples to the laboratory, all waste rock samples were crushed and ground and then passed through 10-mesh (2.000 mm), 100-mesh (0.147 mm), and 200-mesh (0.074 mm) sieves. Then, these samples were sealed in zip-locked bags and stored in 4 °C refrigerator for later characterizations and leaching experiments.

2.3. Characterizations

The morphologies and element distributions of 10–mesh samples were analyzed by using scanning electron microscope (SEM, JEOL, JSK-6490LV, Tokyo, Japan) and X-ray spectrometer (EDS, EDAX, NEPTUNE TEXS HP, CA, USA) at a voltage of 20 kV [35]. Moreover, the element distributions of these samples were further exanimated using TOF-SIMS (ULVAC-PHI Inc., PHI nano TOF II, Kanagawa, Japan) at a voltage of 30 kV and an ion current of 2 nA [36]. Samples of the 200-mesh type were used for mineral composition measurement by using XRD (Bruker, D8 ADVANCE, Karlsruhe, Germany) with CuKa radiation (λ = 1.5406 nm) at a voltage of 40 kV and a current of 50 mA [37] The chemical compositions of samples were measured by using X-ray fluorescence spectrometer (XRF, Bruker, S4 Pioneer, Karlsruhe, Germany) with an accuracy of 0.02% and a limit of detection (LOD) of 0.5 mg/kg [38]. Samples of the 100-mesh type were digested by lead nitrate and acetic acid buffer solution (pH = 5) and then we detected the sulfate–sulfur through ethylene diamine tetra acetic acid titration [39]. The total carbon and sulfur determination of samples were investigated by using carbon sulfur analyzer (LECO, CS600, San Jose, CA, USA); the relative standard deviation (RSD), limit of blank (LOB), and limit of quantitation (LOQ) of total carbon and total sulfur were 0.5%, 0.3 mg/kg, and 0.6 mg/kg, respectively for both [40,41]. The content of sulfide–sulfur in samples was the difference between total sulfur and sulfate–sulfur [42].

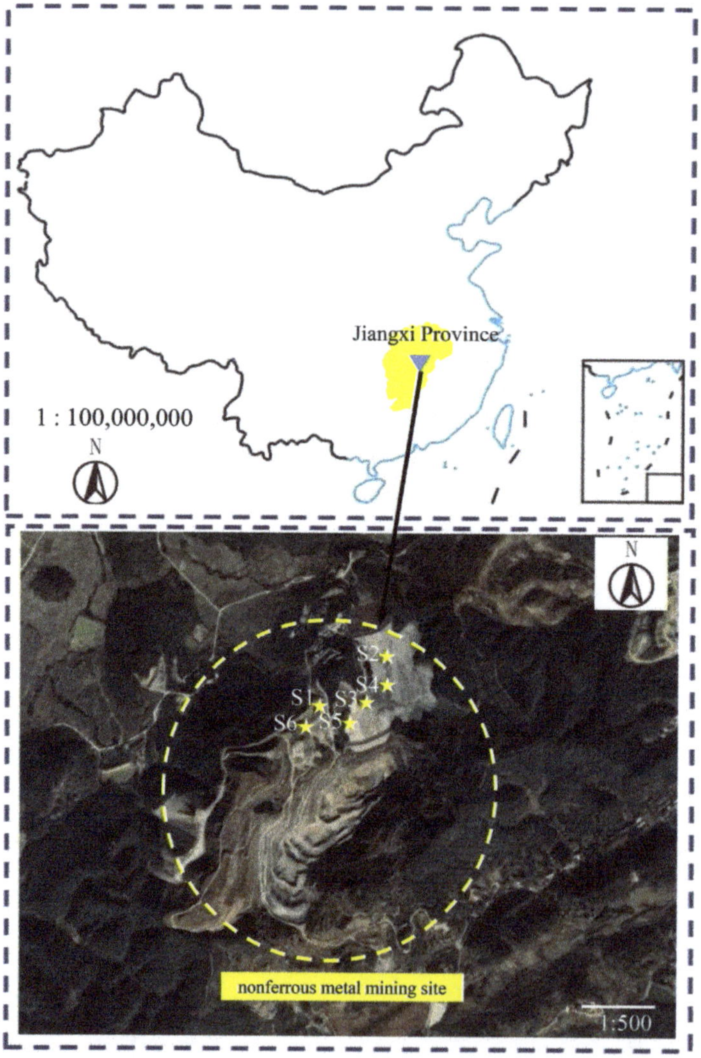

Figure 1. Location of the mining site and sampling overview. S1, S2, S3, S4, S5, and S6 were the six waste rock samples.

2.4. Batch Leaching Tests

The leaching solutions including MD and RW were also collected from the lead–zinc mining area. In order to set up a comparison experiment, an alkaline leaching solution (AS) was prepared by adjusting the pH value of deionized water with NaOH. The solution chemical parameters of MD, RW, and AS are shown in Table 1. The leaching experiment was set as follows: 50 g of waste rocks (10-mesh) was added into MD, RW, and AS in conical flasks, and the liquid–solid ratio (mL/g) was set to 10:1. Then, the conical flasks were shaken at 25 °C and 180 r/min for 15 d in a vertical temperature oscillation incubator (Tianjin Lai Bo Terry Instrument Equipment Co., ZQPL-200, Tianjin, China). Samples were taken at 0.5, 1, 2, 3, 4, 6, 9, and 15 d, respectively. After static sedimentation, the leachates were provided to examine the hydrochemistry parameters including the values of pH, redox potential (Eh), and EC and the total concentrations of As and Pb. The pH value was measured by using a digital pH meter (Leici, PHS-3E, Shanghai, China) according to standard procedure NY/T 1121.2−2006 [43]; the accuracy of this pH meter was 0.01 [43]. The Eh value was determined by using the digital pH meter (Leici, E-301-F, Shanghai, China) and Ag/AgCl electrodes with Pt electrodes according to standard procedure HJ 746−2015 [44]; the accuracy values of this pH meter and Ag/AgCl electrodes with Pt electrodes were 0.01 [44]. The EC value was exanimated by using conductivity meter (Mettler Toledo, FiveEasy Plus FE38, Zurich, Switzerland) based on HJ 802−2016 [45]; the accuracy of this conductivity meter was 0.1 [45]. The measurement of the total concentration of As and Pb was conducted with an ICP-OES (PerkinElmer, Avio500, Waltham, MA, USA); the ICP-OES had an accuracy of 0.001 mg/L. The RSD, LOB, and LOQ of As detected by ICP-OES were 0.5%, 0.053 mg/L, and 0.056 mg/L, respectively; these values for Pb, determined by ICP-OES, were 0.5%, 0.090 mg/L, and 0.091 mg/L, respectively [46].

Table 1. Initial chemical parameters of leaching solutions. MD is 'mine drainage', RW is 'rain water', AS is 'alkaline solution', Eh is 'redox potential', and EC is 'electric conductivity'.

Chemical Parameter	MD	RW	AS
pH	3.24	7.63	12.00
Eh (mV)	495.67	170.00	−54.67
EC (μS/cm)	443.23	780.90	919.63
As	N.D.	N.D.	N.D.
Pb	N.D.	N.D.	N.D.

N.D.: Not detected.

2.5. Statistical Analysis

Statistical analyses were performed using Excel 2021 in Microsoft 365 (Office). All graphs were plotted by using Origin 2021b [43]. RDA analysis was performed and plotted by base R package vegan (v. 2.5.7) from R (v. 3.6.3). The release amount of As and Pb was calculated as follows: $Cm = (C \times V)/m$. Here, Cm is the release amount (mg/kg), C is the concentration of As and Pb in leaching solutions (mg/L), V is the leaching solution volume (L), and m is the mass of waste rocks used for leaching (kg). To ensure the accuracy and precision of the determination, quality assurance (QA) protocol was followed via the utilization of a blank test and repeating the experiment three times. To ensure the quality control (QC), accuracy of ICP-OES determination of As and Pb concentrations was controlled by using standard solutions of As and Pb (0.1, 0.2, 0.5, 1, 5, 10, 20, and 100 mg/L) to recommend standard curve ($R^2 \geq 0.999$). We used 100 mg/L standard solutions of As (GSB 04−1714−2004) and Pb (GSB 04−1742-2004) [47]. The results showed that the concentrations of standard As and Pb solutions were 100.21 and 99.94 mg/L, respectively. To check the pH of the solutions, the 4.00 buffer was determined using a corrected pH meter, which gave a result of 3.99. To check the EC of the solutions, an EC buffer of 1413 μS/cm was determined using a calibrated EC meter, which gave a result of 1413 μS/cm. All data were recorded three times.

3. Results and Discussion

3.1. Mineral Component of Waste Rocks

The XRD patterns of the waste rocks and the proportions of identified minerals are shown in Figure 2. As shown, the mineral component of waste rocks could be classified into five categories: (i) sulfide minerals, including pyrite (FeS_2), chalcopyrite ($CuFeS_2$), sphalerite (ZnS), and galena (PbS); (ii) sulfate minerals such as gypsum ($CaSO_4$); (iii) carbonate minerals, including dolomite [$CaMg(CO_3)_2$], ferro-dolomite [$Ca(Mg,Fe)(CO_3)_2$], siderite ($FeCO_3$), and calcite ($CaCO_3$); (iv) secondary clay minerals, including kaolinite [$Al_4(Si_4O_{10})(OH)_8$] and chlorite [$Y_3(Z_4O_{10}](OH)_2 \cdot Y_3(OH)_6$. Y: Mg^{2+}, Fe^{2+}, Al^{3+}, Fe^{3+}. Z: Si^{4+} or Al^{3+}]; and (v) primary aluminosilicates and silicate minerals, including quartz (SiO_2), feldspar ($KAlSi_3O_8$, $NaAlSi_3O_8$, and $CaAl_2Si_2O_8$), and mica [$KAl_2(AlSi_3O_{10})(OH)_2$].

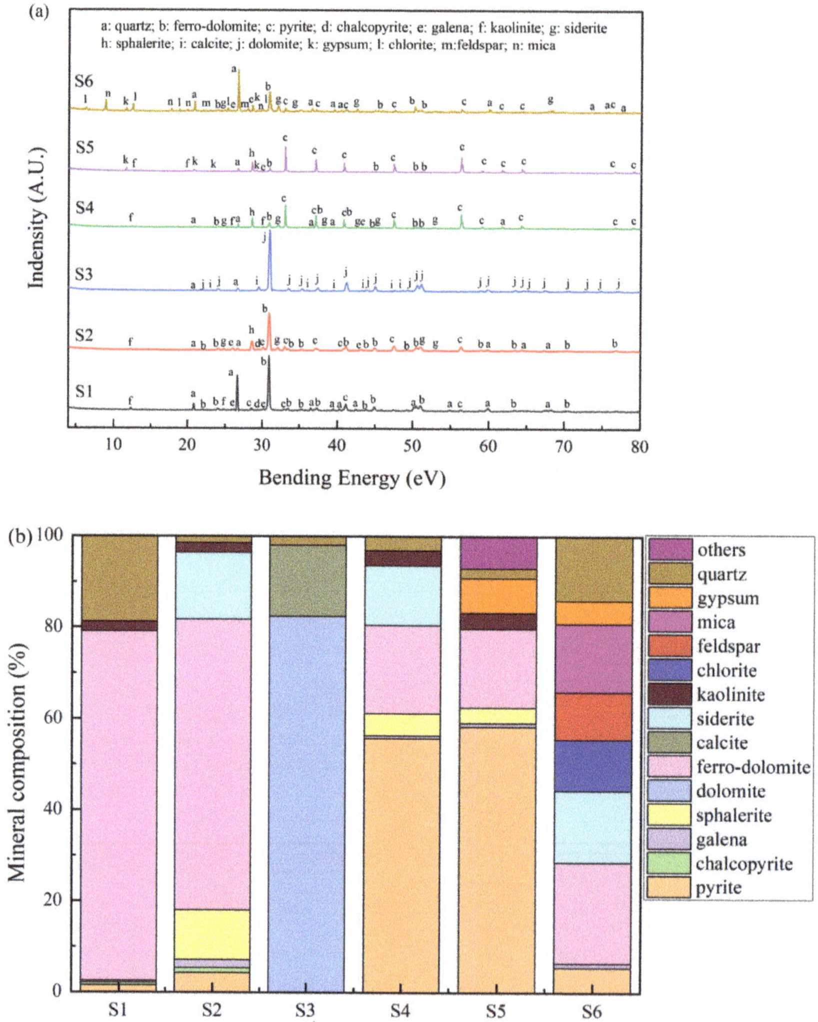

Figure 2. XRD spectra of waste rocks (**a**) and their mineral proportion (**b**).

As and Pb are often associated with sulfide minerals and are easily released in acid MD by the oxidation and weathering of waste rocks [48]. The acid production abilities of sulfide minerals were different. Arsenopyrite, pyrite, and chalcopyrite (leachate pH = 1.68, 1.82, and 2.73, respectively) had higher acid production ability, whereas sphalerite and galena (leachate pH = 4.32, and 4.07, respectively) had lower production ability [49]. As shown in Figure 2b, S4 and S5 were mainly composed of sulfide minerals. Their proportion reached 61.2 and 62.5%, respectively. However, the proportions of sulfide minerals in S1, S2, S3, S6 were only 2.5, 18.0, 0, and 11.5%, respectively. This indicated that S4 and S5 had higher acid production ability than other waste rocks, and As and Pb were more easily released in these waste rocks through oxidation and dissolution.

Carbonate mineral and siliceous minerals in the waste rocks can neutralize acid produced by the sulfide mineral oxidation [50]. It has been reported that the waste rocks without carbonate minerals have a greater potential to produce high concentrations of metal(loid)s than those containing carbonate minerals after acid MD leaching [51]. As shown in Figure 2b, the proportions of carbonate minerals in S1, S2, S3, S4, S5, and S6 were 76.6, 78.4, 98.0, 32.4, 17.2, and 37.9%, respectively. S1, S2, and S3 contained more carbonate minerals than S4, S5, and S6. This suggested that S1, S2, and S3 had stronger acid neutralization ability, potentially resulting in reducing the production of MD and the release of As and Pb. Clay minerals have the ability to absorb or encapsulate metal(loid) ions including As and Pb [52]. It has been reported that clay could absorb As via ligand exchange and electrostatic adsorption and absorb Pb through competitive adsorption onto non-specific sites on the surface of clay [35,53,54]. Figure 2b shows that the contents of clay minerals in all waste rocks were very low (2.1, 2.1, 0, 3.3, and 3.5%) except in S6 (11.1%). This meant that As and Pb released in S6 would be re-adsorbed to waste rock through clay mineral adsorption, leading to a relatively low release amount of S6 [55].

3.2. Chemical Composition of Waste Rocks

The chemical composition of these waste rocks is shown in Table 2. As seen, the proportions of alkaline oxide including CaO and MgO in S1, S2, and S3 were less than that in S4, S5, and S6. This suggested that the acid neutralization ability of S1, S2, and S3 might be lower than that of S4, S5, and S6 due to the hydrogen protons caching the properties of the alkaline oxides [56]. However, the acid neutralization ability of S1, S2, and S3 might have been improved due to the higher carbonate mineral contents as shown in the XRD results (Figure 2). In addition, it had been reported that Fe_2O_3, Al_2O_3, and MnO have adsorption ability to adsorb metal(loid)s including Pb and As [57–59]. The amounts of Fe_2O_3 and Al_2O_3 in S1, S2, and S3 were higher than that in S4, S5, and S6. S1 and S2 contained more Fe_2O_3, reaching 73.50 and 77.43%, respectively; the content of Al_2O_3 in S3 was 22.96%, which was higher than that in the other waste rocks. This indicated that Ca and Mg of the ferro-dolomite in S1, S2, and S3 were replaced by Fe and Al [60]. As a result, S1, S2, and S3 had low CaO and MgO content but high Fe_2O_3 and Al_2O_3 content. This suggested that S1, S2, and S3 might have greater adsorption potential for As and Pb.

Table 2. Chemical composition of waste rocks provided. The oxides, As, and Pb were determined using XRF. Total carbon and total sulfur were determined using a carbon sulfur analyzer. Sulfide–sulfur was determined through titration. Sulfate–sulfur was calculated using the difference between total sulfur and sulfide–sulfur.

Chemical Composition	S1	S2	S3	S4	S5	S6
Na_2O	0.11%	0.78%	1.31%	N.D.	1.51%	N.D.
MgO	2.45%	1.53%	2.88%	17.66%	23.97%	29.44%
Al_2O_3	4.85%	4.44%	22.96%	6.99%	2.96%	0.62%
SiO_2	7.79%	5.70%	32.93%	30.24%	3.87%	3.70%
P_2O_5	0.07%	0.07%	0.26%	0.06%	0.09%	N.D.
K_2O	0.06%	0.02%	2.84%	0.17%	0.06%	0.04%
CaO	4.24%	4.05%	7.36%	30.22%	41.41%	56.84%
MnO	0.43%	0.23%	0.79%	0.70%	1.69%	0.57%
Fe_2O_3	73.50%	77.42%	23.82%	8.82%	1.13%	4.83%
As	0.02%	0.03%	N.D.	0.03%	0.04%	0.06%
Pb	0.33%	1.99%	0.03%	0.51%	0.71%	0.34%
Total carbon	6.75%	8.66%	10.80%	1.55%	0.71%	3.56%
Total sulfur	1.42%	42.54%	0.17%	48.60%	49.92%	3.78%
Sulfide–sulfur	1.01%	41.23%	0.01%	47.40%	46.14%	2.29%
Sulfate–sulfur	0.41%	1.31%	0.16%	1.20%	3.78%	1.49%

N.D.: Not detected.

The total carbon content values of S1, S2, and S3 (6.75–10.80%) were higher than those of S4, S5, and S6 (0.71–3.56%). This indicated that S1, S2, and S3 contained more carbonate minerals, which was consistent with the XRD results as shown in Figure 2. The total sulfur contents of S2, S4, and S5 were 42.54%, 48.60%, and 49.92%, respectively, which were much higher than those of S1, S3, and S6, which reached 1.42, 0.17, and 3.78%, respectively. Most of them were in the form of sulfide–sulfur, except in S3. This sulfide–sulfur might comprise sulfide minerals such as pyrite, chalcopyrite, and galena, according to the results shown in Figure 2. This suggested that the waste rocks of S2, S4, and S5 contained large amounts of sulfide–sulfur and had a stronger ability to produce acid, leading to the release of more As and Pb from these waste rocks [61]. The proportions of As and Pb in these waste rocks were 0.06% and 1.99%, respectively. These values exceeded the national risk control value [62], potentially posing a significant threat to the surrounding environment.

3.3. Morphology and Element Distribution of Waste Rocks

The morphology and element content of the waste rocks are illustrated in Figure 3. All the waste rocks were heterogeneous and contained various mineral components. All of them were quartz and dolomite [63]. S4, S5, and S6 had more irregularly shaped particles than S1, S2, and S3. This indicated that S4, S5, and S6 had larger surface areas in contact with the leaching solution and it was easier for them to oxidize and release As and Pb. EDS results are shown in Figure 3; Pb exists in all waste rocks except S3. The results were similar to those of the XRF (Table 2). As reflected by the EDS results, Pb concentrations in S1, S2, S4, S5 and S6 were 85.80, 84.14, 84.27, 87.64, and 89.36, respectively. Meanwhile, the S concentration were 11.08, 12.20, 11.42, 12.06, and 9.52%, respectively. Pb was associated with S, which indicated that Pb in these waste rocks might have been in the form of galena. In addition, As was only found in S2, S3, and S6, as exhibited by EDS, in concentrations of 19.68%, 1.31% and 31.23%, respectively. EDS results showed that As coexisted with S, Fe, Cu, Al, Ca, and Si. This suggested that As in these waste rocks might be associated with pyrite or chalcopyrite, which was consistent with the XRD results. As(III) and As(-I) would substitute Fe(II) and S_2(–II), respectively in pyrite and chalcopyrite [64]. A small part of As in these waste rocks was adsorbed or coated by clay minerals like kaolinite and chlorite in the form of arsenate [54].

An S2 sample was chosen for TOF-SIMS analysis to further study the distribution characteristics of As and Pb. Figure 4 shows selected secondary ion images acquired for the S2 sample. As shown in Figure 4a–f, As, Pb, S(–), CO_3^{2-}, Si, and Al in S2 were found. CO_3^{2-} were uniformly distributed in S2, while As, Pb S(–), Si, and Al were enriched. As and Pb had low signal intensity, while S(–), CO_3^{2-}, Si, and Al had high signal intensity. This was consistent with the results of the XRF shown in Table 2. The overlap of As and S(–), CO_3^{2-}, Si, and Al is illustrated in Figure 4g–j. The signal of As exhibited a high overlap with S, indicating that As might be associated with sulfide minerals such as pyrite, chalcopyrite, and arsenopyrite [65]. These sulfide minerals conformed with the results of SEM-EDS. The signal of As showed some overlap with Al and Si, and Al highly overlapped with Si, as shown in Figure 4o. This indicated that some parts of As in S2 might have been adsorbed or encased in clay minerals [66,67]. There was little overlap between As and CO_3^{2-}. This suggested that there was a low correlation between As and carbonate minerals in these waste rocks. Figure 4k–n show the overlap of Pb and S(–), CO_3^{2-}, Si, and Al. The overlap of Pb's signal with S(–), CO_3^{2-}, Si, and Al was similar to that of As. The signal of Pb exhibited a high overlap with S(–). This suggested that Pb in S2 was mainly composed of galena [68]. There was some overlap between Pb, Al, and Si. This indicated that a part of Pb was adsorbed or encased with clay minerals [69]. Pb and CO_3^{2-} had little overlap. This indicated a very low correlation between Pb and carbonate minerals.

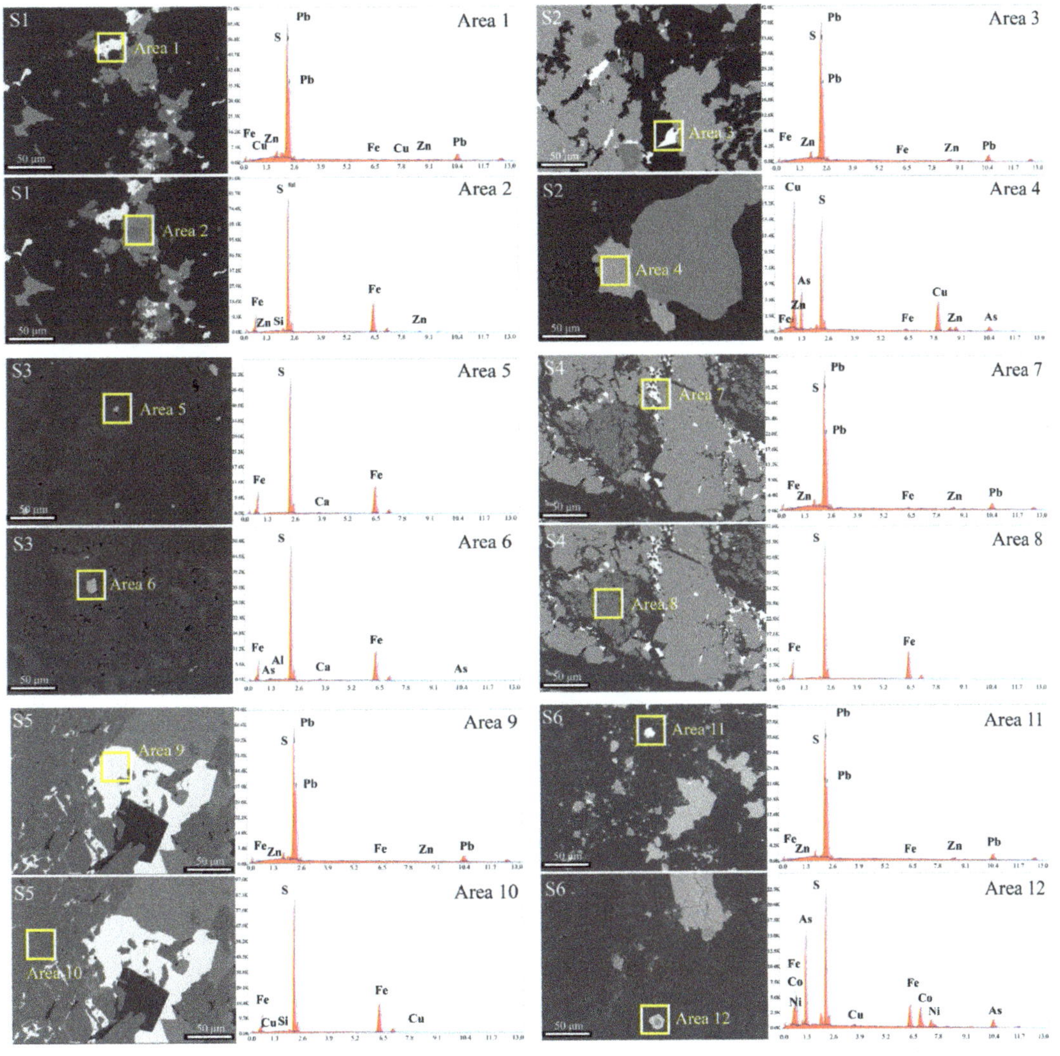

Figure 3. SEM-EDS images of waste rocks. Area 1–12 were the mineral components where As and Pb may exist on six waste rock samples (S1–S6) determined by SEM-EDS.

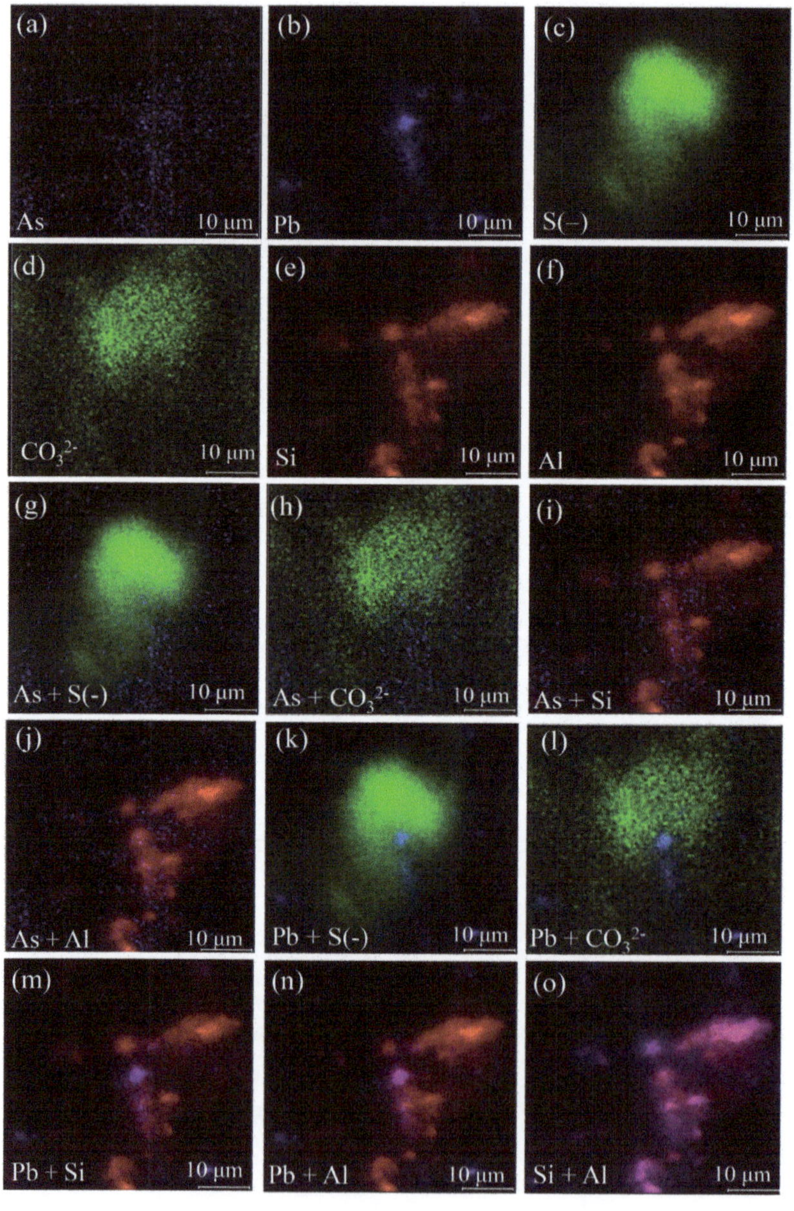

Figure 4. TOF-SIMS images in S2: (**a–f**) distribution of As, Pb, S(−), CO_3^{2-}, Si, and Al; (**g–j**) overlap of As with S(−), CO_3^{2-}, Si, and Al, respectively; (**k–n**) overlap of Pb with S(−), CO_3^{2-}, Si, and Al, respectively; (**o**) overlap of Si with Al. The color is simply to distinguish the distribution of different elements.

3.4. The Changes in pH, Eh, and EC Values of Leaching Solutions

The values of pH, Eh, and EC of the leaching solutions during waste rock leaching by MD, RW, and AS are illustrated in Figure 5. Their values remained stable after leaching by MD, RW, and AS. This might be attributed to the acid–base buffer characteristics of these waste rocks due to the existence of sulfide minerals and carbonate minerals [49,50].

Carbonate mineral dissolution could neutralize the acid produced by sulfide mineral dissolution [70]. As shown in Figure 5a, the pH values of leaching solutions in S1–S3 samples ranged from 7.5 to 8.5 when leached by MD, RW, and AS. This could be attributed to the high acid neutralizing ability of the dolomite, ferro-dolomite, siderite, calcite, CaO, and MgO and the weak acid production ability of the sulfide minerals. Therefore, although a small amount of sulfide minerals was oxidized to produce acid in S1–S3, acid would be neutralized quickly by these carbonate minerals and alkaline oxides. The pH values of S4–S5 in MD, RW, and AS decreased to 3.90–4.57 after 15 d of leaching. The net acid production ability of S4 and S5 was high, due to their having greater amounts of sulfide minerals (reaching 61.2 and 62.5%, respectively). The pH value of the final leachates in S6 was maintained in the range of 7.0–8.0. The content of carbonate minerals was slightly greater than that of sulfide minerals. This indicated that the acid neutralization ability was slightly greater than the acid production ability.

The Eh value of the leaching solution could reflect the redox state in the leaching process [71]. The initial Eh value of MD (495.67 mV) was much higher than those of RW (170.00 mV) and AS (−54.67 mV). This meant that the oxidation potential of MD was much higher than those of RW and AS, resulting in the acceleration of the oxidative dissolution of sulfide minerals in waste rocks when leached by MD. As shown in Figure 5b, Eh decreased in MD leaching in all waste rocks at 1 d; Eh was unchanged in RW leaching at 1 d, while it increased in AS leaching. The Eh values of the three leaching solutions became increasingly similar as leaching time increased. This might be ascribed to the waste rocks having the same redox conditions due to the convergence of pH values during leaching [72]. The Eh of the MD leaching solution was higher than those of RW and AS, indicating that As and Pb were more easily released during MD leaching. Moreover, the Eh values of S1, S2, S3, S4, S5, and S6 in MD decreased from 495.67 to 125.33, 161.33, 122.00, 249.33, 318.67, and 155.00 mV, respectively, after 15 d leaching; the Eh values in RW after 15 d were 109.33, 124.00, 86.67, 245.00, 304.67, and 143.00 mV, respectively. The Eh values of these waste rocks increased from −54.67 to 55.00, 127.33, 54.00, 225.33, 275.33, and 130.67 mV, respectively in AS leaching for 15 d. The Eh values of S4 and S5 were higher than those of the other waste rocks. This suggested that the sulfide minerals of S4 and S5 could be easily oxidized, releasing more As and Pb. The Eh values of these waste rocks in MD leaching were higher than those in RW leaching and AS leaching [73]. This indicated that waste rocks were more easily oxidized in MD leaching than in RW and AS.

EC can indirectly reflect the concentration of metal(loid) ions in leaching solutions [74]. The initial EC values of MD, RW, and AS were 443.23, 780.90, and 919.63 μS/cm, respectively. As shown in Figure 5c, the EC values of the leaching solutions in each waste rock tended to be consistent in the middle and late stages of leaching and showed a rising trend. This could be attributed to the release of the As and Pb in waste rocks [74]. The EC values of S1, S2, S3, S4, S5, and S6 after 15 d reached 739.23, 1197.33, 691.97, 3521.00, 3197.00, and 1534.67 μS/cm, respectively in MD leaching; 870.60, 1084.67, 840.00, 3574.00, 3205.67, and 1837.00 μS/cm, respectively in RW leaching; and 899.93, 1084.67, 979.93, 4176.33, 3309.00, and 1993.00 μS/cm, respectively in AS leaching. The EC values of S1, S2, S3, and S6 were much lower than S4 and S5. This indicated that the As and Pb release abilities of S2 and S6 might be higher than those of S1 and S3 and lower than those of S4 and S5 in these leaching solutions.

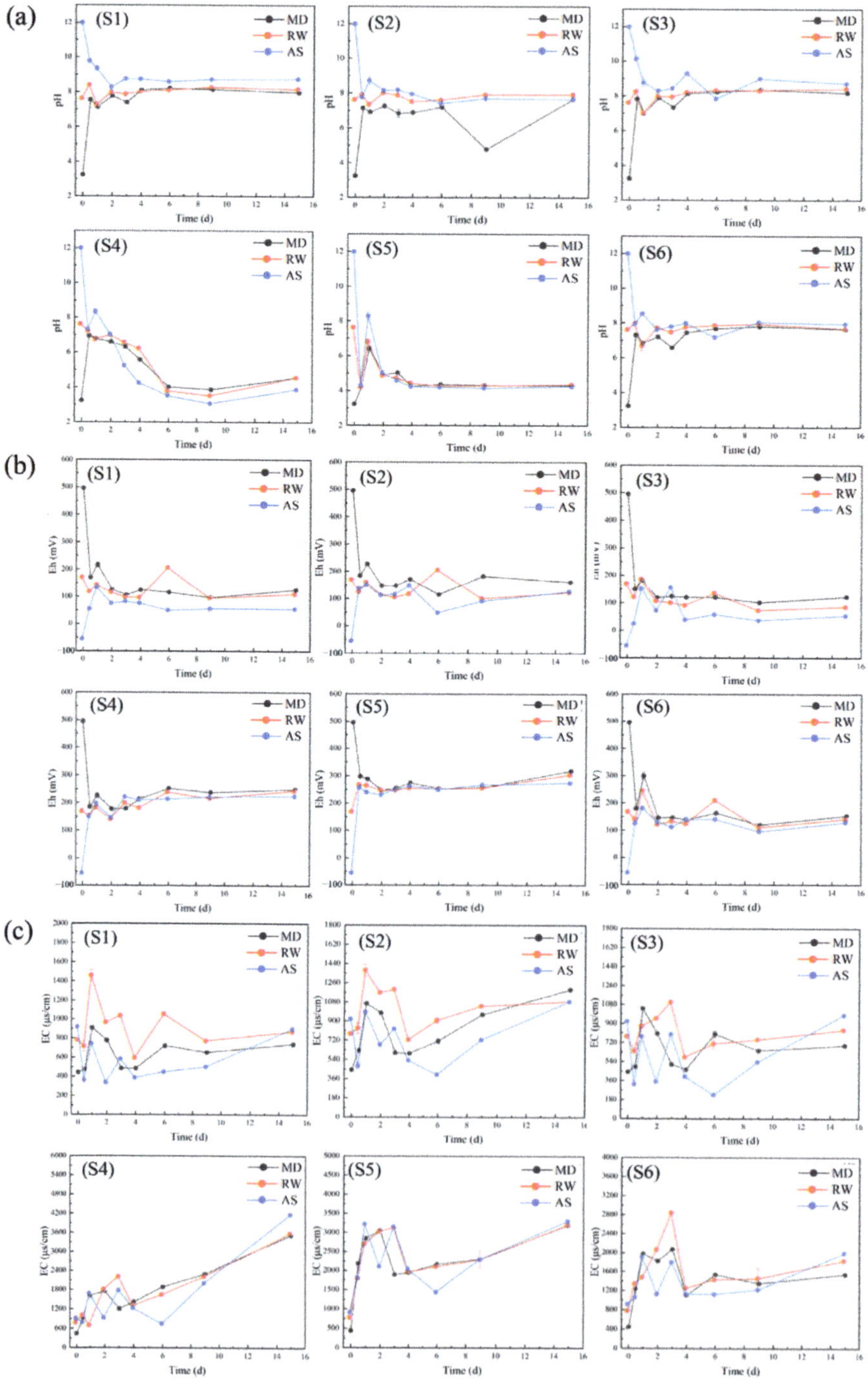

Figure 5. The pH (**a**), Eh (**b**), and EC (**c**) values in leaching solution. S1–S6 are the samples of waste rock.

3.5. The Change in Total Concentrations of As and Pb in Leaching Solutions

Figure 6 shows the release of As and Pb in S1–S6 under MD, RW, and AS. Figure 6a shows the As and Pb released in S1–S6 under MD leaching. As and Pb in S1 and S5 were released the most in these waste rocks, reaching 6.50, 60.32, and 6.57, 47.72 mg/kg, respectively, as shown in Figure 5. The pH of S1 was maintained at about 8 and the Eh was maintained at about 100 mV due to the dissolution of ferro-dolomite, and the solution was in a low-oxidation environment. This indicated that As might have existed in the form of arsenate in S1 [75]. S1 has a high Pb release due to its high total Pb content (0.33%). S5 was mainly composed of pyrite and galena; the pH and Eh of the leaching solution were maintained at 4 and 200 mV, respectively in MD leaching. This meant that sulfide minerals were more easily dissolved and released As and Pb under acidic conditions [76]. The release amounts of As and Pb in S2, S3, S4, and S6 were very low, and values were observed at 0.91 and 0, 0.43 and 0, 1.39 and 0, and 2.46 and 0 mg/kg, respectively. This might have been because the total As and Pb contents in S2, S3, S4, and S6 were very low or the forms of As and Pb in these waste rocks were not easily released [77]. Figure 6b shows the release of As and Pb in S1–S6 under RW leaching. The leaching results of RW were similar to those of MD. S1 and S5 had the highest As release levels, reaching 5.96 and 6.40 mg/kg, respectively under RW leaching. The release of Pb under RW leaching was much lower than that under MD leaching; the concentrations of S1 and S5 under RW leaching were 38.11 and 12.19 mg/kg, respectively. This indicated that Pb was more likely to be released when the initial condition was acidic. The release of As and Pb in these waste rocks under AS leaching is shown in Figure 6c. The release levels of S2, S3, S4, and S6 was much higher than under MD and RW leaching, reaching 1.06, 1.16, 1.85, and 5.43 mg/kg. This indicated that the As in these waste rocks could mainly exist in the form of anions, which were easy to release under alkaline conditions. The release of Pb in AS leaching was also much higher than in RW leaching, but less than in MD leaching. This suggested that the release of Pb could increase under alkaline conditions [78]. In general, both As and Pb were more likely to be released under acidic and alkaline conditions, with the acidic conditions being significantly better.

3.6. Correlation between Release of As and Pb and Mineral Composition

This was selected for analysis, due to the significant release of As and Pb obtained in MD leaching. Figure 7 shows the XRD patterns of waste rocks after MD leaching and the proportions of identified minerals. As shown in Figures 2 and 7, the ferro-dolomite of S1, S2, S4, S5, and S6 was transformed into dolomite by deferrization under MD leaching. This indicated that the oxidation and dissolution of waste rocks would be accelerated due to the presence of Fe(III). Moreover, the content of gypsum in S4 increased from 0 to 4.1%, and anglesite ($PbSO_4$) occurred in S5 after MD leaching. These suggested that sulfide minerals in waste rocks have been oxidized to generate secondary sulfate minerals. Furthermore, the carbonate mineral content in S1–S5 decreased from 76.6%, 78.4%, 98.0%, 32.4%, and 17.2% to 62.2%, 60.9%, 94.8%, 20.2%, and 4.6% whereas the content of sulfur minerals increased in all waste rocks. This indicated that carbonate minerals dissolve under MD leaching, leading to an increase in the proportions of sulfur minerals. The correlation between the release concentrations of As and Pb under MD leaching and the mineral composition of waste rocks was obtained via RDA analysis (Figure 8). The sulfide mineral composition in waste rocks was positively correlated with the release concentration of As and Pb in the leaching solution [79]. This indicated that the release of As and Pb mainly came from the sulfide minerals in these waste rocks. When the waste rock was soaked in acid MD, metal sulfide minerals such As pyrite were dissolved. Fe and As were also dissolved in the leaching solution. Fe(II) was oxidized to Fe(III), which accelerated the dissolution of other sulfide minerals such as galena [80]. It was found that the high pollution of As and Pb in a lead–zinc mining area occurred due to the oxidation dissolution of sulfide minerals in waste rocks [79,81]. Carbonate and clay minerals in waste rocks were negatively correlated with the release concentration of As and Pb. The negative correlations of carbonate minerals with Pb releasing were lower than with As releasing. The dissolution of carbonate minerals would promote the reaction of Fe(II) with the thioarsenite to form arsenopyrite, thereby reducing the release of As [82]. The negative correlation between the dissolution of clay minerals and the release of Pb was greater than that of As. Part of the As was adsorbed or encased in clay minerals in the form of anions [66,67]. This indicated that more Pb than As would be released during the dissolution of clay minerals, but the opposite would be true in the dissolution of clay minerals.

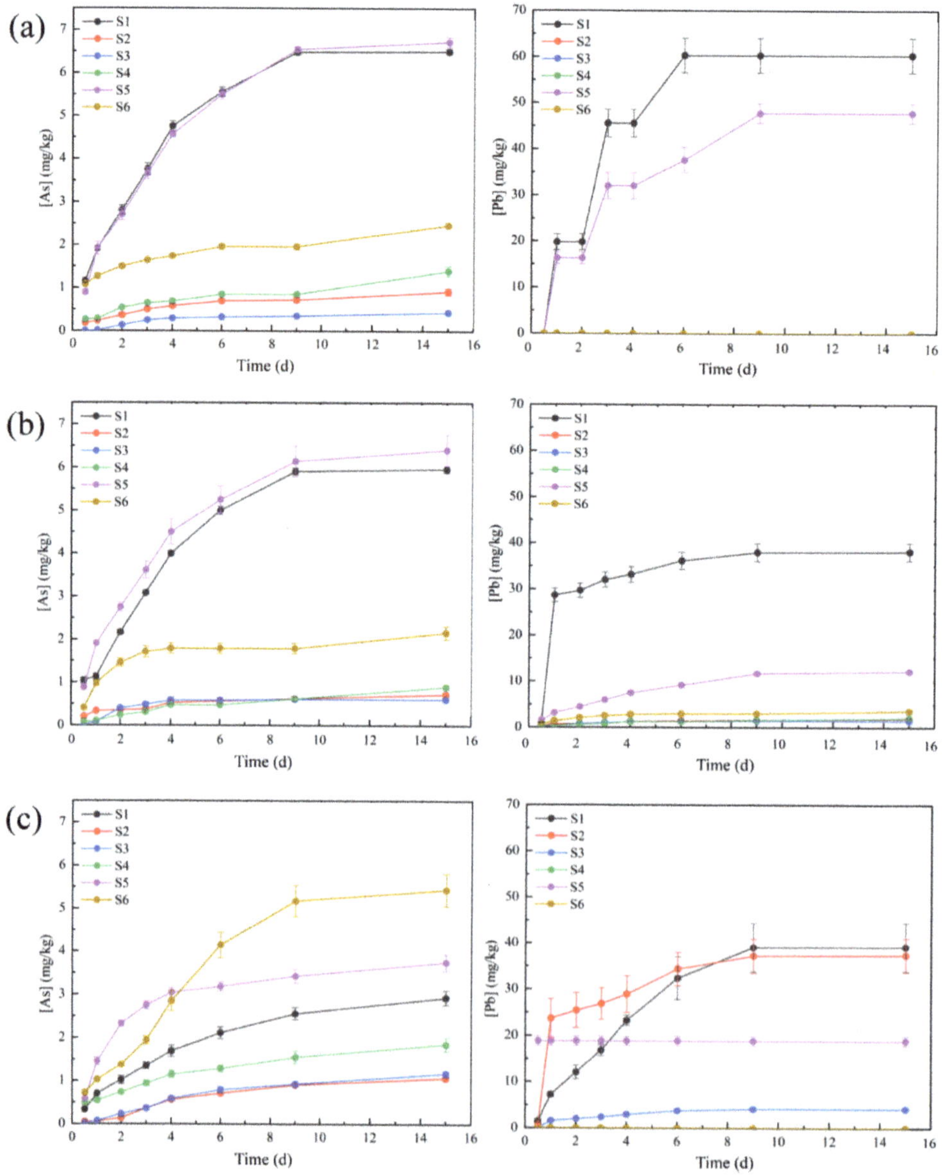

Figure 6. Releasing concentrations of As and Pb under MD (**a**), RW (**b**), and AS (**c**) leaching. On the left is the release of As in waste rock, and on the right is the release of Pb in waste rock.

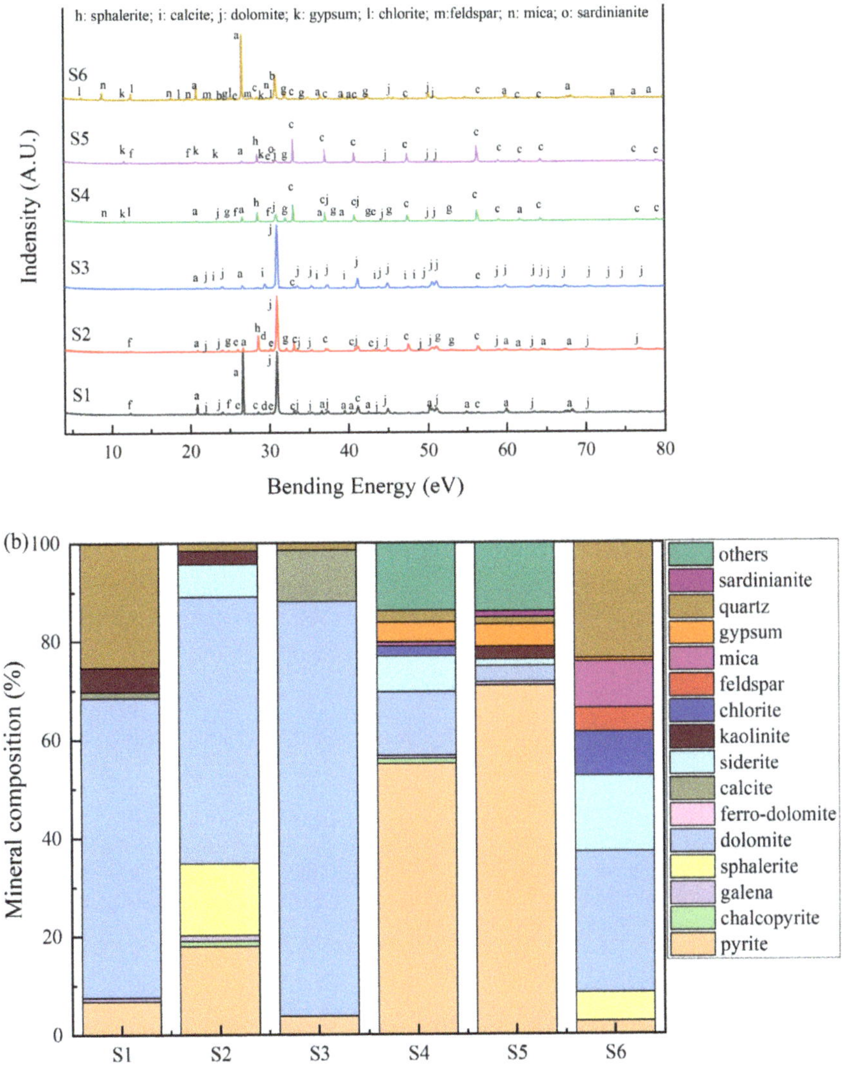

Figure 7. The XRD spectra of waste rocks after MD leaching (**a**) and their mineral proportions (**b**).

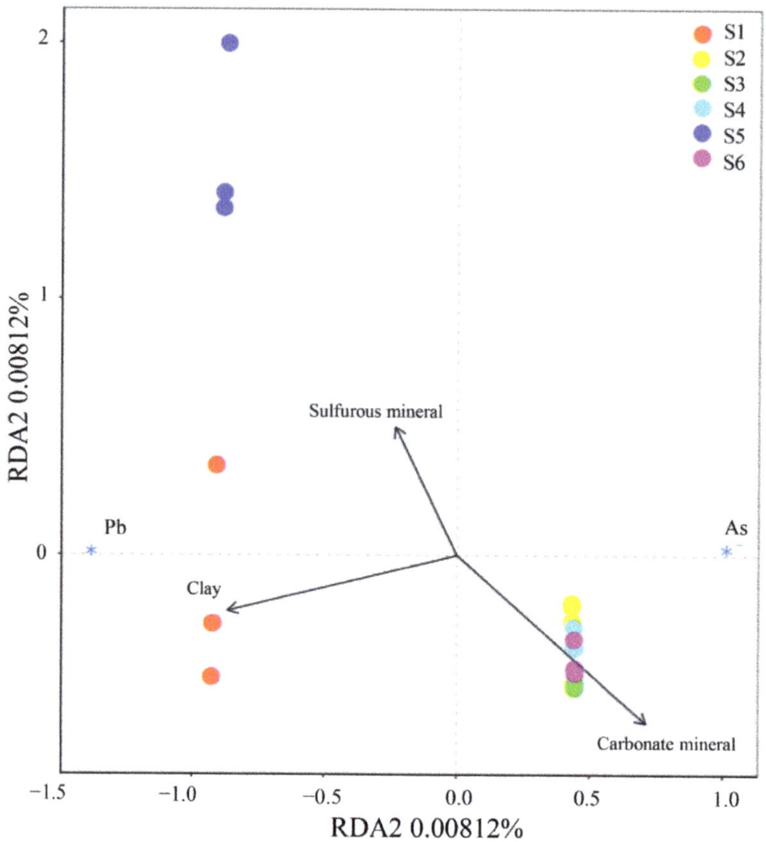

Figure 8. Ordination diagrams from RDA of releasing concentrations of As and Pb and the mineral composition in waste rocks during MD leaching. The * in the figure only indicates the position of As and Pb, and has no other meaning.

4. Conclusions

The findings showed that waste rocks in lead–zinc mine were mainly composed of carbonate minerals and sulfide minerals. The net acid production potential of S4 and S5 was greater due to their having a higher proportion of sulfide minerals and lower content of carbonate minerals. The main occurrence form of As was arsenopyrite, and a small part was associated with pyrite and chalcopyrite. However, the main occurrence form of Pb was galena. In addition, small amounts of As and Pb were adsorbed or encased by clay minerals. During the leaching of MD and RW, the pH, Eh and EC of each waste rock tended to be similar since each of these waste rocks had an acid–base neutralization ability. The leachate pH values of S1, S2, S3, and S6 were significantly higher than those of S4 and S5 due to their having a higher content of carbonate minerals and acid neutralization ability. Because MD had a higher acidity and Eh value, As and Pb were more easily released during MD leaching when compared to RW. Part of the As existed in the form of arsenate and was easier to release under alkaline conditions. Furthermore, the release of As and Pb in MD leaching showed a positive correlation with the change in sulfide minerals in these waste rocks but a negative correlation with clay minerals and carbonate minerals. This suggested that the release of As and Pb in waste rocks was affected by the environment and mineral composition and their physical and chemical properties. At present, As and Pb pollution control is still urgent. This paper has provided the basic data for subsequent pollution control and prevention. Future pollution control should focus on preventing the oxidation and dissolution of sulfide minerals in waste rocks.

Author Contributions: Investigation, Z.G., J.Y., K.L. and J.S.; Resources, B.M., H.L. and X.L.; Data curation, Z.G.; Writing—original draft, Z.G.; Writing—review & editing, Y.P., E.K.S. and L.J.; Supervision, X.L. and L.J. All authors have read and agreed to the published version of the manuscript.

Funding: This research was funded by the National Key Research and Development Program of China (Grants No. 2018YFC1800400), the National Natural Science Foundation of China (Grants No. 51909282), the Natural Science Foundation of Hunan Province of China (Grant No. 2022JJ40583), and Hunan Provincial Key Research and Development Plan (2022WK2017 and 2023NK2030).

Data Availability Statement: The original data presented in the study are included in the article; further inquiries can be directed to the corresponding author.

Acknowledgments: This project is funded under the National Key Research and Development Program of China (Grants No. 2018YFC1800400), the National Natural Science Foundation of China (Grants No. 51909282), the Natural Science Foundation of Hunan Province of China (Grant No. 2022JJ40583), and Hunan Provincial Key Research and Development Plan (2022WK2017 and 2023NK2030).

Conflicts of Interest: The authors declare no conflict of interest.

References

1. Hong, J.; Liu, L.; Zhang, Z.; Xia, X.; Yang, L.; Ning, Z.; Liu, C.; Qiu, G. Sulfate-accelerated photochemical oxidation of arsenopyrite in acidic systems under oxic conditions: Formation and function of schwertmannite. *J. Hazard. Mater.* **2022**, *433*, 128716. [CrossRef]
2. Jain, M.K.; Das, A. Impact of Mine Waste Leachates on Aquatic Environment: A Review. *Curr. Pollut. Rep.* **2017**, *3*, 31–37. [CrossRef]
3. Zhai, M.; Hu, R.; Wang, Y.; Jiang, S.; Wang, R.; Li, J.; Chen, H.; Yang, Z.; Lü, Q.; Qi, T. Mineral resource science in China: Review and perspective. *Geogr. Sustain.* **2021**, *2*, 107–114. [CrossRef]
4. Khaboushan, A.S.; Osanloo, M.; Esfahanipour, A. Optimization of open pit to underground transition depth: An idea for reducing waste rock contamination while maximizing economic benefits. *J. Clean. Prod.* **2020**, *277*, 123530. [CrossRef]
5. Peng, J.-Y.; Zhang, S.; Wang, Y.-J.; Zhao, R.-F.; Zhou, Y.-L.; Zhou, J.-W. Identification of priority pollutants and key factors affecting environmental risks of lead-zinc mine tailing sites. *Sci. Total Environ.* **2023**, *889*, 164039. [CrossRef] [PubMed]
6. Qi, M.; Wu, Y.; Zhang, S.; Li, G.; An, T. Pollution Profiles, Source Identification and Health Risk Assessment of Heavy Metals in Soil near a Non-Ferrous Metal Smelting Plant. *Int. J. Environ. Res. Public Health* **2023**, *20*, 1004. [CrossRef] [PubMed]
7. Jiang, Z.C.; Guo, Z.H.; Peng, C.; Liu, X.; Zhou, Z.R.; Xiao, X.Y. Heavy metals in soils around non-ferrous smelteries in China: Status, health risks and control measures. *Environ. Pollut.* **2021**, *282*, 117038. [CrossRef]
8. Jiang, L.; Sun, H.; Peng, T.; Ding, W.; Liu, B.; Liu, Q. Comprehensive evaluation of environmental availability, pollution level and leaching heavy metals behavior in non-ferrous metal tailings. *J. Environ. Manag.* **2021**, *290*, 112639. [CrossRef]
9. Vodyanitskii, Y.N. Contamination of soils with heavy metals and metalloids and its ecological hazard (analytic review). *Eurasian Soil Sci.* **2013**, *46*, 793–801. [CrossRef]
10. Zeng, J.; Luo, X.; Cheng, Y.; Ke, W.; Hartley, W.; Li, C.; Jiang, J.; Zhu, F.; Xue, S. Spatial distribution of toxic metal(loid)s at an abandoned zinc smelting site, Southern China. *J. Hazard. Mater.* **2022**, *425*, 127970. [CrossRef]
11. Huang, S.; Yuan, C.; Li, Q.; Yang, Y.; Tang, C.; Ouyang, K.; Wang, B. Distribution and risk assessment of heavy metals in soils from a typical Pb-Zn mining area. *Pol. J. Environ. Stud.* **2017**, *26*, 1105–1112. [CrossRef]
12. Song, L.; Qian, J.Z.; Zhang, F.W.; Kong, X.K.; Li, H.; Luan, S.; Zhang, Q.J.; Kang, Z.Q.; Han, Z.T.; Zhang, Z.J. An ecological remediation model combining optimal substrate amelioration and native hyperaccumulator colonization in non-ferrous metal tailings pond. *J. Environ. Manag.* **2022**, *322*, 116141. [CrossRef] [PubMed]
13. Hu, Z.; Wang, C.; Li, K.; Zhu, X. Distribution characteristics and pollution assessment of soil heavy metals over a typical nonferrous metal mine area in Chifeng, Inner Mongolia, China. *Environ. Earth Sci.* **2018**, *77*, 638. [CrossRef]
14. Obiora, S.C.; Chukwu, A.; Chibuike, G.; Nwegbu, A.N. Potentially harmful elements and their health implications in cultivable soils and food crops around lead-zinc mines in Ishiagu, Southeastern Nigeria. *J. Geochem. Explor.* **2019**, *204*, 289–296. [CrossRef]
15. Vural, A. Assessment of metal pollution associated with an alteration area: Old Gumushane, NE Black Sea. *Environ. Sci. Pollut. Res.* **2015**, *22*, 3219–3228. [CrossRef] [PubMed]
16. Kan, X.Q.; Dong, Y.Q.; Feng, L.; Zhou, M.; Hou, H.B. Contamination and health risk assessment of heavy metals in China's lead-zinc mine tailings: A meta-analysis. *Chemosphere* **2021**, *267*, 128909. [CrossRef] [PubMed]
17. Bigot, M.; Guterres, J.; Rossato, L.; Pudmenzky, A.; Doley, D.; Whittaker, M.; Pillai-McGarry, U.; Schmidt, S. Metal-binding hydrogel particles alleviate soil toxicity and facilitate healthy plant establishment of the native metallophyte grass *Astrebla lappacea* in mine waste rock and tailings. *J. Hazard. Mater.* **2013**, *248*, 424–434. [CrossRef] [PubMed]

18. Dong, Y.B.; Chen, D.N.; Lin, H. The behavior of heavy metal release from sulfide waste rock under microbial action and different environmental factors. *Environ. Sci. Pollut. Res.* **2022**, *29*, 75293–75306. [CrossRef] [PubMed]
19. Demers, I.; Bouda, M.; Mbonimpa, M.; Benzaazoua, M.; Bois, D.; Gagnon, M. Valorization of acid mine drainage treatment sludge as remediation component to control acid generation from mine wastes, part 2: Field experimentation. *Miner. Eng.* **2015**, *76*, 117–125. [CrossRef]
20. Wang, P.; Sun, Z.H.; Hu, Y.N.; Cheng, H.F. Leaching of heavy metals from abandoned mine tailings brought by precipitation and the associated environmental impact. *Sci. Total Environ.* **2019**, *695*, 133893. [CrossRef]
21. Yin, T.T.; Lin, H.; Dong, Y.B.; Wei, Z.S.; Li, B.; Liu, C.J.; Chen, X. Inhibition of cadmium releasing from sulfide tailings into the environment by carbonate-mineralized bacteria. *J. Hazard. Mater.* **2021**, *419*, 126479. [CrossRef] [PubMed]
22. Gong, B.N.; Wu, P.X.; Huang, Z.J.; Li, Y.Y.; Yang, S.S.; Dang, Z.; Ruan, B.; Kang, C.X. Efficient inhibition of heavy metal release from mine tailings against acid rain exposure by triethylenetetramine intercalated montmorillonite (TETA-Mt). *J. Hazard. Mater.* **2016**, *318*, 396–406. [CrossRef]
23. Ollson, C.J.; Smith, E.; Scheckel, K.G.; Betts, A.R.; Juhasz, A.L. Assessment of arsenic speciation and bioaccessibility in mine-impacted materials. *J. Hazard. Mater.* **2016**, *313*, 130–137. [CrossRef] [PubMed]
24. Pan, Y.X.; Chen, M.; Wang, X.T.; Chen, Y.D.; Dong, K. Ecological Risk Assessment and Source Analysis of Heavy Metals in the Soils of a Lead-Zinc Mining Watershed Area. *Water* **2023**, *15*, 15010113. [CrossRef]
25. Aykol, A.; Budakoglu, M.; Kumral, M.; Gultekin, A.H.; Turhan, M.; Esenli, V.; Yavuz, F.; Orgun, Y. Heavy metal pollution and acid drainage from the abandoned Balya Pb-Zn sulfide Mine, NW Anatolia, Turkey. *Environ. Geol.* **2003**, *45*, 198–208. [CrossRef]
26. Elghali, A.; Benzaazoua, M.; Taha, Y.; Amar, H.; Ait-khouia, Y.; Bouzahzah, H.; Hakkou, R. Prediction of acid mine drainage: Where we are. *Earth-Sci. Rev.* **2023**, *241*, 104421. [CrossRef]
27. Grathwohl, P.; Susset, B. Comparison of percolation to batch and sequential leaching tests: Theory and data. *Waste Manag.* **2009**, *29*, 2681–2688. [CrossRef]
28. Hama, J.R.; Jorgensen, D.B.G.; Diamantopoulos, E.; Bucheli, T.D.; Hansen, H.C.B.; Strobel, B.W. Indole and quinolizidine alkaloids from blue lupin leach to agricultural drainage water. *Sci. Total Environ.* **2022**, *834*, 155283. [CrossRef]
29. Yin, K.; Chan, W.P.; Dou, X.M.; Ren, F.; Chang, V.W.C. Measurements, factor analysis and modeling of element leaching from incineration bottom ashes for quantitative component effects. *J. Clean. Prod.* **2017**, *165*, 477–490. [CrossRef]
30. Parbhakar-Fox, A.; Lottermoser, B.; Bradshaw, D. Evaluating waste rock mineralogy and microtexture during kinetic testing for improved acid rock drainage prediction. *Miner. Eng.* **2013**, *52*, 111–124. [CrossRef]
31. Chai, J.C.; Onitsuk, K.; Hayashi, S. Cr(VI) concentration from batch contact/tank leaching and column percolation test using fly ash with additives. *J. Hazard. Mater.* **2009**, *166*, 67–73. [CrossRef] [PubMed]
32. Almeida, C.; Grosselli, M.; Gonzalez, P.; Martinez, D.; Gil, R. Batch leaching tests of motherboards to assess environmental contamination by bromine, platinum group elements and other selected heavy metals. *Chemosphere* **2016**, *144*, 1–6. [CrossRef] [PubMed]
33. Li, L.Y.; Ohtsubo, M.; Higashi, T.; Yamaoka, S.; Morishita, T. Leachability of municipal solid waste ashes in simulated landfill conditions. *Waste Manag.* **2007**, *27*, 932–945. [CrossRef] [PubMed]
34. Hosseini, S.A.; Raygan, S.; Rezaei, A.; Jafari, A. Leaching of nickel from a secondary source by sulfuric acid. *J. Environ. Chem. Eng.* **2017**, *5*, 3922–3929. [CrossRef]
35. Lv, J.F.; Zheng, Y.X.; Tong, X.; Li, X. Clean utilization of waste rocks as a novel adsorbent to treat the beneficiation wastewater containing arsenic and fluorine. *J. Clean. Prod.* **2021**, *293*, 126160. [CrossRef]
36. Kappen, P.; Ferrando-Miguel, G.; Reichman, S.M.; Innes, L.; Welter, E.; Pigram, P.J. Antimony leaching and chemical species analyses in an industrial solid waste: Surface and bulk speciation using ToF-SIMS and XANES. *J. Hazard. Mater.* **2017**, *329*, 131–140. [CrossRef] [PubMed]
37. Ullah, S.; Ahmad, F.; Yusoff, P.S.M.M. Effect of boric acid and melamine on the intumescent fire-retardant coating composition for the fire protection of structural steel substrates. *J. Appl. Polym. Sci.* **2013**, *128*, 2983–2993. [CrossRef]
38. Soltan, A.M.M.; Pöhler, K.; Fuchs, F.; Abd El-Raoof, F.; El-Kaliouby, B.A.H.; Koenig, A.; Pöllmann, H. Clay-bricks from recycled rock tailings. *Ceram. Int.* **2016**, *42*, 16685–16696. [CrossRef]
39. Wang, D.F.; Zhang, Y.S.; Li, Z.H.; Shi, J.S.; Liu, C.; Pang, B.; Chen, Y.D.; Liu, G.J.; Sun, G.W. Systemical investigation on the determination of sulfate in cement-based materials based on a promoted conductometric titrator. *Measurement* **2022**, *203*, 111909. [CrossRef]
40. Sonnichsen, C.; Atamanchuk, D.; Hendricks, A.; Morgan, S.; Smith, J.; Grundke, I.; Luy, E.; Sieben, V.J. An Automated Microfluidic Analyzer for In Situ Monitoring of Total Alkalinity. *ACS Sens.* **2023**, *8*, 344–352. [CrossRef]
41. Jia, G.; Wang, Y.; Yang, F.; Ma, Z. Preparation of CFB fly ash/sewage sludge ceramsite and the morphological transformation and release properties of sulfur. *Construction and Building Materials* **2023**, *373*, 130864. [CrossRef]
42. Xu, F.; Chu, M.; Chang, Z.; Gu, Z.; Sun, X. Sulfur release and transformation during the pyrolysis of lignite with different particle sizes. *J. Anal. Appl. Pyrolysis* **2021**, *156*, 105162. [CrossRef]
43. Yang, J.J.; Guo, Z.W.; Jiang, L.H.; Sarkodie, E.K.; Li, K.W.; Shi, J.X.; Deng, Y.; Zhang, Z.C.; Liu, H.W.; Liang, Y.L.; et al. Cadmium, lead and arsenic contamination in an abandoned nonferrous metal smelting site in southern China: Chemical speciation and mobility. *Ecotoxicol. Environ. Saf.* **2022**, *239*, 113617. [CrossRef] [PubMed]

44. Zeng, J.; Tabelin, C.B.; Gao, W.; Tang, L.; Luo, X.; Ke, W.; Jiang, J.; Xue, S. Heterogeneous distributions of heavy metals in the soil-groundwater system empowers the knowledge of the pollution migration at a smelting site. *Chem. Eng. J.* **2023**, *454*, 140307. [CrossRef]
45. Ma, T.; Luo, H.; Huang, K.; Pan, Y.; Tang, T.; Tao, X.; Lu, G. Integrated ecological risk assessment of heavy metals in an oil shale mining area after restoration. *J. Environ. Manag.* **2021**, *300*, 113797. [CrossRef] [PubMed]
46. Panda, S.; Akcil, A.; Mishra, S.; Erust, C. Synergistic effect of biogenic Fe^{3+} coupled to S° oxidation on simultaneous bioleaching of Cu, Co, Zn and as from hazardous Pyrite Ash Waste. *J. Hazard. Mater.* **2017**, *325*, 59–70. [CrossRef] [PubMed]
47. Jin, L.; Le, Z.; Li, Z.; Liu, J.M.; Ren, A. Placental concentrations of mercury, lead, cadmium, and arsenic and the risk of neural tube defects in a Chinese population. *Reprod. Toxicol.* **2013**, *35*, 25–31. [CrossRef] [PubMed]
48. Elghali, A.; Benzaazoua, M.; Bouzahzah, H.; Abdelmoula, M.; Dynes, J.J.; Jamieson, H.E. Role of secondary minerals in the acid generating potential of weathered mine tailings: Crystal-chemistry characterization and closed mine site management involvement. *Sci. Total Environ.* **2021**, *784*, 147105. [CrossRef]
49. Jennings, S.R.; Dollhopf, D.J.; Inskeep, W.P. Acid production from sulfide minerals using hydrogen peroxide weathering. *Appl. Geochem.* **2000**, *15*, 235–243. [CrossRef]
50. Sakala, E.; Fourie, F.; Gomo, M.; Madzivire, G. Natural Attenuation of Acid Mine Drainage by Various Rocks in the Witbank, Ermelo and Highveld Coalfields, South Africa. *Nat. Resour. Res.* **2021**, *30*, 557–570. [CrossRef]
51. Holmstrom, H.; Ljungberg, J.; Ohlander, B. Role of carbonates in mitigation of metal release from mining waste. Evidence from humidity cells tests. *Environ. Geol.* **1999**, *37*, 267–280. [CrossRef]
52. Gu, S.Q.; Kang, X.N.; Wang, L.; Lichtfouse, E.; Wang, C.Y. Clay mineral adsorbents for heavy metal removal from wastewater: A review. *Environ. Chem. Lett.* **2019**, *17*, 629–654. [CrossRef]
53. Du, H.H.; Chen, W.L.; Cai, P.; Rong, X.M.; Feng, X.H.; Huang, Q.Y. Competitive adsorption of Pb and Cd on bacteria-montmorillonite composite. *Environ. Pollut.* **2016**, *218*, 168–175. [CrossRef] [PubMed]
54. Fakhreddine, S.; Fendorf, S. The effect of porewater ionic composition on arsenate adsorption to clay minerals. *Sci. Total Environ.* **2021**, *785*, 147096. [CrossRef] [PubMed]
55. Covelo, E.F.; Vega, F.A.; Andrade, M.L. Heavy metal sorption and desorption capacity of soils containing endogenous contaminants. *J. Hazard. Mater.* **2007**, *143*, 419–430. [CrossRef] [PubMed]
56. Kastyuchik, A.; Karam, A.; Aider, M. Effectiveness of alkaline amendments in acid mine drainage remediation. *Environ. Technol. Innov.* **2016**, *6*, 49–59. [CrossRef]
57. Li, Y.; Yin, H.; Cai, Y.; Luo, H.; Yan, C.; Dang, Z. Regulating the exposed crystal facets of alpha-Fe_2O_3 to promote Fe_2O_3-modified biochar performance in heavy metals adsorption. *Chemosphere* **2023**, *311*, 136976. [CrossRef]
58. Zhang, H.P.; Gu, L.Q.; Zhang, L.; Zheng, S.R.; Wan, H.Q.; Sun, J.Y.; Zhu, D.Q.; Xu, Z.Y. Removal of aqueous Pb(II) by adsorption on Al_2O_3-pillared layered MnO_2. *Appl. Surf. Sci.* **2017**, *406*, 330–338. [CrossRef]
59. Faheem; Yu, H.X.; Liu, J.; Shen, J.Y.; Sun, X.Y.; Li, J.S.; Wang, L.J. Preparation of MnO_x-loaded biochar for Pb^{2+} removal: Adsorption performance and possible mechanism. *J. Taiwan Inst. Chem. Eng.* **2016**, *66*, 313–320. [CrossRef]
60. Mitchell, R.H.; Smith, D.L. Geology and mineralogy of the Ashram Zone carbonatite, Eldor Complex, Quebec. *Ore Geol. Rev.* **2017**, *86*, 784–806. [CrossRef]
61. Mbamba, C.K.; Harrison, S.T.L.; Franzidis, J.P.; Broadhurst, J.L. Mitigating acid rock drainage risks while recovering low-sulfur coal from ultrafine colliery wastes using froth flotation. *Miner. Eng.* **2012**, *29*, 13–21. [CrossRef]
62. Liu, Y.B.; Cui, J.; Peng, Y.; Lu, Y.F.; Yao, D.R.; Yang, J.; He, Y. Atmospheric deposition of hazardous elements and its accumulation in both soil and grain of winter wheat in a lead-zinc smelter contaminated area, Central China. *Sci. Total Environ.* **2020**, *707*, 135789. [CrossRef] [PubMed]
63. Sun, H.R.; Yin, W.Z.; Yang, B.; Han, F. Simultaneous separation of quartz and dolomite from magnesite using monosodium phosphate as a regulator via reverse flotation. *Miner. Eng.* **2021**, *172*, 107185. [CrossRef]
64. Qiu, G.; Gao, T.; Hong, J.; Tan, W.; Liu, F.; Zheng, L. Mechanisms of arsenic-containing pyrite oxidation by aqueous arsenate under anoxic conditions. *Geochim. Cosmochim. Acta* **2017**, *217*, 306–319. [CrossRef]
65. Ramírez-Aldaba, H.; Valles, O.P.; Vazquez-Arenas, J.; Rojas-Contreras, J.A.; Valdez-Pérez, D.; Ruiz-Baca, E.; Meraz-Rodríguez, M.; Sosa-Rodríguez, F.S.; Rodríguez, A.G.; Lara, R.H. Chemical and surface analysis during evolution of arsenopyrite oxidation by *Acidithiobacillus thiooxidans* in the presence and absence of supplementary arsenic. *Sci. Total Environ.* **2016**, *566*, 1106–1119. [CrossRef] [PubMed]
66. Pumure, I.; Renton, J.J.; Smart, R.B. The interstitial location of selenium and arsenic in rocks associated with coal mining using ultrasound extractions and principal component analysis (PCA). *J. Hazard. Mater.* **2011**, *198*, 151–158. [CrossRef] [PubMed]
67. Playter, T.; Konhauser, K.; Owttrim, G.; Hodgson, C.; Warchola, T.; Mloszewska, A.M.; Sutherland, B.; Bekker, A.; Zonneveld, J.P.; Pemberton, S.G.; et al. Microbe-clay interactions as a mechanism for the preservation of organic matter and trace metal biosignatures in black shales. *Chem. Geol.* **2017**, *459*, 75–90. [CrossRef]
68. Deng, M.G.; Zhao, J.X.; Liu, F.X.; Yu, H.J.; Sun, B.D.; Liu, F.; Li, S.B. Discussion on sources of metallogenic fluids and materials of the Shuitoushan Pb-Zn deposit in Zhenkang, western Yunnan: Evidence from H, O, S and Pb isotopes. *Acta Petrol. Sin.* **2017**, *33*, 2001–2017.
69. Su, M.; Han, F.Y.; Wang, M.X.; Ma, J.X.; Wang, X.W.; Wang, Z.J.; Hu, S.J.; Li, Z. Clay-assisted protection of *Enterobacter* sp. from Pb (II) stress. *Ecotoxicol. Environ. Saf.* **2021**, *208*, 111704. [CrossRef]

70. Perez-Lopez, R.; Nieto, J.M.; de Almodovar, G.R. Utilization of fly ash to improve the quality of the acid mine drainage generated by oxidation of a sulphide-rich mining waste: Column experiments. *Chemosphere* **2007**, *67*, 1637–1646. [CrossRef]
71. Minamikawa, K.; Sakai, N. The effect of water management based on soil redox potential on methane emission from two kinds of paddy soils in Japan. *Agric. Ecosyst. Environ.* **2005**, *107*, 397–407. [CrossRef]
72. Yao, B.M.; Wang, S.Q.; Xie, S.T.; Li, G.; Sun, G.X. Optimal soil Eh, pH for simultaneous decrease of bioavailable Cd, as in co-contaminated paddy soil under water management strategies. *Sci. Total Environ.* **2022**, *806*, 151342. [CrossRef] [PubMed]
73. Asael, D.; Matthews, A.; Oszczepalski, S.; Bar-Matthews, M.; Halicz, L. Fluid speciation controls of low temperature copper isotope fractionation applied to the Kupferschiefer and Timna ore deposits. *Chem. Geol.* **2009**, *262*, 147–158. [CrossRef]
74. Kasemodel, M.C.; Sakamoto, I.K.; Varesche, M.B.A.; Rodrigues, V.G.S. Potentially toxic metal contamination and microbial community analysis in an abandoned Pb and Zn mining waste deposit. *Sci. Total Environ.* **2019**, *675*, 367–379. [CrossRef] [PubMed]
75. Opiso, E.M.; Sato, T.; Morimoto, K.; Asai, A.; Anraku, S.; Numako, C.; Yoneda, T. Incorporation of arsenic during the formation of Mg-bearing minerals at alkaline condition. *Miner. Eng.* **2010**, *23*, 230–237. [CrossRef]
76. Kong, L.H.; Hu, X.Y.; Peng, X.J.; Wang, X.L. Specific H_2S Release from Thiosulfate Promoted by UV Irradiation for Removal of Arsenic and Heavy Metals from Strongly Acidic Wastewater. *Environ. Sci. Technol.* **2020**, *54*, 14076–14084. [CrossRef] [PubMed]
77. Mukwaturi, M.; Lin, C.X. Mobilization of heavy metals from urban contaminated soils under water inundation conditions. *J. Hazard. Mater.* **2015**, *285*, 445–452. [CrossRef]
78. Zhang, J.F.; Xie, X.D.; Meng, X.G.; Li, Y.; Zhu, W.H. The critical role of oxidative debris in the adsorption and desorption of Pb(II) to graphene oxides under alkaline groundwater conditions. *Sci. Total Environ.* **2020**, *704*, 135254. [CrossRef]
79. Lee, P.-K.; Kang, M.-J.; Jeong, Y.-J.; Kwon, Y.K.; Yu, S. Lead isotopes combined with geochemical and mineralogical analyses for source identification of arsenic in agricultural soils surrounding a zinc smelter. *J. Hazard. Mater.* **2020**, *382*, 121044. [CrossRef]
80. Shi, S.Y.; Fang, Z.H. Bioleaching of marmatite flotation concentrate by *Acidithiobacillus ferrooxidans*. *Hydrometallurgy* **2004**, *75*, 1–10. [CrossRef]
81. Tum, S.; Matsumoto, S.; Nishikata, M.; Yasutaka, T. Assessment of seasonal changes in groundwater quality of waste rock dump in temperate continental climate, northern Japan. *Chemosphere* **2023**, *327*, 138482. [CrossRef]
82. Kim, M.J.; Nriagu, J.; Haack, S. Carbonate ions and arsenic dissolution by groundwater. *Environ. Sci. Technol.* **2000**, *34*, 3094–3100. [CrossRef]

toxics

MDPI

Article

Mathematical Estimation of Endogenous Proline as a Bioindicator to Regulate the Stress of Trivalent Chromium on Rice Plants Grown in Different Nitrogenous Conditions

Chengzhi Li †, Yuxi Feng †, Peng Tian and Xiaozhang Yu *

College of Environmental Science & Engineering, Guilin University of Technology, Guilin 541004, China; yu-xifeng@foxmail.com (Y.F.)
* Correspondence: xzyu@glut.edu.cn; Tel.: +86-773-5897016
† The author contributed equally to this work.

Abstract: The accumulation of proline impacts the defense mechanisms of plants against the harmful effects of adverse environmental conditions; however, its concentration in plants is associated with the metabolism of N. Therefore, the effects of exogenous organic [glutamate (Glu)/arginine (Arg)] and inorganic [nitrate (NO_3^-)/ammonium (NH_4^+)] N on the accumulation of proline (Pro) in rice plants under trivalent chromium [Cr(III)] stress were studied through using the mass balance matrix model (MBMM). Application of 'NH_4^+' showed the largest contribution to the Pro content in rice shoots under different concentrations of Cr(III), followed by 'NO_3^-', 'Arg', and 'Glu' applications. On the other hand, 'Arg' application displayed the largest contribution to the Pro content in roots under Cr(III) stress, followed by 'NH_4^+', 'Glu', and 'NO_3^-' applications. The combined application of 'NH_4^++Arg' showed the greatest contribution to the Pro content in both roots and shoots of Cr(III)-treated rice seedlings, while the application of 'NO_3^-+Glu' showed the least contribution to the Pro content in rice seedlings. The current study indicated that the endogenous level of Pro in rice seedlings is quite sensitive to Cr(III) stress under different N sources, and the mathematical modeling showed a reliable result while estimating the relationship between Pro content and N source application.

Keywords: proline; rice; trivalent chromium; nitrogen source; mass balance

1. Introduction

Chromium (Cr) is widely distributed in the environment through different activities of industries such as metallurgical and chemical [1]. Due to its high solubility, mobility, and oxidizing potential, Cr is considered as one of the top 20 hazardous materials to be remediated on an early basis [2]. Naturally, Cr can exist in several oxidation states, ranging from "−2" to "+6", in which the hexavalent [Cr(VI)] and trivalent chromium [Cr(III)] are the most stable forms of Cr [1]. Continuous input of Cr into the environment from the industrial sector makes it a serious threat to plants [3,4]. Although plants uptake a small amount of Cr from the soil, its over-accumulation in plants affects their nitrogen (N) metabolism. For example, Cr exposure imbalanced the assimilation of nitrate (NO_3^-) and ammonium (NH_4^+) by modifying the activities of nitrate reductase (NR), nitrite reductase (NiR), glutamine (GS), and glutamate dehydrogenase (GDH) in *Sorghum bicolor* and *Solanum. lycopersicum* (Martins et al. [5]).

Nitrogen is a crucial macroelement for supporting plant growth and development [6], wherein NO_3^- and NH_4^+ are the most available inorganic forms of N for plants. Uptake and subsequent assimilation of NO_3^- or NH_4^+ by plants play a crucial role in improving plant growth and yield [7]. It provides the building blocks for the synthesis of a plethora of biomolecules, such as proteins, nucleic acids, and chlorophyll. Among the biomolecules, amino acids are the major component of plant biomass [8]. Proline (Pro), a common amino acid, plays an important role in plants. For instance, it maintains osmotic balance, protects subcellular structures, scavenges reactive oxygen species (ROS), stabilizes protein and DNA, and provides N sources in responses to stress conditions including drought, high temperature, salinity, UV radiation, pathogens, and chemical exposure [9]. In addition to acting as an excellent osmolyte, proline also plays a major role as a metal chelator, an antioxidative defense molecule, and a signaling molecule [10]. Therefore, the level of Pro accumulation in plant tissues has been suggested as a sensitive indicator to evaluate the overall performance of plant growth in various contaminated sites. It has been reported that the content of amino acids in plants is highly dependent on N nutrition [11]. Our previous study also confirmed that NH_4^+-fed rice seedlings showed a significantly dose-dependent increase in Pro in shoots, while the innate level of Pro in NO_3^--fed rice seedlings is independent of the NO_3^- dose supplied. Additionally, we noticed that accumulation of Pro was observable in rice plants supplied with additional Glu and Arg, where the latter demonstrated much higher potential than the former during the synthesis of Pro in rice plants [12].

Unfavorable environmental conditions inhibit plant growth and development by altering various biological processes including N metabolism. In this case, a change in the N cycle could alter the composition and the synthesis of amino acids [13]. Literature on the effect of chromium toxicity on enzymes of nitrogen metabolism is available [14]. It is also reported that fertilization of inorganic N (NO_3^- and NH_4^+) can influence the bioavailability and toxicity of Cu, Cd, and Cr in plants by altering the synthesis of organic molecules (with N), e.g., Pro, Glu, and Arg [15,16]. Our previous study reported that the innate level of Pro in rice plants is changeable due to the application of different N sources (NO_3^- and NH_4^+) under Cr stress [17]. Rice is one of the world's most produced crops and a major energy source in the world. Research on the synthesis of Pro in rice plants from both inorganic (NO_3^- and NH_4^+) and organic (Arg and Glu) N sources showed a completely different increment pattern [12]. However, no information is available to investigate the endogenous Pro in rice plants fertilized with different nitrogenous chemicals under Cr(III) stress. Therefore, we hypothesized that exogenous N affects the accumulation of proline (Pro) in rice plants under trivalent chromium.

To achieve the objective, we carried out the present study in the following manner: (1) determined the content of Pro in rice tissues under Cr(III) stress with different inorganic (NO_3^- and NH_4^+) and organic (Arg and Glu) N sources alone and in combination; (2) mathematically evaluated the contribution of different N sources alone and in combination to Pro content in rice plants under Cr(III) stress, based on the mass balance matrix model (MBMM); (3) predicted a suitable combination of different N sources for regulation of Cr(III) stress in rice plants using the content of Pro as a bioindicator. Overall, our study provides a new method to elucidate the contribution of different N sources to Pro accumulation in rice plants under Cr(III) stress.

2. Methods and Materials

2.1. Rice Seedlings and Cr Treatment

Rice (*Oryza sativa* L. XZX 45) seedlings were obtained using the method described in our previous study [6]. Briefly, after soaking the seeds with deionized water for 24 h, the seeds were cultivated in sand soils and placed in a growth chamber (temperature: 25 ± 0.5 °C and relative humidity: $60 \pm 2\%$). Rice seedlings were irrigated with a modified ISO8629 nutrient solution during the entire growth period, i.e., 16 days [1]. Seedlings of similar sizes were selected for the following treatments (Figure 1).

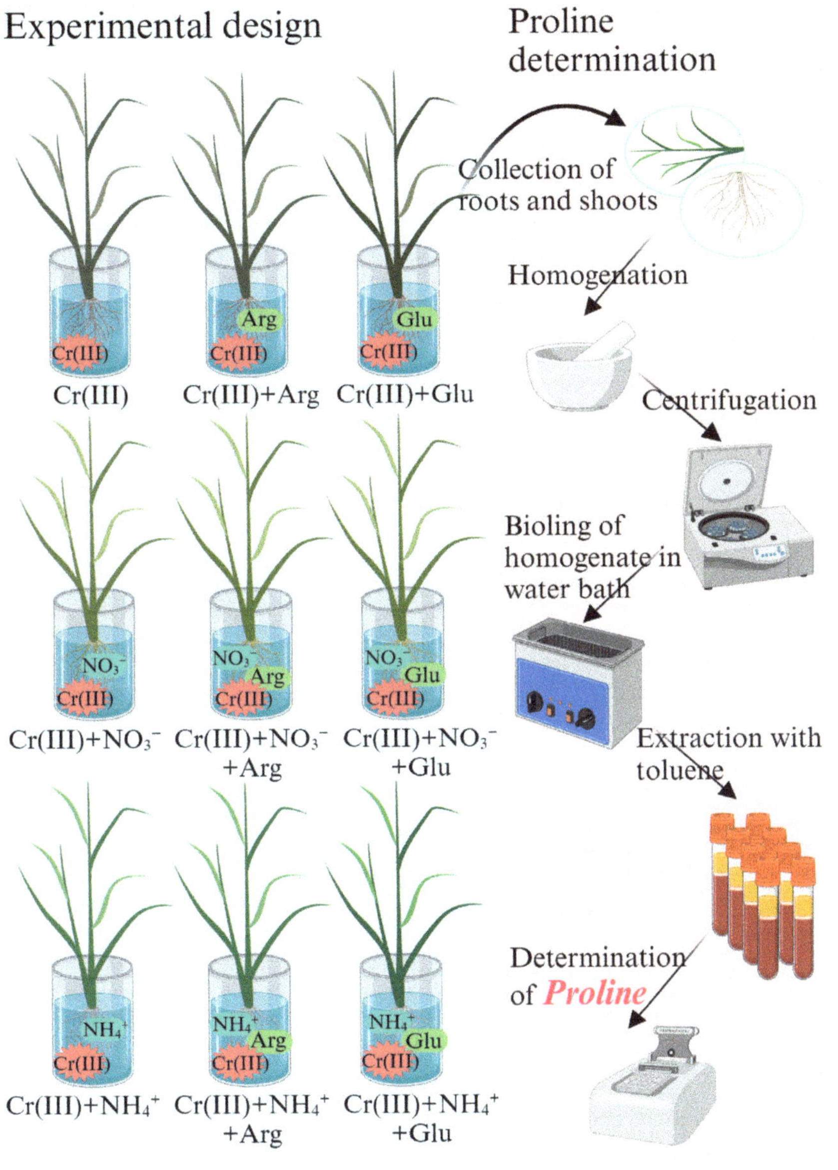

Figure 1. The design of experimental treatments.

(1) 'Cr(III)+($-N_I$)' treatments: Rice seedlings were pre-treated with the nutrition solution without KNO_3/NH_4Cl ('$-N_I$'), but with 3 mM Arg ($+N_{Arg}$) or 10 mM Glu ($+N_{Glu}$) for 12 h [12], and subsequently exposed to Cr(III) solution at 0, 12.0, 24.0 and 40.0 mg Cr/L for 2 days (Figure 1).

(2) 'Cr(III)+($+N_{NO3^-}$)' treatments: Rice seedlings were pre-treated with the KNO_3-containing nutrient solution ($+N_{NO3^-}$) with 3 mM Arg ($+N_{Arg}$) or 10 mM Glu ($+N_{Glu}$) for 12 h, respectively [12], and subsequently exposed to Cr(III) solution at 0, 12.0, 24.0 and 40.0 mg Cr/L for 2 days (Figure 1).

(3) 'Cr(III)+($+N_{NH4^+}$)' treatments: Rice seedlings were pre-treated with the NH_4Cl-containing nutrition solution '$+N_{NH4^+}$', with 3 mM Arg ($+N_{Arg}$) or 10 mM Glu ($+N_{Glu}$) for 12 h [12], and exposed to Cr(III) solution at 0, 12.0, 24.0 and 40.0 mg Cr/L for 2 days (Figure 1).

The symbol '$-N_O$' indicates N treatment without Arg or Glu; '$+N_{Arg}$' indicates N treatment with Arg; '$+N_{Glu}$' indicates N treatment with Glu; '$-N_I$' indicates the N treatment without KNO_3 or NH_4Cl; '$+N_{NO3^-}$' indicates KNO_3 treatment'; '$+N_{NH4^+}$' indicates NH_4Cl treatment. The weight of KNO_3 and NH_4Cl in the nutrient solution is equal to 39.5 mg N/L. All glass containers were wrapped with aluminum foil to minimize water loss and inhibit algae growth. Each treatment was prepared in four biological replicates. All chemicals used were of analytical grade and purchased from Aladdin Chemistry Co., Ltd. (Shanghai, China).

2.2. Measurement of Relative Growth Rate

Rice seedlings were weighed before and after exposure to Cr(III) treatments. The relative growth rate (RGR, %) was calculated based on our previous study [12].

2.3. Measurement of Pro Content in Rice Seedlings

After 2 days of exposure, rice seedlings were divided into roots and shoots. Subsequently, these tissues were homogenized in a pre-chilled mortar with 3% sulfosalicylic acid (5 mL). The homogenate was transferred to a 10 mL tube for centrifuging (4 °C, 11,000 rpm, 15 min). After centrifugation, 2 mL of the supernatant was taken and mixed with equal quantity, i.e., 2.0 mL of glacial acetic acid and 2.5% ninhydrin (glacial acetic acid: 6 mol/L phosphoric acid, 60:40) solution, and boiled in a water bath for 1 h. For instant cooling, the solution was kept on ice for 5 min followed by extraction of desired product using 4 mL toluene. The amount of Pro content was estimated using a spectrophotometer at 520 nm against a toluene reference, as per standard protocol formulated by Li et al. [12]. A standard curve was constructed to measure proline content between the absorbance at 520 nm and L-proline content (Supplementary Materials). Series content of L-proline was 0, 2, 4, 6, 8, and 10 μg/mL of ddH_2O. A linear regression was observed between the absorbance values at 520 nm and L-proline contents at 0–10 μg (R^2 = 0.9992).

2.4. Modeling the "Mass Balance Matrix"

In this study, we developed a "mass balance matrix" model (MBMM), based on the elementary rows (*r*)/columns (*c*) transformation to predict the optimal tolerance strategies for rice seedlings grown in different N sources under Cr(III) stress using the content of Pro as the dependent variable. Accordingly, the contribution of different N sources to Pro content was estimated. The elements of all matrixes are denoted by a_{ij} (i, j = 1, 2, 3).

The fundamental matrix is as follows:

$$\begin{bmatrix} (-N_O)\&(-N_I) & (+N_{Arg})\&(-N_I) & (+N_{Glu})\&(-N_I) \\ (-N_O)\&(+N_{NO_3^-}) & (+N_{Arg})\&(+N_{NO_3^-}) & (+N_{Glu})\&(+N_{NO_3^-}) \\ (-N_O)\&(+N_{NH_4^+}) & (+N_{Arg})\&(+N_{NH_4^+}) & (+N_{Glu})\&(+N_{NH_4^+}) \end{bmatrix} \quad (1)$$

The rows (i) and columns (j) of Matrix (1) are denoted by r_i and c_j (i, j = 1, 2, 3), respectively. The '&' indicated the combination of two different N sources. For example, a_{11} refers to the treatment without organic N, '$-N_O$' and inorganic N (KNO_3/NH_4Cl) '$-N_I$'; a_{12} refers to the treatment with organic N (Arg) '$+N_{Arg}$', but without inorganic N (KNO_3/NH_4Cl) '$-N_I$'; a_{13} refers to the treatment with organic N (Glu) '$+N_{Glu}$', but without inorganic N (KNO_3/NH_4Cl) '$-N_I$'.

To compare the contribution of organic and inorganic N to Pro content, Matrix (1) was performed with the elementary row (r) transformation, i.e., $r_2 - r_1$, $r_3 - r_1$, and $r_3 - r_2$. Therefore, the following three matrices were obtained:

$$\begin{bmatrix} (+N_{NO_3^-})-(-N_I) & (+N_{NO_3^-})-(-N_I) & (+N_{NO_3^-})-(-N_I) \\ (-N_O)+(+N_{NO_3^-}) & (+N_{Arg})+(+N_{NO_3^-}) & (+N_{Arg})+(+N_{NO_3^-}) \\ (-N_O)+(+N_{NH_4^+}) & (+N_{Arg})+(+N_{NH_4^+}) & (+N_{Arg})+(+N_{NH_4^+}) \end{bmatrix} \tag{2}$$

$$\begin{bmatrix} (+N_{NH_4^+})-(-N_I) & (+N_{NH_4^+})-(-N_I) & (+N_{NH_4^+})-(-N_I) \\ (-N_O)+(+N_{NO_3^-}) & (+N_{Arg})+(+N_{NO_3^-}) & (+N_{Glu})+(+N_{NO_3^-}) \\ (-N_O)+(+N_{NH_4^+}) & (+N_{Arg})+(+N_{NH_4^+}) & (+N_{Glu})+(+N_{NH_4^+}) \end{bmatrix} \tag{3}$$

$$\begin{bmatrix} (-N_O)+(-N_I) & (+N_{Arg})+(-N_I) & (+N_{Glu})+(-N_I) \\ (+N_{NH_4^+})-(+N_{NO_3^-}) & (+N_{NH_4^+})-(+N_{NO_3^-}) & (+N_{NH_4^+})-(+N_{NO_3^-}) \\ (-N_O)+(+N_{NH_4^+}) & (+N_{Arg})+(+N_{NH_4^+}) & (+N_{Glu})+(+N_{NH_4^+}) \end{bmatrix} \tag{4}$$

r_1 in Matrix (2), r_1 in Matrix (3) and r_2 in Matrix (4) were all extracted to obtain Matrix (5):

$$\begin{bmatrix} (+N_{NO_3^-})-(-N_I) & (+N_{NO_3^-})-(-N_I) & (+N_{NO_3^-})-(-N_I) \\ (+N_{NH_4^+})-(-N_I) & (+N_{NH_4^+})-(-N_I) & (+N_{NH_4^+})-(-N_I) \\ (+N_{NH_4^+})-(+N_{NO_3^-}) & (+N_{NH_4^+})-(+N_{NO_3^-}) & (+N_{NH_4^+})-(+N_{NO_3^-}) \end{bmatrix} \tag{5}$$

where r_1 is the contribution of inorganic N (KNO_3) '$+N_{NO3}{}^{-}$' to Pro content; r_2 is the contribution of inorganic N (NH_4Cl) '$+N_{NH4}{}^{+}$' to Pro content; r_3 is the difference between the contribution of inorganic N (NH_4Cl) '$+N_{NH4}{}^{+}$' and inorganic N (KNO_3) '$+N_{NO3}{}^{-}$' to Pro content.

Additionally, Matrix (1) was performed with the elementary column (c) transformations, i.e., $c_2 - c_1$, $c_3 - c_1$, and $c_3 - c_2$. Therefore, the following three matrices were obtained:

$$\begin{bmatrix} (+N_{Arg})-(-N_O) & (+N_{Arg})+(-N_I) & (+N_{Glu})+(-N_I) \\ (+N_{Arg})-(-N_O) & (+N_{Arg})+(+N_{NO_3^-}) & (+N_{Glu})+(+N_{NO_3^-}) \\ (+N_{Arg})-(-N_O) & (+N_{Arg})+(+N_{NH_4^+}) & (+N_{Glu})+(+N_{NH_4^+}) \end{bmatrix} \tag{6}$$

$$\begin{bmatrix} (+N_{Glu})-(-N_O) & (+N_{Arg})+(-N_I) & (+N_{Glu})+(-N_I) \\ (+N_{Glu})-(-N_O) & (+N_{Arg})+(+N_{NO_3^-}) & (+N_{Glu})+(+N_{NO_3^-}) \\ (+N_{Glu})-(-N_O) & (+N_{Arg})+(+N_{NH_4^+}) & (+N_{Glu})+(+N_{NH_4^+}) \end{bmatrix} \tag{7}$$

$$\begin{bmatrix} (-N_O)+(-N_I) & (+N_{Glu})-(+N_{Arg}) & (+N_{Glu})+(-N_I) \\ (-N_O)+(+N_{NO_3^-}) & (+N_{Glu})-(+N_{Arg}) & (+N_{Glu})+(+N_{NO_3^-}) \\ (-N_O)+(+N_{NH_4^+}) & (+N_{Glu})-(+N_{Arg}) & (+N_{Glu})+(+N_{NH_4^+}) \end{bmatrix} \tag{8}$$

c_1 in Matrix (6), c_1 in Matrix (7) and c_2 in Matrix (8) were all extracted to obtain Matrix (9):

$$\begin{bmatrix} (+N_{Arg})-(-N_O) & (+N_{Glu})-(-N_O) & (+N_{Glu})-(+N_{Arg}) \\ (+N_{Arg})-(-N_O) & (+N_{Glu})-(-N_O) & (+N_{Glu})-(+N_{Arg}) \\ (+N_{Arg})-(-N_O) & (+N_{Glu})-(-N_O) & (+N_{Glu})-(+N_{Arg}) \end{bmatrix} \tag{9}$$

where c_1 is the contribution of organic N (Arg) '$+N_{Arg}$' to Pro content; c_2 is the contribution of organic N (Glu) '$+N_{Glu}$' to Pro content; c_3 is the difference between the contribution of organic N (Glu) '$+N_{Glu}$' and organic N (Arg) '$+N_{Arg}$' to Pro content.

Next, '$(-N_O)$ and $(-N_I)$' was set to 0, and Matrix (9) was subtracted from Matrix (5) to yield Matrix (10):

$$\begin{bmatrix} (+N_{Arg})-(+N_{NO_3^-}) & (+N_{Glu})-(+N_{NO_3^-}) & (+N_{Glu})-(+N_{Arg})-(+N_{NO_3^-}) \\ (+N_{Arg})-(+N_{NH_4^+}) & (+N_{Glu})-(+N_{NH_4^+}) & (+N_{Glu})-(+N_{Arg}\text{ xFFFD}-(+N_{NH_4^+}) \\ (+N_{Arg})-(+N_{NH_4^+})-(+N_{NO_3^-}) & (+N_{Glu})-(+N_{NH_4^+})-(+N_{NO_3^-}) & (+N_{Glu})-(+N_{Arg})-(+N_{NH_4^+})-(+N_{NO_3^-}) \end{bmatrix} \quad (10)$$

Then, a_{11}, a_{12}, a_{21}, and a_{22} from Matrix (10) were extracted to form Matrix (11):

$$\begin{bmatrix} (+N_{Arg})-(+N_{NO_3^-}) & (+N_{Glu})-(+N_{NO_3^-}) \\ (+N_{Arg})-(+N_{NH_4^+}) & (+N_{Glu})-(+N_{NH_4^+}) \end{bmatrix} \quad (11)$$

where a_{11} represents the difference between the contribution of 'Arg and NO_3^-' to Pro content; a_{12} represents the difference between the contribution of 'Glu and NO_3^-' to Pro content; a_{21} represents the difference between the contribution of 'Arg and NH_4^+' to Pro content; a_{22} represents the difference between the contribution of 'Glu and NH_4^+' to Pro content. In addition, the values of a_{3j} in Matrix (5), a_{i3} in Matrix (9), and a_{11}, a_{12}, a_{21} and a_{22} in Matrix (11) were used to reflect the contribution of organic (Arg/Glu) and inorganic (KNO_3/NH_4Cl) N alone to Pro content.

3. Results

3.1. Pro Content in Rice Tissues under 'Cr(III)+($-N_I$)' Treatments

Under 'Cr(III)+($-N_I$)' treatments, the Pro content in shoots of rice seedlings cultivated with '$-N_O$', '$+N_{Arg}$', and '$+N_{Glu}$' was determined to be "14.30 to 27.28 μg/g FW", "21.96 to 60.85 μg/g FW", and "15.53 to 37.65 μg/g FW", respectively (Figure 2a). These results reveal that Cr(III) treatment increased Pro content in a dose-dependent manner compared with the untreated rice plants. The addition of arginine and glutamate significantly improved Pro content under Cr(III) treatments in rice seedlings in a dose-dependent manner.

The Pro content in roots of rice seedlings cultivated with '$-N_O$', '$+N_{Arg}$', and '$+N_{Glu}$' was "10.59 to 14.82 μg/g FW", "14.23 to 19.35 μg/g FW", and "9.95 to 15.64 μg/g FW", respectively (Figure 2b). This shows that Pro content was lower in roots compared with shoots of rice seedlings under the same treatments. Higher proline content under Cr(III) treatments was observed in arginine-supplemented rice plants.

3.2. Pro Content in Rice Tissues under 'Cr(III)+($+N_{NO3}^-$)' Treatments

Under 'Cr(III)+($+N_{NO3}^-$)' treatments, the Pro content in shoots of rice seedlings cultivated with '$-N_O$', '$+N_{Arg}$', and '$+N_{Glu}$' was determined to be "42.40 to 67.55 μg/g FW", "47.16 to 73.83 μg/g FW", and "34.24 to 47.70 μg/g FW", respectively (Figure 2c). Although the increase in proline content was in a dose-dependent manner in Cr(III)+($-N_O$) treatments, proline content in Cr(III)+($+N_{NO3}^-$) treatments supplemented with arginine and glutamate showed an opposite trend.

The Pro content in roots of rice seedlings cultivated with '$-N_O$', '$+N_{Arg}$', and '$+N_{Glu}$' was "10.21 to 10.78 μg/g FW", "11.78 to 12.24 μg/g FW", and "10.50 to 12.72 μg/g FW", respectively (Figure 2d). There was no significant difference in Pro content of shoots among all the treatments.

3.3. Pro Content in Rice Tissues under 'Cr(III)+($+N_{NH4}^+$)' Treatments

Under 'Cr(III)+($+N_{NH4}^+$)' treatment, the Pro content in shoots of rice seedlings cultivated with '$-N_O$', and '$+N_{Arg}$', '$+N_{Glu}$' was determined to be "80.61 to 140.71 μg/g FW", "63.11 to 104.94 μg/g FW", and "41.23 to 95.38 μg/g FW", respectively (Figure 2e). It was observed that increase in Cr concentration significantly increased Pro content under Cr(III)+($+N_{NH4}^+$) treatments without arginine and glutamate. However, the supplementation of arginine and glutamate showed an opposite trend for Pro accumulation in rice roots under Cr(III)+($+N_{NH4}^+$) treatments.

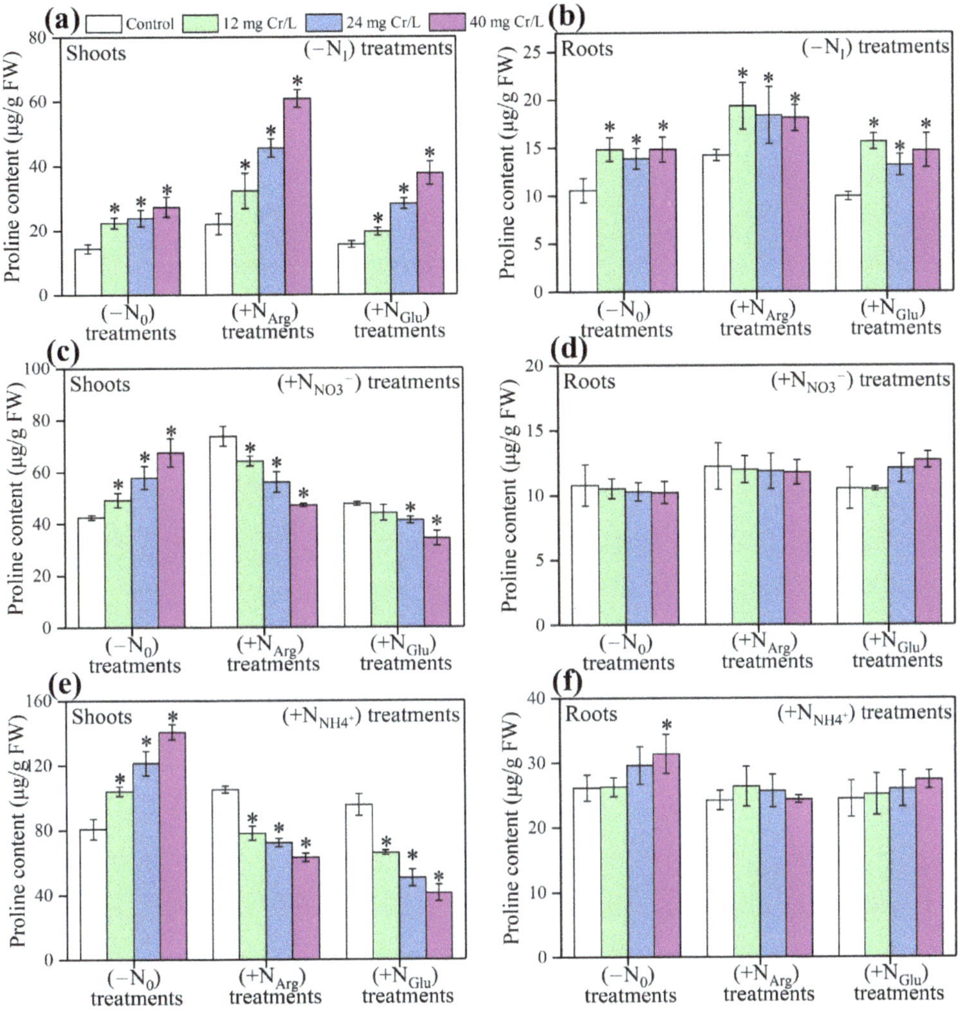

Figure 2. The content of Pro in shoots and roots of rice seedlings under 'Cr(III)+($-N_I$)', 'Cr(III)+($+N_{NO3}^-$)', and 'Cr(III)+($+N_{NH4}^+$)' treatments. The Pro content in (**a**) shoots and (**b**) roots of rice seedlings under 'Cr(III)+($-N_I$)' treatments. The Pro content in (**c**) shoots and (**d**) roots of rice seedlings under 'Cr(III)+($+N_{NO3}^-$)' treatments. The Pro content in (**e**) shoots and (**f**) roots of rice seedlings under 'Cr(III)+($+N_{NH4}^+$)' treatments. "*" indicate significant differences compared with the control group ($p < 0.05$).

The Pro content in roots of rice seedlings cultivated with '$-N_O$', '$+N_{Arg}$', and '$+N_{Glu}$' was "26.12 to 31.39 μg/g FW", "24.18 to 26.34 μg/g FW", and "24.39 to 27.37 μg/g FW", respectively(Figure 2f). There was no significant difference in Pro content of shoots among all the treatments.

3.4. The Contribution of Organic and Inorganic N Application to Pro Content in Rice Seedlings

Herein, we take the treatment with 0.0 mg Cr/L application as an example to predict the contribution of organic and inorganic N alone and in combination with Pro content in rice seedlings (Table 1).

Table 1. The ranking of the contribution of organic and inorganic N to Pro content in rice seedlings.

Treatments	Organic and Inorganic N Treatments Alone		Organic and Inorganic N Treatments in Combination	
	Shoot	Root	Shoot	Root
Control	$N_{NH4}{}^{+} > N_{NO3}{}^{-} > N_{Arg} > N_{Glu}$	$N_{Arg} > N_{NH4}{}^{+} > N_{Glu} > N_{NO3}{}^{-}$	$N_{NH4}{}^{+} + N_{Arg} > N_{NH4}{}^{+} + N_{Glu} > N_{NO3}{}^{-} + N_{Arg} > N_{NO3}{}^{-} + N_{Glu}$	$N_{NH4}{}^{+} + N_{Arg} \approx N_{NH4}{}^{+} + N_{Glu} > N_{NO3}{}^{-} + N_{Arg} \approx N_{NO3}{}^{-} + N_{Glu}$
12 mg Cr/L	$N_{NH4}{}^{+} > N_{NO3}{}^{-} > N_{Arg} > N_{Glu}$	$N_{Arg} > N_{NH4}{}^{+} > N_{Glu} > N_{NO3}{}^{-}$	$N_{NH4}{}^{+} + N_{Arg} > N_{NH4}{}^{+} + N_{Glu} > N_{NO3}{}^{-} + N_{Arg} > N_{NO3}{}^{-} + N_{Glu}$	$N_{NH4}{}^{+} + N_{Arg} \approx N_{NH4}{}^{+} + N_{Glu} > N_{NO3}{}^{-} + N_{Arg} \approx N_{NO3}{}^{-} + N_{Glu}$
24 mg Cr/L	$N_{NH4}{}^{+} > N_{NO3}{}^{-} > N_{Arg} > N_{Glu}$	$N_{Arg} > N_{NH4}{}^{+} > N_{Glu} > N_{NO3}{}^{-}$	$N_{NH4}{}^{+} + N_{Arg} > N_{NO3}{}^{-} + N_{Arg} > N_{NH4}{}^{+} + N_{Glu} > N_{NO3}{}^{-} + N_{Glu}$	$N_{NH4}{}^{+} + N_{Arg} \approx N_{NH4}{}^{+} + N_{Glu} > N_{NO3}{}^{-} + N_{Arg} \approx N_{NO3}{}^{-} + N_{Glu}$
40 mg Cr/L	$N_{NH4}{}^{+} > N_{NO3}{}^{-} > N_{Arg} > N_{Glu}$	$N_{Arg} > N_{NH4}{}^{+} > N_{Glu} > N_{NO3}{}^{-}$	$N_{NH4}{}^{+} + N_{Arg} > N_{NO3}{}^{-} + N_{Arg} > N_{NH4}{}^{+} + N_{Glu} > N_{NO3}{}^{-} + N_{Glu}$	$N_{NH4}{}^{+} + N_{Arg} \approx N_{NH4}{}^{+} + N_{Glu} > N_{NO3}{}^{-} + N_{Arg} \approx N_{NO3}{}^{-} + N_{Glu}$
In summary	$N_{NH4}{}^{+} > N_{NO3}{}^{-} > N_{Arg} > N_{Glu}$	$N_{Arg} > N_{NH4}{}^{+} > N_{Glu} > N_{NO3}{}^{-}$	0 and 12 mg Cr/L treatment: $N_{NH4}{}^{+} + N_{Arg} > N_{NH4}{}^{+} + N_{Glu} > N_{NO3}{}^{-} + N_{Arg} > N_{NO3}{}^{-} + N_{Glu}$ 24 and 40 mg Cr/L treatment: $N_{NH4}{}^{+} + N_{Arg} > N_{NO3}{}^{-} + N_{Arg} > N_{NH4}{}^{+} + N_{Glu} > N_{NO3}{}^{-} + N_{Glu}$	$N_{NH4}{}^{+} + N_{Arg} \approx N_{NH4}{}^{+} + N_{Glu} > N_{NO3}{}^{-} + N_{Arg} \approx N_{NO3}{}^{-} + N_{Glu}$

3.4.1. Pro Content in Shoots of Cr(III)-Treated Rice Seedlings

Using the measured data of Pro content in rice shoots, Matrix (12) was formulated as follows:

$$\begin{bmatrix} 14.30 & 21.96 & 15.53 \\ 42.40 & 73.83 & 47.70 \\ 80.61 & 104.94 & 95.38 \end{bmatrix} \tag{12}$$

As shown in Matrix (12), the contribution of the '$N_I + N_O$' application to Pro content was in the following order: '$N_{NH4}{}^{+} + N_{Arg}$' > '$N_{NH4}{}^{+} + N_{Glu}$' > '$N_{NO3}{}^{-} + N_{Arg}$' > '$N_{NO3}{}^{-} + N_{Glu}$'.

To compare the contribution of inorganic N ($N_{NH4}{}^{+}/N_{NO3}{}^{-}$) application to Pro content, the result is shown in Matrix (13):

$$\begin{bmatrix} 28.10 & 51.87 & 32.17 \\ 66.31 & 82.98 & 79.85 \\ 38.21 & 31.11 & 47.68 \end{bmatrix} \tag{13}$$

According to the calculation presented in the third row (r_3) of Matrix (13), the $N_{NH4}{}^{+}$ application showed a higher contribution to Pro content than the $N_{NO3}{}^{-}$ application.

Then, Matrix (14) can be obtained as follows:

$$\begin{bmatrix} 7.66 & 1.23 & -6.43 \\ 31.43 & 5.30 & -26.13 \\ 24.33 & 14.77 & -9.56 \end{bmatrix} \quad (14)$$

Based on the results presented in the third column (c_3) of Matrix (14), the contribution of the N_{Arg} application to Pro content is higher than that of the N_{Glu} application.

The contribution of each individual organic/inorganic N to Pro content was calculated based on Matrix (15), and the result is shown as follows:

$$\begin{bmatrix} -20.44 & -50.64 \\ -34.88 & -77.68 \end{bmatrix} \quad (15)$$

The contribution of inorganic/organic N application to Pro content was in the following manner: '$N_{NO3}{}^{-}$' > 'N_{Arg}', '$N_{NO3}{}^{-}$' > 'N_{Glu}', '$N_{NH4}{}^{+}$' > 'N_{Arg}', and '$N_{NH4}{}^{+}$' > 'N_{Glu}'. Accordingly, based on the values of a_{3j} in Matrix (13), a_{i3} in Matrix (14), and a_{11}, a_{12}, a_{21}, and a_{22} in Matrix (15), the contribution of inorganic/organic N application to Pro content was in the following order: '$N_{NH4}{}^{+}$' > '$N_{NO3}{}^{-}$' > 'N_{Arg}' > 'N_{Glu}'.

3.4.2. Pro Content in Roots of Cr(III)-Treated Rice Seedlings

Using the measured data of Pro content in rice roots, Matrix (16) was formulated as follows:

$$\begin{bmatrix} 10.59 & 14.23 & 9.95 \\ 10.78 & 12.24 & 10.52 \\ 26.12 & 24.18 & 24.39 \end{bmatrix} \quad (16)$$

It can be seen from Matrix (16) that the contribution of '$N_I + N_O$' application to Pro content was in the following order: '$N_{NH4}{}^{+} + N_{Arg}$' ≈ '$N_{NH4}{}^{+} + N_{Glu}$' > '$N_{NO3}{}^{-} + N_{Arg}$' ≈ '$N_{NO3}{}^{-} + N_{Glu}$'.

Matrix (17) was obtained as follows:

$$\begin{bmatrix} 0.19 & -1.99 & 0.57 \\ 15.53 & 9.95 & 14.44 \\ 15.34 & 11.94 & 13.87 \end{bmatrix} \quad (17)$$

According to the third row of Matrix (17), the contribution of inorganic N sources to Pro content was in the following order: '$N_{NH4}{}^{+}$' > '$N_{NO3}{}^{-}$'.

Then, Matrix (18) was obtained as follows:

$$\begin{bmatrix} 3.64 & -0.64 & -4.28 \\ 1.46 & -0.26 & -1.72 \\ -1.94 & -1.73 & 0.21 \end{bmatrix} \quad (18)$$

Based on the third column of Matrix (18), the contribution of organic N sources to Pro content was in the following order: 'N_{Arg}' > 'N_{Glu}'.

Then, Matrix (19) was generated:

$$\begin{bmatrix} 3.45 & 1.35 \\ -14.07 & -10.21 \end{bmatrix} \quad (19)$$

Matrix (19) shows that the contribution of inorganic or organic N sources to Pro content was in the following manner: 'N_{Arg}' > '$N_{NO3}{}^{-}$', 'N_{Glu}' > '$N_{NO3}{}^{-}$', 'N_{Arg}' > '$N_{NH4}{}^{+}$', and '$N_{NH4}{}^{+}$' > 'N_{Glu}'. Similarly, based on the values of a_{3j} in Matrix (17), a_{i3} in Matrix (18), and a_{11}, a_{12}, a_{21}, and a_{22} in Matrix (19), the contribution of inorganic or organic N sources to Pro content was in the following order: 'N_{Arg}' > '$N_{NH4}{}^{+}$' > 'N_{Glu}' > '$N_{NO3}{}^{-}$'. Moreover, we also calculated the contribution of inorganic and organic N sources alone and in combination with Pro content in both roots and shoots of rice seedlings under 12, 24, and 40 mg Cr/L treatments (Tables 2 and 3). The application of N sources alone, 'N_{NH4^+}', showed the largest contribution to Pro content in rice shoots under different concentrations of Cr(III), followed by 'N_{NO3^-}', 'N_{Arg}', and 'N_{Glu}' application, while 'N_{Arg}' application displayed the largest contribution to the Pro content in rice roots under different concentrations of Cr(III), followed by 'N_{NH4^+}', 'N_{Glu}', and 'N_{NO3^-}' applications. Regarding the application of N sources in combination, '$N_{NH4^+} + N_{Arg}$' application showed the largest contribution to Pro content in both roots and shoots of rice seedlings under different concentrations of Cr(III), while '$N_{NO3^-} + N_{Glu}$' application contributed the least to Pro content in both roots and shoots of Cr(III)-treated rice seedlings. These results suggested that the contribution of different N sources to Pro content in Cr(III)-treated rice seedlings is different.

Table 2. The contribution of inorganic and organic N alone and in combination to Pro content in shoot of rice seedlings under 12, 24, and 40 mg Cr/L treatments.

	Treatments	(12)	(13)	(14)	(15)
Shoot	Control	$\begin{bmatrix} 14.30 & 21.96 & 15.53 \\ 42.40 & 73.83 & 47.70 \\ 80.61 & 104.94 & 95.38 \end{bmatrix}$	$\begin{bmatrix} 28.10 & 51.87 & 32.17 \\ 66.31 & 82.98 & 79.85 \\ 38.21 & 31.11 & 47.68 \end{bmatrix}$	$\begin{bmatrix} 7.66 & 1.23 & -6.43 \\ 31.43 & 5.30 & -26.13 \\ 24.33 & 14.77 & -9.56 \end{bmatrix}$	$\begin{bmatrix} -20.44 & -50.64 \\ -34.88 & -77.68 \end{bmatrix}$
	12 mgCr/L	$\begin{bmatrix} 22.39 & 32.26 & 19.57 \\ 49.22 & 64.22 & 44.15 \\ 103.90 & 78.06 & 66.03 \end{bmatrix}$	$\begin{bmatrix} 26.83 & 31.96 & 24.58 \\ 81.51 & 45.80 & 46.46 \\ 54.68 & 13.84 & 21.88 \end{bmatrix}$	$\begin{bmatrix} 9.87 & -2.82 & -12.69 \\ 15.00 & -5.07 & -20.07 \\ -25.84 & -37.87 & -12.03 \end{bmatrix}$	$\begin{bmatrix} -16.96 & -34.78 \\ -66.51 & -50.87 \end{bmatrix}$
	24 mgCr/L	$\begin{bmatrix} 23.88 & 45.53 & 28.23 \\ 57.87 & 56.14 & 41.24 \\ 121.33 & 72.23 & 50.46 \end{bmatrix}$	$\begin{bmatrix} 33.99 & 10.61 & 13.01 \\ 97.45 & 26.70 & 22.23 \\ 63.46 & 16.09 & 9.22 \end{bmatrix}$	$\begin{bmatrix} 21.65 & 4.35 & -17.30 \\ -1.73 & -16.63 & -14.90 \\ -49.10 & -70.87 & -21.77 \end{bmatrix}$	$\begin{bmatrix} -12.34 & -6.26 \\ -99.18 & -43.33 \end{bmatrix}$
	40 mgCr/L	$\begin{bmatrix} 27.28 & 60.85 & 37.65 \\ 67.55 & 47.16 & 34.24 \\ 140.71 & 63.11 & 41.23 \end{bmatrix}$	$\begin{bmatrix} 40.27 & -13.69 & -3.41 \\ 113.43 & 2.23 & 3.58 \\ 73.16 & 15.95 & 6.99 \end{bmatrix}$	$\begin{bmatrix} 33.57 & 10.37 & -23.20 \\ -20.39 & -33.31 & -12.92 \\ -77.60 & -99.48 & -21.88 \end{bmatrix}$	$\begin{bmatrix} -6.70 & 24.06 \\ -133.82 & -35.57 \end{bmatrix}$

Table 3. The contribution of inorganic and organic N alone and in combination to Pro content in root of rice seedlings under 12, 24, and 40 mg Cr/L treatments.

	Treatments	(16)	(17)	(18)	(19)
Root	Control	$\begin{bmatrix} 10.59 & 14.23 & 9.95 \\ 10.78 & 12.24 & 10.52 \\ 26.12 & 24.18 & 24.39 \end{bmatrix}$	$\begin{bmatrix} 0.19 & -1.99 & 0.57 \\ 15.53 & 9.95 & 14.44 \\ 15.34 & 11.94 & 13.87 \end{bmatrix}$	$\begin{bmatrix} 3.64 & -0.64 & -4.28 \\ 1.46 & -0.26 & -1.72 \\ -1.94 & -1.73 & 0.21 \end{bmatrix}$	$\begin{bmatrix} 3.45 & 1.35 \\ -14.07 & -10.21 \end{bmatrix}$
	12 mg Cr/L	$\begin{bmatrix} 14.84 & 19.35 & 15.64 \\ 10.53 & 12.00 & 10.50 \\ 26.24 & 26.34 & 25.10 \end{bmatrix}$	$\begin{bmatrix} -4.31 & -7.35 & -5.14 \\ 11.40 & 6.99 & 9.46 \\ 15.71 & 14.34 & 14.60 \end{bmatrix}$	$\begin{bmatrix} 4.51 & 0.80 & -3.71 \\ 1.47 & -0.03 & -1.50 \\ 0.10 & -1.14 & -1.24 \end{bmatrix}$	$\begin{bmatrix} 8.82 & 8.15 \\ -9.93 & -7.02 \end{bmatrix}$
	24 mg Cr/L	$\begin{bmatrix} 13.88 & 18.37 & 13.19 \\ 10.28 & 11.86 & 12.09 \\ 29.60 & 25.64 & 25.97 \end{bmatrix}$	$\begin{bmatrix} -3.60 & -6.51 & -1.10 \\ 15.72 & 7.27 & 12.78 \\ 19.32 & 13.78 & 13.88 \end{bmatrix}$	$\begin{bmatrix} 4.49 & -0.69 & -5.18 \\ 1.58 & 1.81 & 0.23 \\ -3.96 & -3.63 & 0.33 \end{bmatrix}$	$\begin{bmatrix} 8.09 & 5.82 \\ -14.14 & -5.46 \end{bmatrix}$
	40 mg Cr/L	$\begin{bmatrix} 14.82 & 18.10 & 14.71 \\ 10.21 & 11.78 & 12.72 \\ 31.39 & 24.34 & 27.37 \end{bmatrix}$	$\begin{bmatrix} -4.61 & -6.32 & -1.99 \\ 16.57 & 6.24 & 12.66 \\ 21.18 & 12.56 & 14.65 \end{bmatrix}$	$\begin{bmatrix} 3.28 & -0.11 & -3.39 \\ 1.57 & 2.51 & 0.94 \\ -7.05 & -4.02 & 3.03 \end{bmatrix}$	$\begin{bmatrix} 7.89 & 6.21 \\ -15.00 & -3.73 \end{bmatrix}$

4. Discussion

The toxic effects of Cr on plants have been extensively reported, such as delaying seed germination, inhibiting root growth, reducing plant height, changing the antioxidative enzyme activities, nutrient elements uptake, and amino acids content [18,19]. In recent years, many strategies have been proposed to curtail the negative effects of Cr pollution on plants, wherein plant growth regulators (PGRs) are considered as one of the most practical and cost-effective methods [20,21]. However, great difficulties are often faced in the selection of PGRs and for evaluating their efficiency in field trials [22]. In this study, we investigated the effect of different nitrogenous compounds as substrates for synthesizing Pro in Cr(III)-treated rice seedlings using MBMM. The endogenous level of Pro in plants is highly dependent on the plant's growth conditions [23]. Previous studies also demonstrated that the aerial part is the major site of Pro synthesis in rice plants [24,25]. Accordingly, we have also observed a higher concentration of Pro in shoots of rice seedlings

(Figure 2). On the other hand, Pro concentration increases with an increase in the intensity of stress until it reaches a threshold level. Pro acts as an excellent osmolyte, a metal chelator, an antioxidative defense molecule, and a signaling molecule which reduces the effects of stress conditions [10,25]. Herein, we observed that the accumulation of Pro in shoots of rice seedlings was dependent on the doses of Cr(III) exposure (Figure 2).

Previously, it has been reported that the content of amino acids, i.e., Pro, in plants is highly dependent on N nutrition [11]. In addition, plant growth highly depends on the forms of N present in the growth media, the amount of N available, and the plant species specifically under stress conditions [26,27]. In our case, we noticed that the response of Pro content was different between organic N-fed and inorganic N-fed seedlings under Cr(III) exposure, indicating that the contribution of these N compounds to Pro content is different [12]. Notably, under 'Cr(III)+($-N_I$)' treatments, relatively higher Pro content was observed in '+N_{Arg}'-fed and '+N_{Glu}'-fed rice seedlings than '$-N_O$'-fed rice seedlings (Figure 2a). The possible reasons behind the higher Pro accumulation might be the supplication of arginine and glutamate amino acids, which are the precursors involved in Pro synthesis [28]. Between Arg and Glu, Arg contributed more to Pro content in rice seedlings under Cr(III) stress, which suggests that Arg is more effective than Glu in the synthesis of Pro specifically under stress conditions. This may be due to the degradation of enzymes responsible for Glu pathway of Pro synthesis under Cr stress. Thus, Arg (ornithine) pathway is considered as an alternative to the Glu pathway for Pro synthesis, which continues under stress conditions [29,30] and may have increased Pro content in Arg-treated seedlings. In addition, Arg plays an important role in the nitrogen (N) cycle because it has the highest ratio of N to carbon among amino acids [31]. Supplementation of N sources specifically ammonium increased Pro content in rice seedlings under Cr(III) stress mainly in '$-N_O$' treatments (Figure 2) because Cr stress affects N metabolism which ultimately reduces Pro synthesis [13]. In addition, ammonium is mostly favored over nitrate by rice plants which increases Pro over nitrate supplementation. The decreasing trend of Pro in Arg and Glu supplemented 'Cr(III)+(+$N_{NO3}{}^-$)' and 'Cr(III)+(+$N_{NH4}{}^+$)' treatments (Figure 2c,e) is complex and needs to be studied further. There could be several possible reasons for this trend; one of the possible reasons is that chromium may react with ammonium to produce precipitates of chromium hydroxide which limit N uptake and subsequently reduce Pro content. The formation of chromium hydroxide led to a further increase in Cr stress in the solution while plants were already in Cr stress [32]. Moreover, the application of $NO_3{}^-$ and $NH_4{}^+$ increased the Pro content in '$-N_O$'-fed rice seedlings, but reduced the Pro content in '+N_{Arg}' and '+N_{Glu}'-fed seedlings, suggesting that the utilization and conversion of these inorganic and organic N sources into Pro synthesis in Cr(III)-treated rice seedlings is different. The content of Pro in both roots and shoots of rice seedlings under 'Cr(III)+(+$N_{NH4}{}^+$)' treatments was significantly higher ('$N_{NH4}{}^+$' > '$N_{NO3}{}^-$' > 'N_{Arg}' > 'N_{Glu}') than other treatments, suggesting that '$NH_4{}^+$' is the preferred N source that contributes to Pro synthesis in rice plants under Cr(III) stress. Therefore, Cr(III) exposure did not change the N preference in the rice, because rice is an $NH_4{}^+$-like crop. The uptake and assimilation rate of $NH_4{}^+$ by rice plants is more rapid than other N sources [33,34]. Moreover, $NH_4{}^+$ is mainly assimilated in rice roots [6], while $NO_3{}^-$ is chiefly assimilated in rice shoots [35].

In addition, we found that the content of Pro increased in rice shoots under Cr(III) treatments with '+$N_{NO3}{}^-$' or '+$N_{NH4}{}^+$' application. However, the content of Pro decreased in rice shoots under "Cr(III)+(+$N_{NO3}{}^-$)" and "Cr(III)+(+$N_{NH4}{}^+$)" treatments with '+N_{Arg}' or '+N_{Glu}' application. MBMM also suggested that the relatively higher content of Pro in rice shoots treated with N source alone was higher than that of treated with N source in combination. This may be due to the (1) application of organic N sources, influencing the uptake and metabolism of inorganic N in rice seedlings; (2) application of organic N sources, promoting the catabolism of Pro in rice seedlings. That is to say, the limiting steps of Pro metabolism in plants are different under Cr(III) stress with different N sources supplied [10,36]. To maintain the homeostasis of the internal environment, plants should rapidly metabolize the Pro to maintain normal N levels against Cr(III) stress [36]. In the future, more physiological, biochemical, and molecular studies should be carried out to reveal the effects of different N sources on Pro synthesis under Cr(III) stress.

5. Conclusions

The present study was conducted to analyze the role of endogenous proline as a bioindicator in the regulation of Cr stress in rice plants grown in different nitrogenous conditions. The content of Pro in rice seedlings depended on the doses of Cr(III) exposure and different N sources. The application of '$N_{NH4}{}^{+}$' had the largest contribution to the content of Pro in rice shoots under Cr(III) stress, while 'N_{Arg}' application had the largest contribution to the content of Pro in rice roots under Cr(III) stress. The combined application of '$N_{NH4}{}^{+} + N_{Arg}$' had the largest contribution to the content of Pro in both roots and shoots of rice seedlings under Cr(III) stress. In contrast, '$N_{NO3}{}^{-} + N_{Glu}$' application contributed the least to the content of Pro in Cr(III)-treated rice seedlings. In conclusion, plants can regulate the content of Pro in plant tissues to cope with the potential threat induced by Cr(III) exposure under different nutritional N sources.

Supplementary Materials: The following supporting information can be downloaded at: https://www.mdpi.com/article/10.3390/toxics11100803/s1, Figure S1. The standard curve between the absorbance at 520 nm and L-proline content. Series content of L-proline (0, 2, 4, 6, 8, and 10 μg/mL).

Author Contributions: Conceptualization, X.Y.; Methodology, X.Y.; Software, Y.F.; Investigation, C.L. and P.T.; Data curation, C.L.; Writing—original draft, Y.F.; Writing—review & editing, X.Y.; Supervision, X.Y.; Funding acquisition, X.Y. All authors have read and agreed to the published version of the manuscript.

Funding: This work is financially supported by the Natural Science Foundation of Guangxi (No. 2018GXNSFDA281024).

Institutional Review Board Statement: Not applicable.

Informed Consent Statement: Not applicable.

Data Availability Statement: The authors acknowledge that the data presented in this study must be deposited and made publicly available in an acceptable repository before publication.

Conflicts of Interest: The authors declare that the research was conducted without any commercial or financial relationships that could be construed as a potential conflict of interest.

References

1. Pan, X.; Ullah, A.; Feng, Y.X.; Tian, P.; Yu, X.Z. Proline-mediated activation of glyoxalase II improve methylglyoxal detoxification in *Oryza sativa* L. under chromium injury: Clarification via vector analysis of enzymatic activities and gene expression. *Plant Physiol. Biochem.* **2023**, *201*, 107867. [CrossRef] [PubMed]
2. Feng, Y.X.; Yu, X.Z.; Mo, C.H.; Lu, C.J. Regulation network of sucrose metabolism in response to trivalent and hexavalent chromium in *Oryza sativa*. *J. Agric. Food Chem.* **2019**, *67*, 9738–9748. [CrossRef] [PubMed]
3. Fan, W.J.; Feng, Y.X.; Li, Y.H.; Lin, Y.J.; Yu, X.Z. Unraveling genes promoting ROS metabolism in subcellular organelles of *Oryza sativa* in response to trivalent and hexavalent chromium. *Sci. Total Environ.* **2020**, *744*, 140951. [CrossRef] [PubMed]
4. Zhang, Q.; Feng, Y.X.; Lin, Y.J.; Yu, X.Z. Mathematical quantification of interactive complexity of transcription factors involved in proline-mediated regulative strategies in *Oryza sativa* under chromium stress. *Plant Physiol. Biochem.* **2022**, *182*, 36–44. [CrossRef] [PubMed]
5. Martins, M.; Lopes, J.; Sousa, B.; Soares, C.; Valente, I.M.; Rodrigues, J.A.; Fidalgo, F.; Teixeira, J. Cr (VI)-induced oxidative damage impairs ammonia assimilation into organic forms in *Solanum lycopersicum* L. *Plant Stress* **2021**, *2*, 100034. [CrossRef]
6. Li, C.Z.; Yang, L.; Lin, Y.J.; Zhang, H.; Rad, S.; Yu, X.Z. Assimilation of exogenous cyanide cross talk in *Oryza sativa* L. to the key nodes in nitrogen metabolism. *Ecotoxicology* **2020**, *29*, 1552–1564. [CrossRef]
7. Chen, M.; Zhu, K.; Xie, J.; Liu, J.; Qiao, Z.; Tan, P.; Peng, F. Ammonium-nitrate mixtures dominated by $NH_4{}^{+}$-N promote the growth of pecan (*Carya illinoinensis*) through enhanced N uptake and assimilation. *Front. Plant Sci.* **2023**, *14*, 1186818. [CrossRef] [PubMed]

8. Han, M.; Xu, M.Y.; Wang, S.Z.; Wu, L.D.; Sun, S.Y.; Su, T. Effects of exogenous L-Glutamine as a sole nitrogen source on physiological characteristics and nitrogen use efficiency of poplar. *Plant Physiol. Biochem.* **2022**, *172*, 1–13. [CrossRef]
9. Hosseinifard, M.; Stefaniak, S.; Ghorbani Javid, M.; Soltani, E.; Wojtyla, Ł.; Garnczarska, M. Contribution of exogenous proline to abiotic stresses tolerance in plants: A review. *Int. J. Mol. Sci.* **2022**, *23*, 5186. [CrossRef]
10. Meena, M.; Divyanshu, K.; Kumar, S.; Swapnil, P.; Zehra, A.; Shukla, V.; Upadhyay, R.S. Regulation of L-proline biosynthesis, signal transduction, transport, accumulation and its vital role in plants during variable environmental conditions. *Heliyon* **2019**, *5*, e02952. [CrossRef]
11. Spormann, S.; Nadais, P.; Sousa, F.; Pinto, M.; Martins, M.; Sousa, B.; Soares, C. Accumulation of proline in plants under contaminated soils-are we on the same page? *Antioxidants* **2023**, *12*, 666. [CrossRef]
12. Li, Y.H.; Tian, P.; Li, C.Z.; Yu, X.Z. Elucidating comportment of the glutamate and ornithine pathway on proline accumulation in rice under different nitrogenous nutrition. *Int. J. Environ. Sci. Technol.* **2022**, *19*, 2993–3000. [CrossRef]
13. Luo, L.; Zhang, Y.; Xu, G. How does nitrogen shape plant architecture? *J. Exp. Bot.* **2020**, *71*, 4415–4427. [CrossRef] [PubMed]
14. Sangwan, P.; Kumar, V.; Joshi, U.N. Effect of chromium (VI) toxicity on enzymes of nitrogen metabolism in clusterbean (*Cyamopsis tetragonoloba* L.). *Enzym. Res.* **2014**, *2014*, 784036. [CrossRef]
15. Huo, K.; Shangguan, X.; Xia, Y.; Shen, Z.; Chen, C. Excess copper inhibits the growth of rice seedlings by decreasing uptake of nitrate. *Ecotoxicol. Environ. Saf.* **2020**, *190*, 110105. [CrossRef]
16. Chai, M.; Li, R.; Shen, X.; Tam, N.F.Y.; Zan, Q.; Li, R. Does ammonium nitrogen affect accumulation, subcellular distribution and chemical forms of cadmium in *Kandelia obovata*? *Ecotoxicol. Environ. Saf.* **2018**, *162*, 430–437. [CrossRef] [PubMed]
17. Yu, X.Z.; Lin, Y.J.; Fan, W.J.; Lu, M.R. The role of exogenous proline in amelioration of lipid peroxidation in rice seedlings exposed to Cr (VI). *Int. Biodeter. Biodegrad.* **2017**, *123*, 106–112. [CrossRef]
18. López-Bucio, J.S.; Ravelo-Ortega, G.; López-Bucio, J. Chromium in plant growth and development: Toxicity, tolerance and hormesis. *Environ. Pollut.* **2022**, *312*, 120084. [CrossRef] [PubMed]
19. Danyal, Y.; Mahmood, K.; Ullah, S.; Rahim, A.; Raheem, G.; Khan, A.H.; Ullah, A. Phytoremediation of industrial effluents assisted by plant growth promoting bacteria. *Environ. Sci. Pollut. Res.* **2023**, *30*, 5296–5311. [CrossRef]
20. Choudhary, S.P.; Kanwar, M.; Bhardwaj, R.; Gupta, B.D.; Gupta, R.K. Epibrassinolide ameliorates Cr (VI) stress via influencing the levels of indole-3-acetic acid, abscisic acid, polyamines and antioxidant system of radish seedlings. *Chemosphere* **2011**, *84*, 592–600. [CrossRef]
21. Ullah, A.; Heng, S.; Munis, M.F.H.; Fahad, S.; Yang, X. Phytoremediation of heavy metals assisted by plant growth promoting (PGP) bacteria: A review. *Environ. Exp. Bot.* **2015**, *117*, 28–40. [CrossRef]
22. Lin, Y.J.; Feng, Y.X.; Li, Y.H.; Yu, G.; Yu, X.Z. Fuzzy synthetic evaluation of the impact of plant growth regulators on the root phenotype traits of rice seedlings under thiocyanate stress. *Plant Physiol. Biochem.* **2021**, *158*, 182–189. [CrossRef] [PubMed]
23. Szabados, L.; Savouré, A. Proline: A multifunctional amino acid. *Trends Plant Sci.* **2010**, *15*, 89–97. [CrossRef] [PubMed]
24. Stein, H.; Honig, A.; Miller, G.; Erster, O.; Eilenberg, H.; Csonka, L.N.; Sazbados, L.; Koncz, C.; Zilberstein, A. Elevation of free proline and proline-rich protein levels by simultaneous manipulations of proline biosynthesis and degradation in plants. *Plant Sci.* **2011**, *181*, 140–150. [CrossRef] [PubMed]
25. Cao, X.; Wu, L.; Wu, M.; Zhu, C.; Jin, Q.; Zhang, J. Abscisic acid mediated proline biosynthesis and antioxidant ability in roots of two different rice genotypes under hypoxic stress. *BMC Plant Biol.* **2020**, *20*, 198. [CrossRef] [PubMed]
26. Cambui, C.A.; Svennerstam, H.; Gruffman, L.; Nordin, A.; Ganeteg, U.; Näsholm, T. Patterns of plant biomass partitioning depend on nitrogen source. *PLoS ONE* **2011**, *6*, e19211. [CrossRef] [PubMed]
27. Carlisle, E.; Myers, S.; Raboy, V.; Bloom, A. The effects of inorganic nitrogen form and CO_2 concentration on wheat yield and nutrient accumulation and distribution. *Front. Plant Sci.* **2012**, *3*, 195. [CrossRef]
28. Trovato, M.; Funck, D.; Forlani, G.; Okumoto, S.; Amir, R. Amino acids in plants: Regulation and functions in development and stress defense. *Front. Plant Sci.* **2021**, *12*, 772810. [CrossRef]
29. Hayat, S.; Hayat, Q.; Alyemeni, M.N.; Wani, A.S.; Pichtel, J.; Ahmad, A. Role of proline under changing environments: A review. *Plant Signal. Behav.* **2012**, *7*, 1456–1466. [CrossRef]
30. Winter, G.; Todd, C.D.; Trovato, M.; Forlani, G.; Funck, D. Physiological implications of arginine metabolism in plants. *Front. Plant Sci.* **2015**, *6*, 534. [CrossRef]
31. Chen, Q.; Wang, Y.; Zhang, Z.; Liu, X.; Li, C.; Ma, F. Arginine Increases Tolerance to Nitrogen Deficiency in Malus hupehensis via Alterations in Photosynthetic Capacity and Amino Acids Metabolism. *Front. Plant Sci.* **2022**, *12*, 772086. [CrossRef] [PubMed]
32. Mondal, M.H.; Begum, W.; Nasrollahzadeh, M.; Ghorbannezhad, F.; Antoniadis, V.; Levizou, E.; Saha, B. A comprehensive review on chromium chemistry along with detection, speciation, extraction and remediation of hexavalent chromium in contemporary science and technology. *Vietnam J. Chem.* **2021**, *59*, 711–732.
33. Gaur, V.S.; Singh, U.S.; Gupta, A.K.; Kumar, A. Understanding the differential nitrogen sensing mechanism in rice genotypes through expression analysis of high and low affinity ammonium transporter genes. *Mol. Biol. Rep.* **2012**, *39*, 2233–2241. [CrossRef] [PubMed]
34. Thornton, B.; Robinson, D. Uptake and assimilation of nitrogen from solutions containing multiple N sources. *Plant Cell Environ.* **2005**, *28*, 813–821. [CrossRef]

35. Cao, X.C.; Zhong, C.; Zhu, C.Q.; Zhu, L.F.; Zhang, J.H.; Wu, L.H.; Jin, Q.Y. Ammonium uptake and metabolism alleviate PEG-induced water stress in rice seedlings. *Plant Physiol. Biochem.* **2018**, *132*, 128–137. [CrossRef] [PubMed]
36. Iqbal, N.; Umar, S.; Khan, N.A. Nitrogen availability regulates proline and ethylene production and alleviates salinity stress in mustard (*Brassica juncea*). *J. Plant Physiol.* **2015**, *178*, 84–91. [CrossRef]

toxics

Review

Advances in Soil Amendments for Remediation of Heavy Metal-Contaminated Soils: Mechanisms, Impact, and Future Prospects

Xinyi Nie [1,2], Xianhuai Huang [1,2,*], Man Li [3], Zhaochi Lu [3] and Xinhe Ling [4]

1 School of Environment and Energy Engineering, Anhui Jianzhu University, Hefei 230601, China
2 Anhui Provincial Key Laboratory of Environmental Pollution Control and Resource Reuse, Hefei 230601, China
3 Institute of Geotechnical Engineering, Southeast University, Nanjing 211189, China
4 College of Civil Engineering, Anhui Jianzhu University, Hefei 230601, China
* Correspondence: xianhuai_h@126.com

Abstract: Heavy metal contamination is a critical factor contributing to soil degradation and poses significant environmental threats with profound implications for ecosystems and human health. Soil amendments have become an effective strategy to address these challenges by reducing heavy metal hazards and remediating contaminated soils. This review offers a comprehensive analysis of recent advancements in soil amendments for heavy metal-contaminated soils, with a focus on natural, synthetic, natural-synthetic copolymer, and biological amendments. By thoroughly examining and contrasting their remediation mechanisms and effects, this study provides a detailed evaluation of their influence on soil physicochemical properties, leachable heavy metal content, and microbial communities. Through bibliometric analysis, current research priorities and trends are highlighted, offering a multidimensional comparison of these amendments and clarifying their varying applicability and limitations. Furthermore, this review explores future prospects and the inherent challenges in soil amendments for heavy metal contamination, aiming to offer valuable insights and theoretical references for the development and selection of novel, efficient, multifunctional, environmentally friendly amendments.

Keywords: heavy metal; soil degradation; contaminated soils; soil amendments; bibliometric analysis

1. Introduction

Heavy metals are released into the soil through both natural and anthropogenic sources [1]. With increasing levels of industrialization and the continuous development of transportation networks worldwide, industries such as metallurgy, mining, construction materials manufacturing, and transportation generate a large amount of industrial waste containing heavy metals. This has led to the entry of heavy metals such as lead (Pb), mercury (Hg), chromium (Cr), and cadmium (Cd) into the atmosphere, water, and soil, causing serious environmental pollution [2]. Industrial waste, agricultural fertilizers, spills from petroleum products, coal combustion, livestock manure, wastewater irrigation, and urban sewage sludge are all important sources of heavy metals in contaminated soils [3].

Unlike other organic compounds, heavy metals are not easily purified through natural processes, leading to their accumulation in the environment. This accumulation can have adverse effects on the biological community, including changes in microbial communities, accumulation in plants and soil organisms, and risks to animal, plant, and human health [4]. The results of cancer risk studies show that approximately 76% and 15.7% of cancer risks are caused by high levels of Cd and arsenic (As), respectively [5]. Consuming rice grown in As, Cd, and Cr-contaminated fields significantly raises the risk of cancer. Mercury deposited in land and water can form more toxic methylmercury, damaging the central nervous system and causing blindness, deafness, and other symptoms [6]. The gradual accumulation of heavy metals in the soil leads to increasingly severe soil degradation phenomena, such as weak soil structure, susceptibility to erosion, nutrient loss in plants, and decreased soil biodiversity [7].

The excessive accumulation of heavy metals in the soil can alter the physical and chemical properties of the soil. Pearson correlation analysis demonstrated that soil pH, $CaCO_3$, and clay content are critical factors influencing the binding forms and types of heavy metals in soil fractions, thereby significantly affecting their bioavailability and chemical reactivity [8,9]. Heavy metal toxicity disrupts microbial communities by inhibiting their growth and metabolic activities, reducing diversity and ecological function. This disruption and altered enzyme activities due to heavy metal interference impair organic matter decomposition and nutrient cycling. Consequently, these changes adversely affect soil structure, leading to reduced porosity, compressive strength, and water-holding capacity, further degrading soil health [3].

Soil contaminated with heavy metals has a detrimental impact on human health and the ecological environment [1]. Remediating heavy metal-contaminated soil is a growing area of research. Techniques involve physical, chemical, and biological methods to reduce pollutant concentrations, immobilize them, and convert them into less harmful forms, preventing their spread in the ecosystem [1,3,10–13]. Approaches for remediating heavy metal-contaminated sites are generally categorized into in situ and ex situ technologies. In situ methods include electrokinetic separation, soil amendment, thermal treatment, in situ chemical leaching, chemical oxidation-reduction, solidification/stabilization, phytoremediation, nanotechnology, and microbial remediation [1,3,14–17]. Ex situ technologies comprise attrition scrubbing, landfilling, chemical leaching, thermal desorption, co-disposal in cement kilns, and the use of prefabricated beds [1,11,18]. The United States Environmental Protection Agency (USEPA) analyzed the remediation technologies utilized at 106 sites, which were documented in 172 source decision documents from the Superfund program from 2018 to 2020 [19]. Table 1 summarizes the specific types of soil remediation techniques selected in these documents. Of the 48 (37%) documents designated for treatment, in situ remediation technologies accounted for 28% of the total projects, while ex situ remediation technologies made up 13%. The most selected in situ technologies for soil remediation in recent years include in situ thermal treatment (ISTT), soil vapor extraction (SVE), solidification/stabilization (S/S), and chemical treatment methods, which encompass in situ chemical oxidation (ISCO) and in situ chemical reduction (ISCR). For ex situ soil treatment, S/S and physical separation are the most frequently selected remedies.

When selecting soil remediation techniques for different types of soil pollution, factors such as remediation objectives, socio-economic conditions, pollution types, pollution levels, remediation time, and feasibility of the techniques must be considered [11]. Additionally, soil remediation techniques for heavy metal-contaminated sites can be categorized into physical, chemical, and biological techniques based on their operational principles [1,3,11–13]. Some of these techniques require the use of soil amendments, such as weak acids and acid salts, to remove heavy metals [14,20–22].

Soil amendments can be categorized into four main groups based on the source of raw materials—natural, synthetic, natural-synthetic copolymers, and biological amendments. Representative examples in each category are summarized as shown in Figure 1 [10]. Table 2 presents a selection of typical soil amendments used for heavy metal soil remediation. The soil amendments are organized by classification, experiment types, types of heavy metals in contaminated soils, and remediation effectiveness, along with their corresponding references. However, both the nature of the amendment and its application have limitations in remediating heavy metal-contaminated soils. For instance, industrial solid wastes such as red mud, fly ash, and calcium carbide slag may contribute to secondary pollution in these soils. Organic amendments can either release nitrogen, reducing the soil's capacity to retain organic carbon, or contain high levels of pathogens, parasites, and toxic substances, which can adversely affect plants and soil organisms to varying degrees. Excessive application of organic amendments, particularly biochar, can reduce soil nitrogen availability by increasing microbial nitrogen demand, raising the carbon–nitrogen ratio, and temporarily fixing nitrogen, ultimately leading to nitrogen deficiency [23]. However, more research is needed on the types and amounts of these combinations [24].

Table 1. Soil remedies selected most frequently in recent decision documents (FY 2018–2020) [19].

Selected Remedy	Number	Percent
Treatment	48	37%
In Situ Treatment	37	28%
Thermal Treatment	14	11%
Soil Vapor Extraction	13	10%
Solidification/Stabilization	8	6%
Chemical Treatment	5	4%
Bioremediation	3	2%
Flushing	2	2%
Multi-phase Extraction	2	2%
Soil Amendments	2	2%
Ex Situ Treatment	17	13%
Solidification/Stabilization	7	5%
Physical Separation	6	5%
Thermal Treatment	2	2%
Containment/Disposal	90	69%
Monitored Natural Attenuation	1	1%
Institutional Controls	98	75%
Other	21	16%

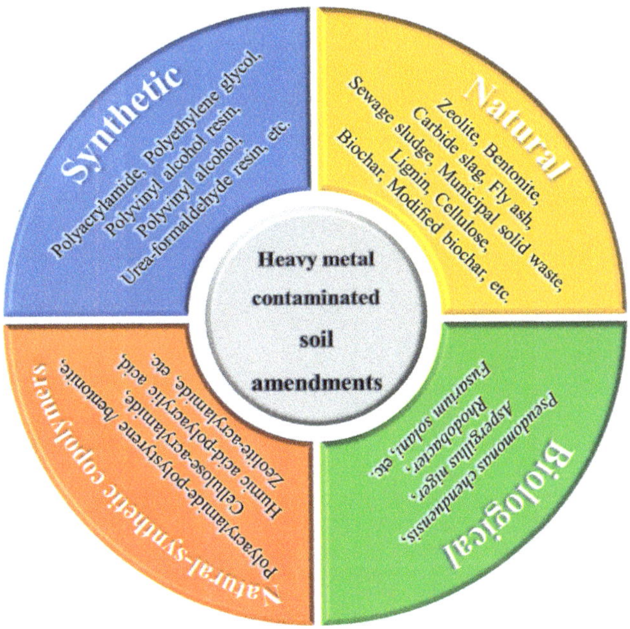

Figure 1. Classification system chart of heavy metal contaminated soil amendments [10].

In this review, we aim to: (1) comprehensively analyze recent advances in soil amendments for heavy metal-contaminated soils, focusing on natural, synthetic, natural-synthetic copolymers, and biological amendments; (2) investigate and compare their remediation mechanisms thoroughly, as well as their impacts on soil physicochemical properties, leachable heavy metal contents, and microbial communities; (3) highlight current research focuses and trends through bibliometric analyses relevant to the soil heavy metal remediation, providing multidimensional comparisons of these amendments and elucidating their varying applications and limitations; and (4) explore future prospects and challenges for the development and selection of novel, highly efficient, and multifunctional environmentally friendly soil amendments.

Table 2. Some typical soil amendment samples are used for the remediation of heavy metal-contaminated soils.

Soil Amendments	Classification	Experiment Types	Heavy Metals	Remediation Effectiveness	Reference
Vermiculite	Natural mineral	Pot experiments	Cu, Cr, Ni	Significantly reduce the absorption of metal pollutants by mustard and spinach plants.	[25]
Cement, Fly ash, Desulfurization gypsum	Inorganic solid waste	Solidification/ Stabilization	Cu, Ni	Significantly increase the compressive strength and permeability of contaminated soils.	[26]
Vermicompost, Leaf compost, Spent mushroom compost	Organic solid waste	Greenhouse experiments	Cd, Cr, Pb, Mn	Decrease the absorption of Cd, Cr, Pb, and Mn by plants, promoting plant growth.	[27]
Lignin, Chitin	Naturally extracted polymer compounds	Kinetics experiments, Adsorption experiments	Cu, Fe	Show high adsorption capacity for metal ions, especially at low concentrations.	[28]
Biochar	Organic material	Field experiments	Cd, Pb	Significantly increase the pH value and total organic carbon content of the soil and effectively immobilize Cd and Pb in the soil.	[29,30]
Polyacrylamide	Synthetic	Adsorption experiments	Pb, Cr	Immobilize Pb and Cr on the surface of clay minerals.	[31]
Chitosan-grafted poly(acrylamide-co-acrylic acid)/biochar	Natural-synthetic copolymer	Kinetics experiments, Adsorption experiments	Cu	Effectively increase the adsorption capacity of soils for heavy metals and improve the water retention of soils.	[32]
Pseudomonas chenduensis	Biological	Pot experiments	Cu, Cd, Pb, Zn	Enhance the role of the microbial community in transforming Cd components and reduce Cd accumulation in rice grains and roots.	[33]

2. Bibliometric Analysis of Hotspots and Frontiers on Remediation Materials for Heavy Metal-Contaminated Soils

Using the Thomson Reuters Web of Science (WOS) database, the search period was 2013–2023, and the search formula was Topic = (heavy metal soil) and (soil remediation) and (soil amendments) and (materials), and a total of 317 articles were retained in the Web of Science Core Collection as a data sample for systematic analysis through screening and de-duplication to remove irrelevant and duplicated literature. Using CiteSpace software (version 6.16) and Origin 2022 software, the relevant literature published in the last 11 years was analyzed econometrically.

Figure 2 shows publication number and citation count trends. The number of publications is a key indicator of research interest in a field. The citations in Figure 2 represent the total number of citations to articles on heavy metal-contaminated soil amendments from 2013 to that year. The greater the slope of the line, the more citations there were that year. The bar chart in Figure 2 illustrates the number of publications each year. From 2013 to 2016, the number of publications and citations in the relevant fields was low. During the period 2017–2022, the publication and citation number increased rapidly, indicating the increasing demand for remediation of heavy metal-contaminated soils.

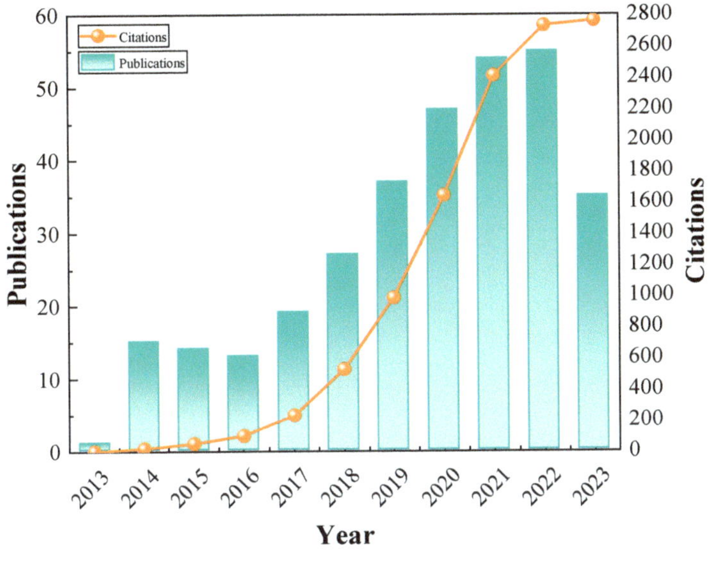

Figure 2. The distribution of citations and publication number by year.

2.1. Country Co-Authorship Analysis

Figure 3 depicts the collaborative network among 61 countries researching soil remediation for heavy metals, with 133 collaborative links. Table 3 ranks the top 10 countries by publication count. China leads with 155 papers, followed by Spain (35), USA (25), France (21), and Australia (20), accounting for 78.29% of the total. These countries have made significant contributions to the field. China's high publication count suggests a strong focus on this area. Spain, China, the USA, and Australia have higher centrality, indicating greater influence. Extensive interconnections in the visualization suggest active communication and collaboration between researchers worldwide.

Table 3. The top 10 countries in terms of publication volume.

Ranking	Count	Centrality	Countries
1	155	0.57	People's Republic of China
2	35	0.64	Spain
3	25	0.27	USA
4	21	0.17	France
5	20	0.25	Australia
6	18	0.1	Pakistan
7	16	0.06	South Korea
8	15	0.01	Italy
9	11	0.21	Czech Republic
10	11	0.01	Poland

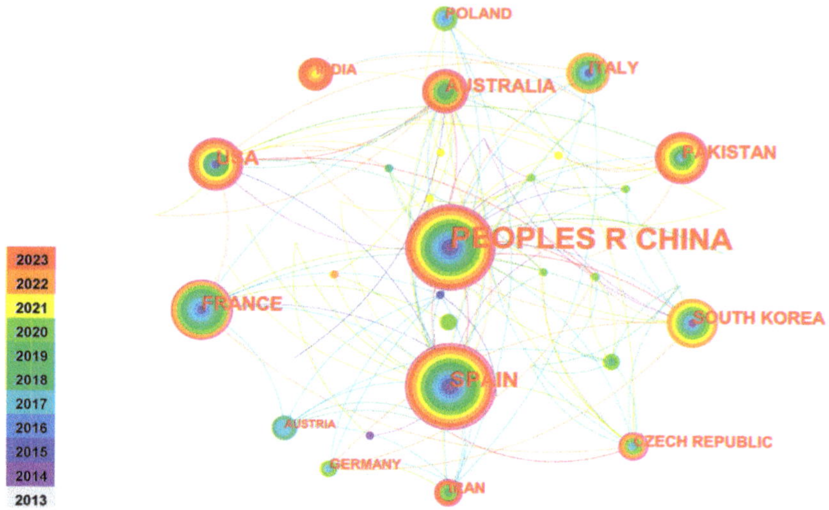

Figure 3. Cooperation network of productive countries from 2013 to 2023.

2.2. Keyword Co-Occurrence Analysis

The frequency of keywords in a specific period reflects research hotspots. Emergence analysis with a 1-year time slice using CiteSpace software reveals research progress and cutting-edge hotspots. Figure 4 displays keyword co-occurrence in heavy metal-contaminated soil amendments from 2013 to 2023, while Table 4 ranks the top 10 keywords based on centrality. "Stabilization" has the highest occurrence frequency after "Heavy metal", with related terms like "Immobilization", "Stabilization/solidification", and "Solidification/stabilization" also in the top ten, indicating a focus on solidification remediation.

Figure 4. A network of keyword co-occurrence from 2013 to 2023.

Table 5 presents the burst intensity, start and end times, and duration of these keywords. The interval time period is marked by blue lines, while the burst time period of the keywords is indicated by red lines. The red timeline highlights bursts corresponding to highly cited keywords [34]. Table 5 highlights the emergence analysis of keywords in heavy metal-contaminated soil remediation materials from 2013 to 2023. From 2013 to 2018, initial interest in "organic amendment", "mine tailing", and "black carbon" waned, possibly due to limited effectiveness and concerns about secondary pollution. Between 2021 and 2023, "pollution" and "black carbon" became the most popular amendments, reflecting a growing concern about secondary pollution and health risks.

Table 4. The top 10 most occurring keywords.

Ranking	Count	Centrality	Keywords
1	191	0.02	Heavy metal
2	105	0.07	Stabilization
3	79	0.04	Fly ash
4	78	0.11	Immobilization
5	59	0.07	MSWI fly ash
6	57	0.13	Stabilization/solidification
7	53	0.15	Solidification/stabilization
8	49	0.11	Cement
9	44	0.15	Portland cement
10	40	0.09	Behavior

Table 5. The top 12 keywords with the strongest citation bursts.

Keywords	Year	Strength	Begin	End	2013–2023
organic amendment	2013	4.34	2013	2018	
mine tailing	2013	3.03	2013	2017	
black carbon	2014	2.93	2014	2019	
pollution	2021	2.48	2021	2023	
health risk	2021	2.38	2021	2023	
trace element	2017	2.37	2017	2018	
impact	2019	2.22	2019	2020	
plant	2019	2.22	2019	2020	
risk assessment	2017	2.1	2017	2019	
fractionation	2014	2.05	2018	2019	
paddy soil	2020	2.01	2020	2023	
copper	2014	2	2014	2017	

Keyword clustering analysis is a useful technique for presenting important research findings. Table 6 and Figure 5 provide information about the keyword clusters and their associated features. Ten clusters of basic keywords are shown based on graphical and chronological order, where #0 is the largest clustering. The mean silhouette (S) and modularity (Q) are two crucial metrics for evaluating the quality of co-cited node clusters, reflecting the "overall structural characteristics" of the clusters within the co-cited network, which can range from 0 to 1. Higher values of Q signify more effective node clusters, with a Q value greater than 0.3 suggesting a notable community structure in the network. Similarly, increased S values indicate greater uniformity among the nodes in a cluster, and typically, an S value exceeding 0.7 implies that the cluster is highly credible. In this research, Q and S are 0.498 and 0.739, respectively, which indicates that our network structure has a significant community structure and node reliability. The year represents the median year of all the references within this cluster. Log-likelihood ratio (LLR) is a statistical method used to identify and analyze relationships between nodes in a network, and CiteSpace divided these subject categories into 10 clusters using the LLR algorithm [35].

Table 6. Keyword clusters analysis and feature information.

Cluster-ID	Count	Silhouette	Year	Top Terms (LLR)
0	47	0.685	2018	lead-zinc smelting slag; heavy metal; solid waste (SW); material characteristics; cementitious property
1	45	0.635	2017	electrolytic manganese residue; calorimetry; building materials; fly ash; APC residues
2	40	0.685	2020	red mud; sewage sludge; cotreatment NBSP; arsenic-laden spent media; environmental risk assessment
3	34	0.866	2014	calcining pretreatment; sulfates; pickling liquor; hazard-free treatment; Cu/Zn
4	34	0.714	2018	potentially toxic elements; cement; leaching pattern; hazardous waste management; immobilization mechanisms
5	31	0.783	2018	municipal solid waste; fly ash; MSWI fly ash; hazardous waste; arsenic contaminated soil
6	30	0.885	2015	tobermorite; air pollution control residues; heavy metal speciation; reconstructed slag; vitrification
7	24	0.639	2020	pre-treatment; cementitious materials; solidification and stabilization; cement kiln co-processing; XPS
8	23	0.782	2016	MSWI fly ash; compressive strength; stabilization; hydration products; uncertainty
9	18	0.793	2016	heavy metal immobilization; alkali-activated technology; nano-alumina; synergistic effect; gelation

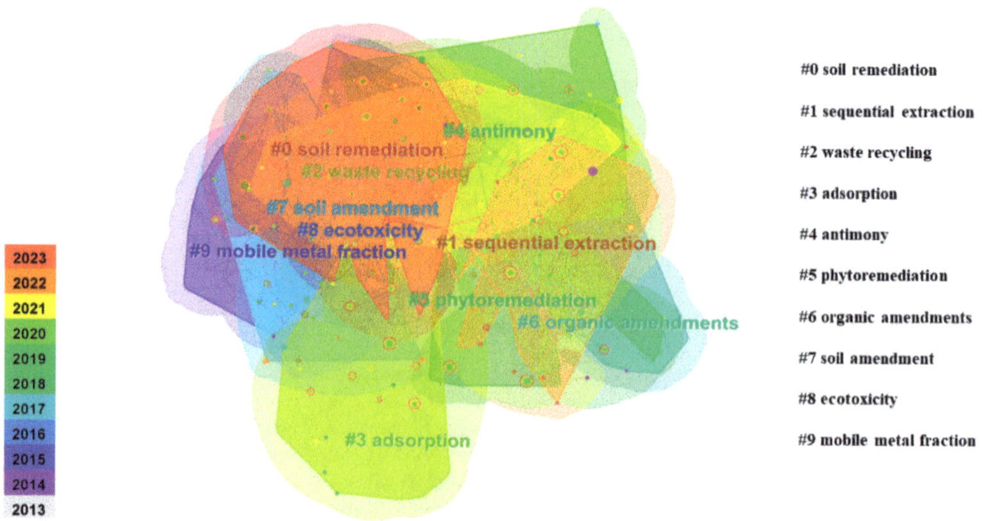

Figure 5. A cluster map of co-cited literature on heavy metal-contaminated soil remediation from 2013 to 2023.

The keyword co-occurrence network diagram for the remediation of heavy metal-contaminated soil is shown in Figure 5. The size of each node corresponds to the frequency of the associated cluster, with larger nodes representing higher frequency [34]. As shown in Figure 5 and Table 4, there is a noticeable variation in the size of the clusters. The largest cluster, #0, contains 47 nodes, accounting for approximately 14.4% of the total nodes of the co-occurrence network. In contrast, cluster #9 is the smallest, with only 18 nodes representing just 5.5% of the co-citation network. From Figure 5 and Table 4, it can be observed that the five largest clusters in the heavy metal-contaminated soil remediation research network are as follows: cluster #0 "soil remediation" (47 occurrences), cluster #1 "sequential extraction" (45 occurrences), cluster #2 "waste recycling" (40 occurrences), cluster #3 "adsorption" (34 occurrences), and cluster #4 "antimony" (40 occurrences). The remaining clusters are detailed in Table 4. These terms represent the current research hotspots in the field, reflecting the prominent themes from 2013 to 2023.

Figure 6 presents the keyword timeline view, offering insights into the temporal distribution of keywords. The size of each node is proportional to the frequency of keyword co-occurrence, while nodes in different colors represent different years. The color of each cluster label corresponds to the average year of that cluster [35]. From Figure 6, the following conclusions can be drawn. Studies conducted between 2013 and 2016 primarily focused on organic amendments, chemical leaching, and chemical oxidation-reduction processes (clusters #3, #6, #8, #9). However, following the promulgation of policies related to soil remediation and the recognition of limitations in some chemical methods, research from 2016 to 2023 gradually shifted towards solid waste reuse and stabilization/solidification (clusters #0, #1, #2, #4, #7). Phytoremediation (cluster #5) has garnered significant attention due to its low cost and minimal environmental impact.

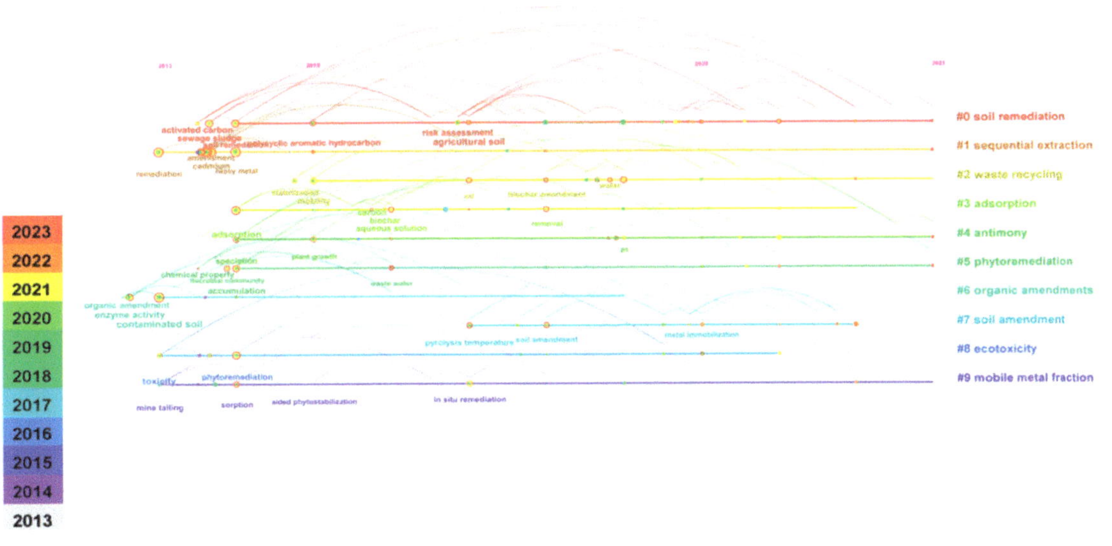

Figure 6. Timeline visualization of the 10 clusters of document co-citations from 2013 to 2023 based on 1-year slices.

3. Natural Soil Amendments

3.1. Inorganic Soil Amendments

3.1.1. Natural Mineral

Common natural minerals used for remediating heavy metal-contaminated soils primarily include clay minerals and natural zeolites. Due to their ion exchange properties, natural zeolites are highly effective at adsorbing metal cations [36]. Among natural zeolites, clinoptilolite is most commonly used for heavy metal adsorption, followed by mordenite and chabazite. Clinoptilolite, the most widely utilized, is particularly effective in removing metals such as Pb, copper (Cu), zinc (Zn), Cd, and nickel (Ni) [37]. Zeolites are porous hydrated aluminosilicates characterized by their molecular sieve properties, which arise from their well-defined network structure [38].

Clay minerals, broadly categorized as hydrous aluminosilicates, form the colloidal portion of soils, sediments, rocks, and water. Known for their strong ion exchange and adsorption capacities, these minerals are key natural adsorbents for heavy metals, playing a crucial role in the remediation of soils contaminated with such pollutants. In agricultural soil remediation, sepiolite has been the most frequently used amendment, followed by palygorskite and bentonite [39]. Bentonite minerals consist of montmorillonite units arranged in a 2:1 layered structure of silica sheets. They have a large surface area, high cation exchange capacity, and a negative charge, making them effective adsorbents [40]. Palygorskite, also known as attapulgite, is a magnesium aluminum phyllosilicate. Its surface hydroxyl groups can form inner-sphere complexes with Cd^{2+} ions, while deprotonated hydroxyl groups can create outer-sphere complexes through electrostatic binding, effectively reducing Cd availability in the soil [41].

Based on the natural mineral substance structure and adsorption principle, the remediation of heavy metal-contaminated soil mainly includes the following methods: (1) improving soil structure. Bentonite, mainly composed of montmorillonite, has certain expansion, dispersion, and adhesiveness. Adding it to the soil can increase the number of aggregates, increase soil porosity, and reduce soil bulk density [42]. (2) adjusting soil pH value. Bentonite is a porous, fibrous, hydrated magnesium silicate. Its structure consists of two tetrahedral silica sheets sandwiching an octahedral magnesium/hydroxide sheet [39]. Bentonite alters soil pH and heavy metal retention through adsorption and ion exchange mechanisms. At low pH, metal retention is mainly due to surface deprotonation and ion exchange, while at higher pH, metal ion precipitation further enhances retention, especially for Cu and Zn [41,43]. Lime, as an alkaline calcium compound, can raise the pH value of acidic soil contaminated with heavy metals through neutralization reactions and reduce the solubility and mobility of heavy metals in the soil [44]. Zeolite has the capacity to adsorb Na^{+} and Cl^{-} from the soil, which effectively reduces soil salinity and alleviates the impacts of salinization. Additionally, zeolite can buffer the pH levels of saline-alkali soils. At high pH, the primary effect is the conversion of toxic inorganic compounds into poorly soluble hydrated oxides and carbonates, thereby mitigating their potential toxicity and improving soil quality [45]. (3) adsorbing heavy metals. Additionally, zeolite is a potent natural adsorbent for heavy metals like Cu^{2+}, Zn^{2+}, and Ni^{2+}, with adsorption efficiency influenced by soil pH. The equilibrium adsorption tests were conducted by mixing sieved and dried zeolite particles with solutions containing various heavy metals at different pH levels. The flasks were shaken at a constant temperature for a duration of four days. The residual concentration of heavy metals in the solution is measured at the end of the reaction. The difference between the initial heavy metal concentration and the residual concentration represents the quantity of heavy metals that have been adsorbed, which indicates that adjusting pH can optimize the adsorption process for practical applications. Tests indicate that zeolite's removal of these metals depends on pH, with higher pH (around 2–3) enhancing adsorption capacity, as seen in Figure 7 [45].

Bentonite and kaolin are widely used for heavy metal adsorption, with varying effectiveness for Pb^{2+}, Cu^{2+}, and Zn^{2+}. Higher H^{+} on clay surfaces and reduced metal ion precipitation at high pH levels decrease heavy metal retention in clay-metal systems. Clay suspension samples were prepared by conducting batch equilibrium tests, in which kaolin or bentonite was mixed with heavy metal solutions at a specified metal ion concentration. After shaking for a predetermined duration, the suspension samples were centrifuged and filtered to separate the liquid phase from the solid phase. Bentonite's adsorption of heavy metals follows Temkin and Langmuir models over a wide concentration range, while kaolin follows Freundlich and Langmuir models. The Temkin isotherm is highly effective for predicting gas-phase equilibrium but may be less suitable for more complex adsorption systems. The isotherm can be described as Equation (1).

$$\text{Temkin: } S = B\ln(A_T) + B\ln C_e \quad (1)$$

where A_T denotes the Temkin isotherm equilibrium binding constant (L/mM) and B (J/mol) represents the constant related to the heat of sorption.

The Freundlich isotherm is commonly applied to multilayer adsorption on heterogeneous surfaces and can be expressed in its linearized form by Equation (2).

$$\text{Freundlich: } \ln S = \ln K + n\ln C_e \quad (2)$$

In the equation, S (mmol/g) represents the specific adsorption coefficient of heavy metal ions per unit mass of soil. K ($mmol^{1-n}$ g^{-1} Ln) represents the Freundlich adsorption capacity coefficient. n (dimensionless) is the adsorption constant, which measures the adsorption strength or surface heterogeneity when $0 < n < 1$. When $n < 1$, it represents a chemical adsorption process, and when $n > 1$, it represents a cooperative adsorption process. C_e (mmol/L) represents the final concentration of heavy metal ions after 48 h of solid/liquid equilibrium.

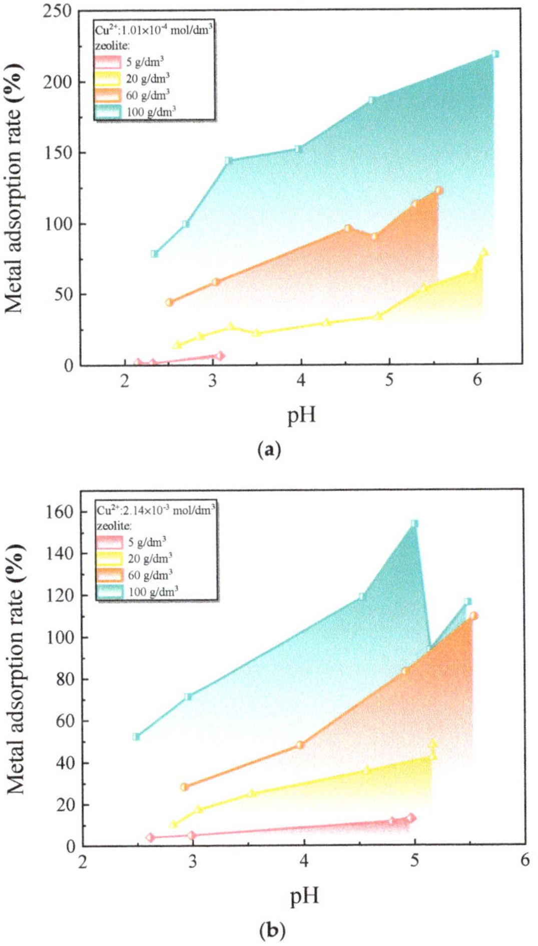

Figure 7. *Cont.*

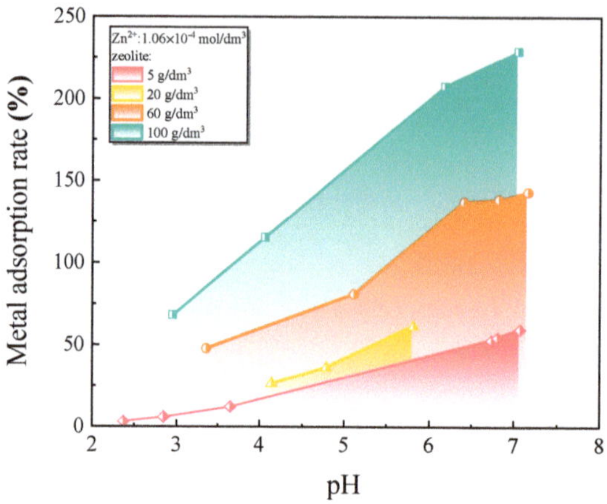

(c)

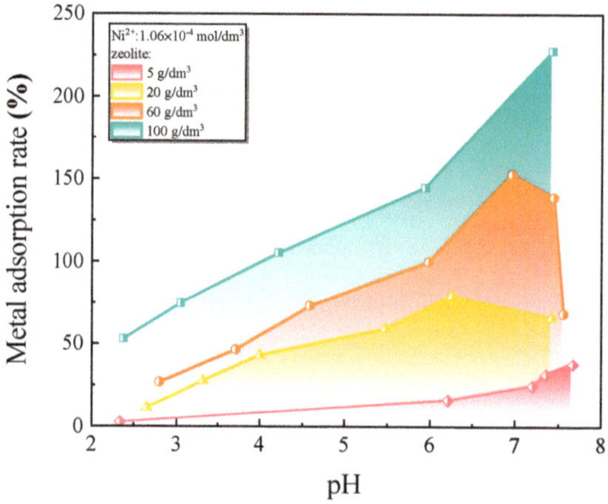

(d)

Figure 7. Experimental metal pH-adsorption edges at various zeolite concentrations: (**a**,**b**) Cu^{2+}; (**c**) Zn^{2+}; (**d**) Ni^{2+} [45].

Contrary to the Freundlich model, the Langmuir adsorption isotherm is typically used to describe monolayer adsorption and homogeneous adsorption. Its mathematical expression is shown in Equation (3) [43].

$$\text{Langmuir: } C_e/S = 1/(bS_{max}) + (1/S_{max})C_e \quad (3)$$

In the equation, S_{max} (mM g^{-1}) represents the maximum Langmuir adsorption capacity, and b (1/Mm) represents the Langmuir model constant.

Vermiculite is a widely available natural clay. Zhao et al. [46] conducted column immersion tests to study the effects of coexisting cations and evaluate the practical performance of vermiculite in heavy metal removal. They found that vermiculite had a good effect on removing heavy metals. Malandrino et al. [25] conducted pot experiments and demonstrated that adding vermiculite significantly reduced the absorption of metal pollutants by lettuce and spinach plants, confirming its potential for the remediation of metal-contaminated soils. Furthermore, zeolite can retain water. Adding zeolite to soil can improve its water retention capacity. It can also enhance the physicochemical properties and hydraulic parameters of the soil, thereby reducing nutrient loss [47].

Natural minerals encounter theoretical and technical challenges in practical use, including managing application rates, methods, and timing and limitations in available reserves for large-scale applications. Studying the ideal application rates for single amendments and the optimal ratios for composite amendments can significantly enhance the efficiency of remediating heavy metal-contaminated soil.

3.1.2. Inorganic Solid Waste

Inorganic solid waste can be categorized into six main groups based on their origin, such as the electric power industry, non-ferrous metal mining industry, thermal industry, metal processing and smelting industry, paper printing industry, and other industries. Table 7 shows the classification of inorganic industrial solid waste.

Table 7. The classification of inorganic industrial solid waste.

Industry Source	Name of Solid Waste	References
Electric power industry	Slag, desulfurization gypsum, fly ash, etc.	[15,26,48,49]
Non-ferrous metal mining industry	Tailings, coal gangue, limestone, gypsum, etc.	[50–53]
Thermal industry	Fly ash, slag, dust, etc.	[15,48,52,54,55]
Metal processing and smelting industry	Blast furnace slag, steel slag, dust, sludge, etc.	[56–59]
Paper printing industry	Deinking residue, plastic debris, tailings, etc.	[50,60,61]
Other industries	Waste clay, nuclear waste residue, etc.	[62,63]

Inorganic solid waste is primarily used in soil as an amendment or stabilizer, offering solidification and stabilization effects. These wastes include stabilization and solidification agents. Stabilization agents reduce heavy metal mobility and solubility in contaminated soil without altering soil particle strength or permeability significantly. Solidification agents bind contaminated soil particles, forming a strong, low-permeability solid mass, effectively fixing heavy metals [64].

Inorganic solid waste amendments commonly used for soil improvement include carbide slag, fly ash, and cement. The remedial effects of inorganic solid waste amendments on heavy metal-contaminated soil are primarily evident in the following three aspects: (1) Improving the physical and chemical properties of the soil. Sun et al. [41] found that incorporating carbide slag into copper (Cu)-contaminated soil can increase the unconfined compressive strength (UCS) and pH value, which helps alleviate soil acidification. Wang et al. [26] investigated the effect of a new composite cementitious material consisting of cement, fly ash, and desulfurization gypsum (CFG) on the compressive strength and impermeability of contaminated soil containing different concentrations of nickel (Ni) and Cu. The ratios of total Ni content to dry soil mass used in this study were 0%, 0.02%, 0.4%, and 1%, while the ratios of total Cu content to dry soil mass were 0% and 1%, resulting in six heavy metal levels: Ni 0 Cu 0, Ni 0.02, Ni 0.4, Ni 1, Cu 1, and Ni 1 Cu 1. Figures 8 and 9 show that unconfined compressive strength (UCS) decreases with increasing heavy metal concentration. Figure 8 indicates that CFG content between 8% and 15% enhances soil strength by forming gel-like hydrates that fill soil pores. Figure 9 illustrates the long-term benefits of CFG for the solidification of Ni-Cu-contaminated soils. The study conducted by Chang et al. [64] utilized three different leaching tests to determine the concentration of Cr in the leachate of cement powder and cement mortar at various substitution levels. The results of the experiment showed that cement mortar can reduce the leaching of Cr through the hydration process and enhance its encapsulation. Co-processing in cement kilns is an effective alternative method for the remediation of heavy metal-contaminated soil.

(2) Increasing the number of soil microorganisms and enhancing enzyme activity. Fly ash, a byproduct of coal-fired power plants, has traditionally been considered a problematic solid waste. However, some studies have suggested that fly ash can be used as a soil amendment to improve the physical, chemical, and biological properties of degraded soil and serve as a source of both micro and macro-nutrients for plants [65]. Fly ash contains trace amounts of toxic elements and heavy metals, with aluminum (Al) mostly bound in

insoluble aluminosilicate structures, which limits its biological toxicity. Additionally, fly ash contains macro and micronutrients, making it an effective soil amendment for improving soil health and crop yield [55]. For instance, Nayak et al. [55] conducted a pot experiment to investigate the effects of fly ash application on microbial reactions, soil enzyme activity, and heavy metal accumulation in soil and rice. The experiment found varying microbial responses to fly ash application rates. Fungi and actinomycetes decreased with higher fly ash rates, while aerobic heterotrophic bacteria remained stable until a 40% fly ash application. Total microbial activity and denitrifies increased initially but plateaued after 40%. Alkaline and acid phosphatase activities decreased with fly ash. Applying 10–20% fly ash by soil volume can boost trace elements, microbial activity, and crop yield. However, high fly ash levels may induce oxidative stress in rice plants, mitigated by antioxidants and enzymes. Lower fly ash doses show promise in enhancing methane-producing bacteria and rice yields in nutrient-poor soils [66]. (3) The application of ash-fly ash mixture (AFAM) improves soil fertility by improving soil structure and increasing porosity and water retention. These mixtures also act as liming agents, neutralizing acidic soils and making essential nutrients more accessible. In addition, they provide plant-friendly micronutrients while immobilizing heavy metals, thereby reducing their bioavailability and potential toxicity in the soil [67].

However, there are limitations in the use of inorganic solid waste in the process of remediating heavy metal-contaminated soil. For example, fly ash has a strong adsorption and fixation effect on the added phosphorus (P), which leads to low effectiveness of P in fly ash [68]. Additionally, the high effectiveness of boron (B) in fly ash is detrimental to crop growth. Applying high doses of fly ash can worsen heavy metal pollution and hinder microbial activity. Therefore, adjusting the proportion of fly ash application can help avoid adverse effects on crops [65]. The proportion of inorganic solid waste application significantly affects the remediation effect. Applying high doses of fly ash can lead to the accumulation of toxic metals in the soil and inhibit the growth and activity of microorganisms [55].

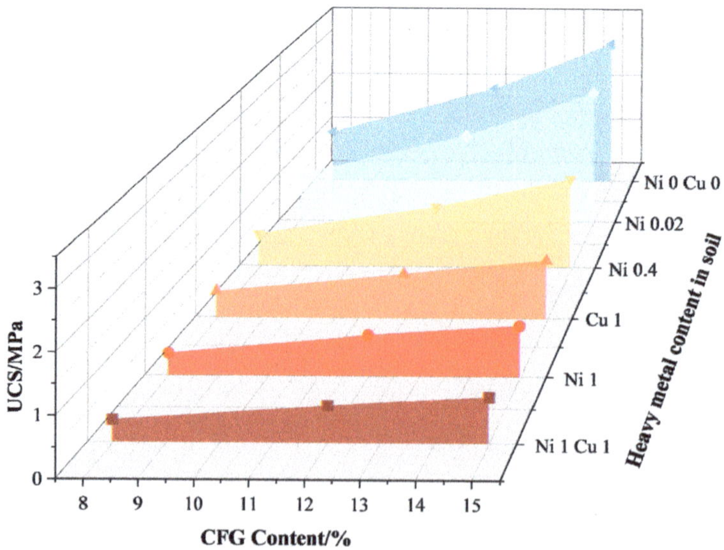

Figure 8. The relationship between curing agent content and UCS [26].

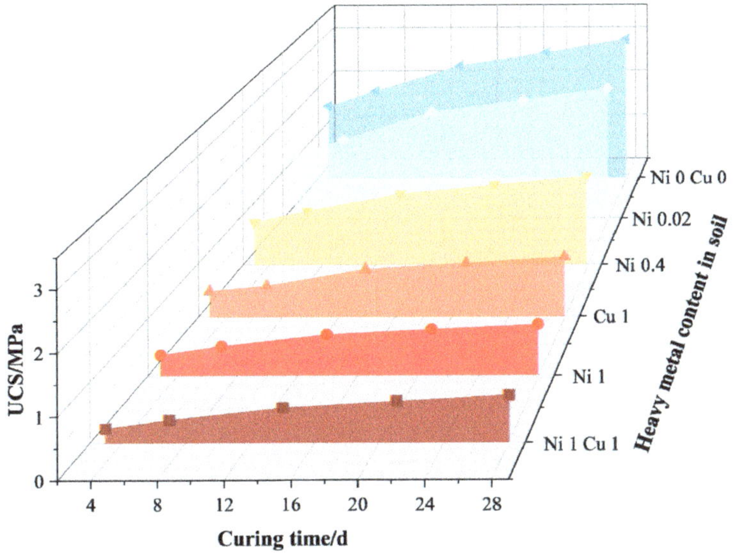

Figure 9. The relationship between curing time and UCS [26].

3.2. Organic Soil Amendments

3.2.1. Organic Solid Waste

Organic solid waste can include sewage sludge, municipal solid waste, food waste, kitchen waste, garden waste, agricultural waste, and animal waste [69]. Organic solid waste consists of biodegradable organic matter with a water content below 85–90%. Waste like paper sludge, sewage sludge, urban solid waste, crop straw, green manure, and animal manure can serve as organic soil amendments, particularly useful in addressing heavy metal pollution. The role of organic solid waste in improving the remediation of heavy metal-contaminated soil can be observed in several aspects. (1) It improves the physical and chemical properties of the soil. Municipal solid waste compost and cow manure are commonly used as organic fertilizers in agricultural production and horticultural cultivation. Studies have shown that under greenhouse conditions, these two types of organic solid waste significantly decrease the soil pH and increase the electrical conductivity (EC), which depends on the raw materials and ion concentrations of the compost [70,71]. (2) It reduces the bioavailability of heavy metals in the soil. Research conducted by Alam et al. found that a mixture of vermicompost (VC), leaf compost (LC), and spent mushroom compost (SMC) in equal proportions can raise the pH of heavy metal-contaminated soil, thereby reducing the bioavailability and mobility of heavy metals in the soil [27,72]. This effectively decreases the absorption of Cd, Cr, Pb, and Mn by plants, promoting plant growth [27]. (3) It improves the biological characteristics of contaminated soil. Incorporating straw into the soil enhances its biological characteristics by altering the bacterial and fungal community structure and increasing the biomass of bacteria and fungi. This, in turn, accelerates straw decomposition [73]. Additionally, the activities of soil phosphatase, urease, and transformation enzymes increase with the application of straw incorporation [74]. Furthermore, the application of biogas residue and compost improves several soil microbial characteristics, such as substrate-induced respiration, potential ammonium oxidation, and nitrogen mineralization [75]. Research has also shown that sewage sludge can serve as a barrier for heavy metal pollutants in solid waste landfill sites, hindering the ion migration of heavy metal pollutants [58]. The relationship between ion strength and electrical conductivity has been summarized by Griffin and Jurinak [76] with a correlation coefficient of 0.996.

$$I = 0.013EC \tag{4}$$

In the equation, I (mol/L) represents ionic strength, and EC (mmhos/cm) represents conductivity.

The compacted sewage sludge hinders the migration of heavy metal ions due to the anaerobic neutral to weakly alkaline conditions caused by microbial respiration [58]. (4) Furthermore, it enhances soil fertility and promotes plant growth. Green manure, as a main organic fertilizer, has been widely proposed as a sustainable method for ecological restoration and improving soil fertility in degraded environments [77]. Leguminous green manure is an important source of carbon (C) and nitrogen (N) in the cropping system, and the biochar derived from leguminous green manure can be used as a soil amendment to enhance soil fertility [23]. In addition, the organic acids released during straw decomposition can activate inorganic phosphorus, making it a new source of nutrients [78].

Although organic solid waste can improve heavy metal-contaminated soil, it also has some negative impacts. Chen et al. [23] found that leguminous green manure can quickly release N through rapid turnover, but it limits the effectiveness of maintaining organic carbon in the soil. In addition, the high carbon-to-nitrogen ratio of sewage sludge can increase the extractable content of Zn, Cu, Cr, and Cu in the soil. Moreover, untreated sludge contains numerous pathogens, parasites, and toxic substances, which can have varying degrees of toxic effects on plants and organisms in the soil. When applying straw, attention should be paid to the dosage, as excessive straw can lead to nitrogen deficiency in crops; hence, it is best to use it in combination with nitrogen fertilizer.

3.2.2. Naturally Extracted Polymer Compounds

Naturally extracted polymer compounds (NEPCs) are high molecular weight substances obtained from natural sources or minerals by chemical processes involving biochemical or photosynthetic mechanisms. Common NEPCs used to amend heavy metal-contaminated soils include lignin, cellulose, and chitin.

Derived from plant biomass, particularly fibers and wood pulp, lignin is a byproduct of industries such as paper and wood. It shows significant potential in improving heavy clay and sandy soils. Lignin, a major component alongside cellulose in plant tissues, consists of hydrophilic and hydrophobic groups, contributing to its high organic molecular weight. Studies by Cai et al. [79] have investigated lignin-treated sludge, evaluating parameters such as particle size distribution, Atterberg limits, compaction characteristics, unconfined compressive strength, pH levels, and resistivity. Enhanced soil performance attributed to lignin treatment is linked to the formation of a more stable soil structure through cation exchange and lignin bonding. Lignin's ability to fill soil pores and improve strength, permeability, and durability is facilitated by hydrolysis reactions, ion exchange mechanisms, and electrostatic forces [79–82].

Similarly, cellulose, a polysaccharide found in plant cell walls, enhances soil structure and moisture retention, benefiting plant growth [83–86]. Chitin, a polymer found in fungal cell walls and exoskeletons of insects and crustaceans, contributes to soil fertility and biological activity, offering potential in sustainable agriculture practices [28,87–89].

These NEPCs collectively contribute to (1) improvement in soil properties—enhancing physical, mechanical, and resistivity characteristics; (2) reduction in heavy metal ion concentrations—immobilizing heavy metal ions in the soil, thereby reducing their availability and toxicity; (3) additionally, it can change the soil's microbial communities, enhancing their ability to suppress harmful microorganisms; (4) improving the carbon-to-nitrogen ratio in the soil enhances plant growth and increases crop yield. However, lignin, cellulose, and chitin can also have different negative effects on contaminated soils. For example, lignin and cellulose have complex structures and are more difficult for microorganisms to degrade, leading to their persistence in the soil [81,82,85,86]. Decomposition of chitin may increase soil acidity, interfering with plant growth and soil microbial activity [28,88]. Therefore, further research and trials are needed on the dosage and ratios of amendments to be used.

3.2.3. Organic Material

Biochar is a material rich in organic carbon that is produced by the pyrolysis of agricultural biomass waste, such as wood chips or crop straw, in a limited oxygen environment [29]. Numerous studies have confirmed that biochar can immobilize heavy metals in the soil and reduce their accumulation in plants. This is mainly attributed to the high porosity, active functional groups, high pH value, and cation exchange capacity (CEC) of biochar [29,30,90,91]. Additionally, the application of biochar and modified biochar in the remediation of heavy metal-contaminated soil has shown good economic benefits [92,93]. For example, biochar made from sludge can not only reduce the huge treatment costs of urban sludge but also recover energy through further processing (such as sludge pyrolysis), promoting sustainable development [94]. Biochar can serve as an organic amendment for heavy metal-contaminated soil, with the following specific functions: (1) improving the physical and chemical properties of the soil. Biochar application can positively influence soil thermal properties by enhancing soil water retention and increasing soil moisture content [95]. Furthermore, biochar is an alkaline material that can increase the pH value of acid-contaminated soil [96]. Wang et al. [97] conducted pot experiments and found that wood-based biochar, bamboo-based biochar, straw-based biochar, and walnut shell-based biochar all significantly improved soil electrical conductivity. The pH and cation exchange capacity (CEC) of soil are interrelated. Biochar composites raise the soil pH, increasing CEC and improving binding with the negatively charged functional parts of organic matter [94]. Githinji and Berihun [98,99] found that the application of biochar to the soil significantly reduced soil bulk density and increased total porosity, soil pH value, total nitrogen, soil organic carbon, and available phosphorus and potassium (K) through experiments. (2) Adsorbing heavy metals in the soil and reducing their accumulation in plants. Biochar's complex humic substances and mineral oxides co-precipitate, creating inner-sphere complexes that exchange heavy metals with cations like Ca^{2+} and Mg^{2+}. Biochar's adsorption on heavy metal-contaminated soil may vary depending on soil type and cation presence [91]. Bian et al. [29,30] conducted a study on rice fields contaminated with heavy metals, where they applied a soil amendment called biochar. After using biochar, the soil's pH and total organic carbon content increased significantly and stayed higher. This helped immobilize Cd and Pb in the soil, as Al, iron (Fe), and phosphorus (P) minerals combined with the contaminated biochar particles. Kim et al. [100] investigated the effect of biochar on heavy metal content in paddy soil and the accumulation of heavy metals in rice through a pot experiment. Their research demonstrated that mineral components such as phosphate and carbonate in biochar can precipitate together with heavy metals. Park et al. [101,102] found that peat moss biochar reduced the mobility and bioavailability of Cu, Cd, and Pb through coordination between metal electrons and chemical bonds. The presence of functional groups in the biochar and their affinity to bind with heavy metal ions lowered their bioavailability. (3) Biochar has an impact on soil microbial populations and activity. Some studies have shown that the dosage of biochar affects the abundance and activity of soil microbial species. At a low concentration of 1% biochar, the relative abundance of bacteria and fungi species increases, while at a higher application rate of 5%, their abundance decreases [103]. The application of biochar to soil can further alter the composition of microbial communities and increase diversity, thereby stimulating specific microbial processes [104]. This enhances soil biochemical cycling through the interaction between rhizobia and bacteria, promoting nutrient absorption and crop yield [93]. (4) Moreover, biochar has been found to promote plant growth and increase crop yield. Wang et al. [97] conducted a pot experiment and found that, apart from bamboo biochar, all other types of biochar increased the dry weight of bamboo compared to the control group. Abulaiti et al. [105] applied biochar to rice fields and observed a significant increase in the net photosynthesis rate, transpiration rate, and grain yield of rice.

Biological charcoal makes significant contributions to improving heavy metal-contaminated soil. However, it is important to consider the application dosage. Research shows that as the dosage of biological charcoal increases, the adsorption of nitrogen by biological charcoal can lead to a decrease in available nitrogen content in the soil, affecting the soil's nutritional status and thus reducing plant growth rate [90].

4. Synthetic Soil Amendments

Synthetic soil amendments are large organic polymers designed to mimic natural amendments. Research on these synthetics, like polyacrylamide and polyvinyl alcohol resin, is a focal point in both domestic and international studies [10]. Among them, polyacrylamide (referred to as PAM) is a widely used artificial synthetic amendment in recent years. Its main mechanisms for soil improvement are as follows: (1) improving the physical properties of the soil. PAM is a water-soluble synthetic polymer composed of repeat units with amide functional groups. The amide groups are non-ionizable, but by introducing anionic or cationic functional groups, the PAM macromolecules can carry charges and form the ionic form of the polymer [106]. Rabiee et al. [107] used free radical polymerization to produce anionic PAM with metal cations and found that anionic PAM can form a composite material with soil structure, enhancing the adhesive force between soil and polymer functional groups, agglomerating soil components, and increasing soil strength. Sandy soils contaminated with heavy metals have poor water-holding capacity, and water and nutrients are easily lost [108]. PAM notably boosted the slope of the soil moisture curve's inflection point, enhancing water retention and decreasing hydraulic conductivity in sandy soil, thus improving its physical properties. Additionally, PAM reduces soil salinity and bulk density while increasing organic matter, water-stable aggregates, and macroaggregate content [105]. (2) Adsorb heavy metals and reduce the bioavailability of heavy metals to organisms. Wiśniewska and Fijałkowska et al. [106] demonstrated through electrokinetic and potentiometric titration measurements by analyzing the solution pH, mineral type, different ionic types of PAM, properties of heavy metal ions, and the order of adsorbent addition, that anionic PAM and cationic PAM have different adsorption capacities at different pH values. Both anionic PAM and cationic PAM can immobilize Pb and Cr on the surface of clay minerals [31]. Superabsorbent polymers are loosely cross-linked polymer chain networks with high hydrophilicity, which can absorb and retain water or aqueous solutions up to several hundred times their weight [109]. Applying PAM to soil can reduce the uptake of Cr, Cu, Pb, and Zn by potato (*Solanum tuberosum* L.) plants irrigated with wastewater [109,110]. (3) Influence the activity of microorganisms and enzymes in the soil. PAM can affect the growth and respiration of microorganisms by lowering the soil pH and altering the chemical properties of the soil. PAM can be decomposed by some microorganisms in the soil, such as Bacillus subtilis, as a source of carbon and nitrogen, promoting microbial reproduction and increasing their population [111]. (4) Reduce the loss of nitrogen in the soil and increase crop yield. PAM can also increase the net photosynthetic rate, transpiration rate, and stomatal conductance of rice, positively impacting rice yield [105].

However, PAM has limitations as a soil amendment for heavy metal pollution. Its intermediate product, acrylamide, is toxic, and excessive use may lead to secondary pollution in heavy metal-contaminated soil. Further research is needed to determine the appropriate application rate of PAM.

5. Natural-Synthetic Copolymer Soil Amendments

Natural-synthetic copolymers like humic acid-polyacrylic acid, cellulose-polyacrylamide, and others can be used to remediate heavy metal-contaminated soil [32,85,112–115]. Due to their high molecular weight and numerous functional groups, natural-synthetic copolymer amendments play an important role in absorbing heavy metal elements. Thus, polymer-based nanocomposites can be used to immobilize metal ions in contaminated soil [32,85,112–115].

The main improvements of natural-synthetic copolymer soil amendments on heavy metal-polluted soil can be summarized as follows: (1) adsorption and fixation of heavy metals in the soil, reducing the effective accumulation of heavy metals in the soil and plants. Mahdavi et al. [116] developed a novel and inexpensive polymer nanocomposite material called polyacrylamide-polyvinyl styrene/montmorillonite. This modifier is prepared by in situ polymerization of acrylamide and styrene between the interlayers of montmorillonite through free radical polymerization [116]. Guleria et al. [85] developed a cellulose-g-poly(acrylamide-co-acrylic acid) copolymer and tested its strong adsorption capacity for Cd, Cu, Pb, and Zn by varying pH, contact time, temperature, and metal ion concentration. The reaction steps for synthesizing cellulose-grafted copolymers are shown in Figure 10.

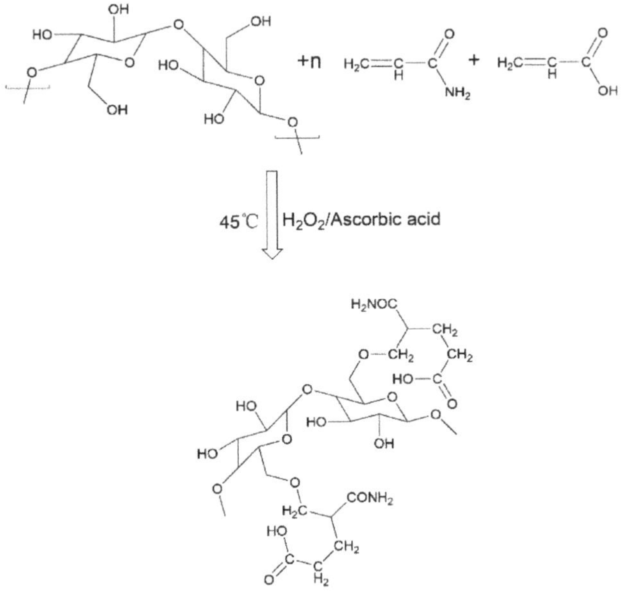

Figure 10. Reaction steps for the synthesis of cellulose graft copolymers [85].

Zhou et al. [113] prepared polyacrylamide/montmorillonite by solution polymerization of acrylamide on γ-methacryloxypropyltrimethoxysilane-modified montmorillonite. Chemical leaching is a remediation technique for heavy metal-polluted soil using chelating agents to leach the contaminated soil [1]. However, chelating agents used in chemical leaching may cause secondary pollution to the soil. Therefore, Zhao et al. [112] studied a biodegradable lignin-based poly(acrylic acid) composite material (LBPAA). It can transfer Cd^{2+} and Pb^{2+} ions from the soil to the eluent, making it a new environmentally friendly composite material. (2) Improve the physical properties of soil to enhance its water retention and soil conservation ability. Li et al. [32] developed a multifunctional microsphere heavy metal-contaminated soil amendment called chitosan-grafted poly(acrylamide-co-acrylic acid)/biochar (CPB). CPB greatly enhances the soil's ability to adsorb heavy metals. Its dense porous structure, which includes N-methylene bisacrylamide and hydrophilic biochar, reduces water loss, improves water retention, and enhances soil's physical properties. Basuki et al. [114] used gamma rays to synthesize soil amendments by grafting chitosan-acrylamide copolymer, and the research showed that this amendment reduced the bulk density of soil, enhanced its water retention capacity, and increased porosity. (3) Control and release nitrogen sources in soil. Niu and Li [115] prepared starch-g-poly(vinyl acetate) (St-g-PVAc), which is a biodegradable carrier material through graft copolymerization of starch and vinyl acetate. It exhibits relatively low swelling, high encapsulation capacity, and a slow release rate.

6. Biological Soil Amendments

Bioremediation is a great option for fixing heavy metal-contaminated soil. It uses microbial amendments to remove heavy metals, offering economic and ecological benefits with minimal negative impacts [1,117]. Microbial remediation of heavy metal-contaminated soil occurs through the biological activity of microorganisms, which adsorb or transform heavy metals into less toxic products [117]. This process primarily involves mechanisms such as biosorption, biovolatilization, bioleaching, bioaccumulation, and biomineralization [118].

Bioremediation agents are suitable for repairing large areas with relatively low pollutant concentrations in shallow sites. The main effects of bioremediation are: (1) improving the physical and chemical properties of soil. Liu et al. [119] added feces to oil-contaminated soil to biostimulate the local microorganisms for bioremediation. The use of the *Pseudomonas chenduensis* strain MBR for microbial remediation of heavy metal-contaminated soil demonstrated that adding MBR can improve the total phosphorus, total carbon, total nitrogen, and pH value of the soil while also increasing the dry weight of grains [33]. (2) To reduce the overall bioavailable fraction of heavy metals in contaminated soil and decrease their toxicity, microorganisms can transform heavy metals from one oxidation state or organic compound to another [12]. *Aspergillus niger* produces metal-solubilizing organic acids, which can leach heavy metals like Cu, Cd, Pb, and Zn from the soil, removing their exchangeable forms and other compounds. Microbial leaching for bioremediation is more effective and cost-efficient than chemical leaching [120]. Peng et al. [13] used *Rhodobacter* bacteria to decrease the bioavailable fractions of Cd and Zn in their exchangeable and carbonate-bound phases while increasing the more stable forms, such as iron/manganese oxides and organic complexes. After wheat seedling experiments with *Fusarium solani*, Cd uptake by plants decreased, and exchangeable Cd and Zn in contaminated soil dropped by 30.7% and 100.0%, respectively. Li et al. [33] used *Pseudomonas chenduensis* strain MBR (referred to as strain MBR) to bioremediate Cd-contaminated paddy soil, and the results showed that the addition of strain MBR enhanced the role of the microbial community in transforming Cd components, significantly reducing Cd accumulation in rice grains and roots. (3) Increasing the quantity of microorganisms in the soil promotes the diversification of soil microbial communities. Liu et al. [119] conducted experiments to demonstrate that soil treated with microbial remediation had significantly increased numbers of heterotrophic bacteria, polycyclic aromatic hydrocarbon-degrading bacteria, and total petroleum hydrocarbon-degrading bacteria. The microbial activity, species, and diversity were all enhanced. (4) Enhancing plant growth and making plants more robust can increase their resistance to pathogenic bacteria. Basyal and Emery [121] conducted in-depth research on the plant-arbuscular mycorrhizal fungi (AMF) symbiotic system and found that AMF can increase leaf number, root biomass, cellulose, and hemicellulose content in plants. This promotes plant growth and improves key indicators related to plant growth and cell wall chemistry, with the most significant effects observed under low soil moisture conditions.

Biological agents have limits in remediation. Microbes can change heavy metals into less toxic forms but cannot remove them from the soil, requiring additional extraction methods [12]. Additionally, microbial remediation is influenced by environmental conditions, requires extended remediation periods, and is not suitable for soils with high contamination levels or low permeability [117].

7. Conclusions and Recommendations

This review synthesizes recent research on soil amendments for mitigating heavy metal contamination, emphasizing their impacts on soil properties, heavy metal concentrations, microbial communities, and crop yields. The selection criteria for amendments are based on their efficacy, application timing, cost-effectiveness, and ecological advantages. Soil amendments for heavy metal-contaminated soils exhibit several key effects: (1) They enhance soil structural integrity, improve water retention and nutrient holding capacity, modify pH levels, and ameliorate other physicochemical properties. (2) They diminish the availability and mobility of heavy metals such as Cu, Ni, Cd, Cr, Pb, Zn, and Co by converting them into less toxic forms, thereby reducing their uptake by plants. (3) They stimulate soil microbial populations, modify microbial community composition and structure, increase community diversity, and boost enzyme activity. (4) They enhance soil fertility, adjust carbon-to-nitrogen ratios, promote plant growth, and subsequently increase crop yields.

There is a notable deficiency of multifunctional, cost-effective, and environmentally benign soil amendments for heavy metal-contaminated soils. The limitations of current approaches are as follows: (1) Natural mineral amendments are constrained by their limited effectiveness and availability, resulting in suboptimal efficiency. (2) Amendments derived from inorganic and organic solid wastes may introduce secondary pollution and inhibit microbial and enzymatic activities due to challenges in dosage control. (3) Artificially synthesized amendments are expensive to produce, and some intermediate products present potential pollution risks. (4) Biological amendments are unable to completely eliminate heavy metals and exhibit delayed effectiveness over extended testing periods. (5) While natural-synthetic copolymer amendments demonstrate high efficiency, further research is required to elucidate their synthesis, underlying mechanisms, and economic feasibility.

Based on the applicability and limitations of the aforementioned soil amendments, we propose the following recommendations: (1) Investigate the mechanisms and impacts of various soil amendments on heavy metal-contaminated soils, with consideration given to pollution control requirements and cost factors associated with industrial production, transportation, and agricultural application. Emphasize the development of cost-effective and environmentally friendly amendments while prioritizing both cost reduction and environmental protection. (2) Relying on a single amendment may yield limited effectiveness. It is advisable to explore combinations of amendments, such as mixing inorganic-organic solid wastes or integrating biological amendments with agricultural and industrial byproducts. Combining different types of amendments can address the limitations of individual treatments. Employing orthogonal experimental designs can assist in identifying optimal combinations and proportions to enhance effectiveness and functionality. (3) While current research predominantly focuses on soil physicochemical properties, heavy metal removal, biological characteristics, and crop yield improvements, further investigation is warranted into the changes in soil properties and biological dynamics before and after remediation, as well as the mechanisms of soil–plant interactions. (4) The effectiveness of soil amendments for remediating heavy metal-contaminated soils may vary across different climates and terrains. Most studies are conducted in laboratory settings, small plots, or greenhouses, which may not accurately predict performance in larger outdoor environments. It is crucial to undertake more large-scale field experiments to better understand the real-world effects of these amendments. (5) Ensure the maintenance of remediated soils, establish comprehensive soil assessment systems, and preserve ecosystem functions to promote sustainable ecological, economic, and social benefits.

Author Contributions: All authors commented on previous versions of the manuscript. Conceptualization: X.N., M.L. and Z.L.; Methodology: X.N., M.L. and Z.L.; Formal analysis and investigation: X.N., M.L., Z.L. and X.L.; Writing—original draft preparation: X.N.; Writing—review and editing: M.L., Z.L., X.L. and X.H.; Funding acquisition: X.H. and M.L.; Resources: X.H.; Supervision: X.H. All authors have read and agreed to the published version of the manuscript.

Funding: This work was supported by the National Natural Science Foundation of China [grant number 52370001].

Institutional Review Board Statement: Not applicable.

Informed Consent Statement: Not applicable.

Data Availability Statement: The authors confirm that the data supporting the study is available within the article.

Acknowledgments: The authors would like to thank the editor and anonymous reviewers for their comments and suggestions. We confirm that the manuscript is original, has not been reviewed by multiple journals at the same time, and has not divided a study into several parts to increase the number of submissions. The results are clear and true, and there is no improper manipulation of the data.

Conflicts of Interest: The authors declare that they have no known competing financial interests or personal relationships that could have appeared to influence the work reported in this paper.

References

1. Khalid, S.; Shahid, M.; Niazi, N.K.; Murtaza, B.; Bibi, I.; Dumat, C. A Comparison of Technologies for Remediation of Heavy Metal Contaminated Soils. *J. Geochem. Explor.* **2017**, *182*, 247–268. [CrossRef]
2. Liu, S.; Zhan, L.; Hu, L.; Du, Y. Environmental Geotechnics: State-of-the-art of Theory, Testing and Application to Practice. *Chin. Civ. Eng. J.* **2016**, *49*, 6–30. [CrossRef]
3. Manzoor, M.M.; Goyal, P.; Gupta, A.P.; Gupta, S. Heavy Metal Soil Contamination and Bioremediation. In *Bioremediation and Biotechnology*; Bhat, R.A., Hakeem, K.R., Dervash, M.A., Eds.; Springer International Publishing: Cham, Switzerland, 2020; Volume 2, pp. 221–239, ISBN 978-3-030-40332-4.
4. Vareda, J.P.; Valente, A.J.M.; Durães, L. Heavy Metals in Iberian Soils: Removal by Current Adsorbents/Amendments and Prospective for Aerogels. *Adv. Colloid Interface Sci.* **2016**, *237*, 28–42. [CrossRef] [PubMed]
5. Mao, C.; Song, Y.; Chen, L.; Ji, J.; Li, J.; Yuan, X.; Yang, Z.; Ayoko, G.A.; Frost, R.L.; Theiss, F. Human Health Risks of Heavy Metals in Paddy Rice Based on Transfer Characteristics of Heavy Metals from Soil to Rice. *CATENA* **2019**, *175*, 339–348. [CrossRef]
6. He, F.; Gao, J.; Pierce, E.; Strong, P.J.; Wang, H.; Liang, L. In Situ Remediation Technologies for Mercury-Contaminated Soil. *Environ. Sci. Pollut. Res.* **2015**, *22*, 8124–8147. [CrossRef]
7. Obalum, S.E.; Chibuike, G.U.; Peth, S.; Ouyang, Y. Soil Organic Matter as Sole Indicator of Soil Degradation. *Environ. Monit. Assess.* **2017**, *189*, 176. [CrossRef]
8. Sungur, A.; Soylak, M.; Yilmaz, E.; Yilmaz, S.; Ozcan, H. Characterization of Heavy Metal Fractions in Agricultural Soils by Sequential Extraction Procedure: The Relationship Between Soil Properties and Heavy Metal Fractions. *Soil Sediment Contam. Int. J.* **2015**, *24*, 1–15. [CrossRef]
9. Minnikova, T.V.; Denisova, T.V.; Mandzhieva, S.S.; Kolesnikov, S.I.; Minkina, T.M.; Chaplygin, V.A.; Burachevskaya, M.V.; Sushkova, S.N.; Bauer, T.V. Assessing the Effect of Heavy Metals from the Novocherkassk Power Station Emissions on the Biological Activity of Soils in the Adjacent Areas. *J. Geochem. Explor.* **2017**, *174*, 70–78. [CrossRef]
10. Chen, Y.; Dong, Y. Progress of Research and Utilization of Soil Amendments. *Ecol. Environ.* **2008**, 1282–1289. [CrossRef]
11. Liu, L.; Li, W.; Song, W.; Guo, M. Remediation Techniques for Heavy Metal-Contaminated Soils: Principles and Applicability. *Sci. Total Environ.* **2018**, *633*, 206–219. [CrossRef]
12. Garbisu, C.; Alkorta, I. Basic Concepts on Heavy Metal Soil Bioremediation. *Eur. J. Miner. Process. Environ. Prot.* **2003**, *3*, 58–66.
13. Peng, W.; Li, X.; Song, J.; Jiang, W.; Liu, Y.; Fan, W. Bioremediation of Cadmium- and Zinc-Contaminated Soil Using Rhodobacter Sphaeroides. *Chemosphere* **2018**, *197*, 33–41. [CrossRef] [PubMed]
14. Wang, Y.; Li, A.; Cui, C. Remediation of Heavy Metal-Contaminated Soils by Electrokinetic Technology: Mechanisms and Applicability. *Chemosphere* **2021**, *265*, 129071. [CrossRef] [PubMed]
15. Jiang, Q.; He, Y.; Wu, Y.; Dian, B.; Zhang, J.; Li, T.; Jiang, M. Solidification/Stabilization of Soil Heavy Metals by Alkaline Industrial Wastes: A Critical Review. *Environ. Pollut.* **2022**, *312*, 120094. [CrossRef] [PubMed]
16. Shah, V.; Daverey, A. Phytoremediation: A Multidisciplinary Approach to Clean up Heavy Metal Contaminated Soil. *Environ. Technol. Innov.* **2020**, *18*, 100774. [CrossRef]
17. Thomé, A.; Reddy, K.R.; Reginatto, C.; Cecchin, I. Review of Nanotechnology for Soil and Groundwater Remediation: Brazilian Perspectives. *Water Air Soil Pollut.* **2015**, *226*, 121. [CrossRef]
18. O'Brien, P.L.; DeSutter, T.M.; Casey, F.X.M.; Khan, E.; Wick, A.F. Thermal Remediation Alters Soil Properties—A Review. *J. Environ. Manag.* **2018**, *206*, 826–835. [CrossRef]
19. Superfund Remedy Report 17th Edition—EPA 542-R-23-001. 2023. Available online: https://www.epa.gov/remedytech/superfund-remedy-report (accessed on 1 November 2024).
20. Desogus, P.; Manca, P.P.; Orrù, G. Heavy Metal Leaching of Contaminated Soils from a Metallurgical Plant. *Int. J. Min. Reclam. Environ.* **2013**, *27*, 202–214. [CrossRef]

21. Huang, Q.; Yu, Z.; Pang, Y.; Wang, Y.; Cai, Z. Coupling Bioleaching and Electrokinetics to Remediate Heavy Metal Contaminated Soils. *Bull. Environ. Contam. Toxicol.* **2015**, *94*, 519–524. [CrossRef]
22. Moon, D.H.; Chang, Y.-Y.; Lee, M.; Koutsospyros, A.; Koh, I.-H.; Ji, W.H.; Park, J.-H. Assessment of Soil Washing for Heavy Metal Contaminated Paddy Soil Using $FeCl_3$ Washing Solutions. *Environ. Geochem. Health* **2021**, *43*, 3343–3350. [CrossRef]
23. Chen, C.-P.; Cheng, C.-H.; Huang, Y.-H.; Chen, C.-T.; Lai, C.-M.; Menyailo, O.V.; Fan, L.-J.; Yang, Y.-W. Converting Leguminous Green Manure into Biochar: Changes in Chemical Composition and C and N Mineralization. *Geoderma* **2014**, *232–234*, 581–588. [CrossRef]
24. Zhang, T.; Yang, Y.-L.; Liu, S.-Y. Application of Biomass By-Product Lignin Stabilized Soils as Sustainable Geomaterials: A Review. *Sci. Total Environ.* **2020**, *728*, 138830. [CrossRef] [PubMed]
25. Malandrino, M.; Abollino, O.; Buoso, S.; Giacomino, A.; La Gioia, C.; Mentasti, E. Accumulation of Heavy Metals from Contaminated Soil to Plants and Evaluation of Soil Remediation by Vermiculite. *Chemosphere* **2011**, *82*, 169–178. [CrossRef] [PubMed]
26. Wang, Q.; Li, M.; Yang, J.; Cui, J.; Zhou, W.; Guo, X. Study on Mechanical and Permeability Characteristics of Nickel-Copper-Contaminated Soil Solidified by CFG. *Environ. Sci. Pollut. Res.* **2020**, *27*, 18577–18591. [CrossRef] [PubMed]
27. Alam, M.; Hussain, Z.; Khan, A.; Khan, M.A.; Rab, A.; Asif, M.; Shah, M.A.; Muhammad, A. The Effects of Organic Amendments on Heavy Metals Bioavailability in Mine Impacted Soil and Associated Human Health Risk. *Sci. Hortic.* **2020**, *262*, 109067. [CrossRef]
28. Saravanan, D.; Gomathi, T.; Sudha, P.N. Sorption Studies on Heavy Metal Removal Using Chitin/Bentonite Biocomposite. *Int. J. Biol. Macromol.* **2013**, *53*, 67–71. [CrossRef]
29. Bian, R.; Joseph, S.; Cui, L.; Pan, G.; Li, L.; Liu, X.; Zhang, A.; Rutlidge, H.; Wong, S.; Chia, C.; et al. A Three-Year Experiment Confirms Continuous Immobilization of Cadmium and Lead in Contaminated Paddy Field with Biochar Amendment. *J. Hazard. Mater.* **2014**, *272*, 121–128. [CrossRef]
30. Bian, R.; Chen, D.; Liu, X.; Cui, L.; Li, L.; Pan, G.; Xie, D.; Zheng, J.; Zhang, X.; Zheng, J.; et al. Biochar Soil Amendment as a Solution to Prevent Cd-Tainted Rice from China: Results from a Cross-Site Field Experiment. *Ecol. Eng.* **2013**, *58*, 378–383. [CrossRef]
31. Fijałkowska, G.; Szewczuk-Karpisz, K.; Wiśniewska, M. Anionic Polyacrylamide Influence on the Lead(II) Ion Accumulation in Soil—The Study on Montmorillonite. *J. Environ. Health Sci. Eng.* **2020**, *18*, 599–607. [CrossRef]
32. Li, S.; Yang, F.; Xiang, K.; Chen, J.; Zhang, Y.; Wang, J.; Sun, J.; Li, Y. A Multifunctional Microspheric Soil Conditioner Based on Chitosan-Grafted Poly(acrylamide-*co*-acrylic acid)/Biochar. *Langmuir* **2022**, *38*, 5717–5729. [CrossRef]
33. Li, L.; Lin, Q.; Li, X.; Li, T.; He, X.; Li, D.; Tao, Y. Dynamics and Potential Roles of Abundant and Rare Subcommunities in the Bioremediation of Cadmium-Contaminated Paddy Soil by Pseudomonas Chenduensis. *Appl. Microbiol. Biotechnol.* **2019**, *103*, 8203–8214. [CrossRef]
34. Zhang, Y.; He, H.; Khandelwal, M.; Du, K.; Zhou, J. Knowledge Mapping of Research Progress in Blast-Induced Ground Vibration from 1990 to 2022 Using CiteSpace-Based Scientometric Analysis. *Environ. Sci. Pollut. Res.* **2023**, *30*, 103534–103555. [CrossRef] [PubMed]
35. Zhang, Y.; Li, C.; Ji, X.; Yun, C.; Wang, M.; Luo, X. The Knowledge Domain and Emerging Trends in Phytoremediation: A Scientometric Analysis with CiteSpace. *Environ. Sci. Pollut. Res.* **2020**, *27*, 15515–15536. [CrossRef] [PubMed]
36. Misaelides, P. Application of Natural Zeolites in Environmental Remediation: A Short Review. *Microporous Mesoporous Mater.* **2011**, *144*, 15–18. [CrossRef]
37. Kuldeyev, E.; Seitzhanova, M.; Tanirbergenova, S.; Tazhu, K.; Doszhanov, E.; Mansurov, Z.; Azat, S.; Nurlybaev, R.; Berndtsson, R. Modifying Natural Zeolites to Improve Heavy Metal Adsorption. *Water* **2023**, *15*, 2215. [CrossRef]
38. Kordala, N.; Wyszkowski, M. Zeolite Properties, Methods of Synthesis, and Selected Applications. *Molecules* **2024**, *29*, 1069. [CrossRef]
39. Xu, Y.; Liang, X.; Xu, Y.; Qin, X.; Huang, Q.; Wang, L.; Sun, Y. Remediation of Heavy Metal-Polluted Agricultural Soils Using Clay Minerals: A Review. *Pedosphere* **2017**, *27*, 193–204. [CrossRef]
40. Majumder, S.; Jha, A.K. Removal of Cr and Mn from Aqueous Medium Using Bentonites and Their Derivatives. *J. Chem. Sci.* **2020**, *132*, 135. [CrossRef]
41. Sun, Y.; Li, Y.; Xu, Y.; Liang, X.; Wang, L. In Situ Stabilization Remediation of Cadmium (Cd) and Lead (Pb) Co-Contaminated Paddy Soil Using Bentonite. *Appl. Clay Sci.* **2015**, *105–106*, 200–206. [CrossRef]
42. Kumar Jha, A. Application of Bentonite Mineral in Removal of Heavy Metals. *Acta Scie Agric.* **2020**, *4*, 63–64. [CrossRef]
43. Liu, X.; Hicher, P.; Muresan, B.; Saiyouri, N.; Hicher, P.-Y. Heavy Metal Retention Properties of Kaolin and Bentonite in a Wide Range of Concentration and Different pH Conditions. *Appl. Clay Sci.* **2016**, *119*, 365–374. [CrossRef]
44. Li, W.; Ni, P.; Yi, Y. Comparison of Reactive Magnesia, Quick Lime, and Ordinary Portland Cement for Stabilization/Solidification of Heavy Metal-Contaminated Soils. *Sci. Total Environ.* **2019**, *671*, 741–753. [CrossRef] [PubMed]
45. Cabrera, C.; Gabaldón, C.; Marzal, P. Sorption Characteristics of Heavy Metal Ions by a Natural Zeolite: Sorption of Heavy Metal Ions by Zeolite. *J. Chem. Technol. Biotechnol.* **2005**, *80*, 477–481. [CrossRef]
46. Zhao, S.; Meng, Z.; Fan, X.; Jing, R.; Yang, J.; Shao, Y.; Liu, X.; Wu, M.; Zhang, Q.; Liu, A. Removal of Heavy Metals from Soil by Vermiculite Supported Layered Double Hydroxides with Three-Dimensional Hierarchical Structure. *Chem. Eng. J.* **2020**, *390*, 124554. [CrossRef]

47. Colombani, N.; Mastrocicco, M.; Di Giuseppe, D.; Faccini, B.; Coltorti, M. Batch and Column Experiments on Nutrient Leaching in Soils Amended with Italian Natural Zeolitites. *CATENA* **2015**, *127*, 64–71. [CrossRef]
48. Miraki, H.; Shariatmadari, N.; Ghadir, P.; Jahandari, S.; Tao, Z.; Siddique, R. Clayey Soil Stabilization Using Alkali-Activated Volcanic Ash and Slag. *J. Rock Mech. Geotech. Eng.* **2022**, *14*, 576–591. [CrossRef]
49. Bai, Y.; Guo, W.; Wang, X.; Pan, H.; Zhao, Q.; Wang, D. Utilization of Municipal Solid Waste Incineration Fly Ash with Red Mud-Carbide Slag for Eco-Friendly Geopolymer Preparation. *J. Clean. Prod.* **2022**, *340*, 130820. [CrossRef]
50. Wu, C.; Li, J.; Lu, Y.; Zhu, D. The Influence of Industrial Solid Waste in Conjunction with Lepidolite Tailings on the Mechanical Properties and Microstructure of Cemented Backfill Materials. *Constr. Build. Mater.* **2024**, *419*, 135422. [CrossRef]
51. Zhou, X.; Zhang, T.; Wan, S.; Hu, B.; Tong, J.; Sun, H.; Chen, Y.; Zhang, J.; Hou, H. Immobilizatiaon of Heavy Metals in Municipal Solid Waste Incineration Fly Ash with Red Mud-Coal Gangue. *J. Mater. Cycles Waste Manag.* **2020**, *22*, 1953–1964. [CrossRef]
52. Gu, Y.; Li, J.-L.; Peng, J.-K.; Xing, F.; Long, W.-J.; Khayat, K.H. Immobilization of Hazardous Ferronickel Slag Treated Using Ternary Limestone Calcined Clay Cement. *Constr. Build. Mater.* **2020**, *250*, 118837. [CrossRef]
53. Toksöz Hozatlıoğlu, D.; Yılmaz, I. Shallow Mixing and Column Performances of Lime, Fly Ash and Gypsum on the Stabilization of Swelling Soils. *Eng. Geol.* **2021**, *280*, 105931. [CrossRef]
54. Wang, G.; Zhang, X.; Liu, X.; Gao, H.; Song, S.; Lu, J. Enhancing Mechanical Properties of Micaceous Weathered Granitic Soil: A Synergetic Approach Using Natural Fibers and Fly Ash. *Bull. Eng. Geol. Environ.* **2024**, *83*, 257. [CrossRef]
55. Nayak, A.K.; Raja, R.; Rao, K.S.; Shukla, A.K.; Mohanty, S.; Shahid, M.; Tripathi, R.; Panda, B.B.; Bhattacharyya, P.; Kumar, A.; et al. Effect of Fly Ash Application on Soil Microbial Response and Heavy Metal Accumulation in Soil and Rice Plant. *Ecotoxicol. Environ. Saf.* **2015**, *114*, 257–262. [CrossRef] [PubMed]
56. Yi, Y.; Zheng, X.; Liu, S.; Al-Tabbaa, A. Comparison of Reactive Magnesia- and Carbide Slag-Activated Ground Granulated Blastfurnace Slag and Portland Cement for Stabilisation of a Natural Soil. *Appl. Clay Sci.* **2015**, *111*, 21–26. [CrossRef]
57. Li, M.; Wang, Q.; Yang, J.; Guo, X.; Zhou, W. Strength and Mechanism of Carbonated Solidified Clay with Steel Slag Curing Agent. *KSCE J. Civ. Eng.* **2021**, *25*, 805–821. [CrossRef]
58. Zhang, H.; Yang, B.; Zhang, G.; Zhang, X. Sewage Sludge as Barrier Material for Heavy Metals in Waste Landfill. *Arch. Environ. Prot.* **2016**, *42*, 52–58. [CrossRef]
59. Goodarzi, A.R.; Movahedrad, M. Stabilization/Solidification of Zinc-Contaminated Kaolin Clay Using Ground Granulated Blast-Furnace Slag and Different Types of Activators. *Appl. Geochem.* **2017**, *81*, 155–165. [CrossRef]
60. Nguyen Phan, B.; Sekine, R.; Hayano, K.; Yamauchi, H. Assessment of Consistency and Strength Properties of Clays Treated with Paper Sludge Ash-Based Stabilizers Using the Water Absorption and Retention Rate. *Constr. Build. Mater.* **2022**, *351*, 128936. [CrossRef]
61. Gómez, M.; Peisino, L.E.; Kreiker, J.; Gaggino, R.; Cappelletti, A.L.; Martín, S.E.; Uberman, P.M.; Positieri, M.; Raggiotti, B.B. Stabilization of Hazardous Compounds from WEEE Plastic: Development of a Novel Core-Shell Recycled Plastic Aggregate for Use in Building Materials. *Constr. Build. Mater.* **2020**, *230*, 116977. [CrossRef]
62. Chen, R.; Cai, G.; Dong, X.; Pu, S.; Dai, X.; Duan, W. Green Utilization of Modified Biomass By-Product Rice Husk Ash: A Novel Eco-Friendly Binder for Stabilizing Waste Clay as Road Material. *J. Clean. Prod.* **2022**, *376*, 134303. [CrossRef]
63. Sanito, R.C.; Bernuy-Zumaeta, M.; You, S.-J.; Wang, Y.-F. A Review on Vitrification Technologies of Hazardous Waste. *J. Environ. Manag.* **2022**, *316*, 115243. [CrossRef] [PubMed]
64. Chang, J.-S.; Cho, Y.-C.; Lin, Y.-P. Regeneration of Heavy Metal Contaminated Soils for Cement Production by Cement Kiln Co-Processing. *Resour. Conserv. Recycl.* **2022**, *176*, 105909. [CrossRef]
65. Pandey, V.C.; Singh, N. Impact of Fly Ash Incorporation in Soil Systems. *Agric. Ecosyst. Environ.* **2010**, *136*, 16–27. [CrossRef]
66. Singh, J.S.; Pandey, V.C. Fly Ash Application in Nutrient Poor Agriculture Soils: Impact on Methanotrophs Population Dynamics and Paddy Yields. *Ecotoxicol. Environ. Saf.* **2013**, *89*, 43–51. [CrossRef]
67. Kováčik, P.; Macák, M.; Ducsay, L.; Halčínová, M.; Jančich, M. Effect of Ash-Fly Ash Mixture Application on Soil Fertility. *J. Elem.* **2011**, *16*, 215–225. [CrossRef]
68. Tripathi, R.C.; Jha, S.K.; Ram, L.C. Impact of Heavy Metals in Indian Fly Ashes on Its Application as Soil Ameliorant. *Energy Sources Part A Recovery Util. Environ. Eff.* **2016**, *38*, 2568–2574. [CrossRef]
69. Chen, T.; Zhang, S.; Yuan, Z. Adoption of Solid Organic Waste Composting Products: A Critical Review. *J. Clean. Prod.* **2020**, *272*, 122712. [CrossRef]
70. Atiyeh, R.; Lee, S.; Edwards, C.; Arancon, N.; Metzger, J. The Influence of Humic Acids Derived from Earthworm-Processed Organic Wastes on Plant Growth. *Bioresour. Technol.* **2002**, *84*, 7–14. [CrossRef]
71. Fekri, M.; Kaveh, S. Heavy Metal Accumulation in Soil after Application of Organic Wastes. *Arab. J. Geosci.* **2013**, *6*, 463–467. [CrossRef]
72. Liu, K. Major Factors Influencing Cadmium Uptake from the Soil into Wheat Plants. *Ecotoxicol. Environ. Saf.* **2015**, *113*, 207–213. [CrossRef]
73. Zhao, S.; Qiu, S.; Xu, X.; Ciampitti, I.A.; Zhang, S.; He, P. Change in Straw Decomposition Rate and Soil Microbial Community Composition after Straw Addition in Different Long-Term Fertilization Soils. *Appl. Soil. Ecol.* **2019**, *138*, 123–133. [CrossRef]
74. Wu, L.; Ma, H.; Zhao, Q.; Zhang, S.; Wei, W.; Ding, X. Changes in Soil Bacterial Community and Enzyme Activity Under Five Years Straw Returning in Paddy Soil. *Eur. J. Soil. Biol.* **2020**, *100*, 103215. [CrossRef]

75. Odlare, M.; Arthurson, V.; Pell, M.; Svensson, K.; Nehrenheim, E.; Abubaker, J. Land Application of Organic Waste—Effects on the Soil Ecosystem. *Appl. Energy* **2011**, *88*, 2210–2218. [CrossRef]
76. Griffin, R.A.; Jurinak, J.J. Test of a New Model for the Kinetics of Adsorption-Desorption Processes. *Soil Sci. Soc. Am. J.* **1973**, *37*, 869–872. [CrossRef]
77. Zhang, Y.; Wang, L.; Guo, Z.; Xu, L.; Zhao, H.; Zhao, P.; Ma, C.; Yi, K.; Jia, X. Revealing the Underlying Molecular Basis of Phosphorus Recycling in the Green Manure Crop Astragalus Sinicus. *J. Clean. Prod.* **2022**, *341*, 130924. [CrossRef]
78. Su, Y.; Lv, J.L.; Yu, M.; Ma, Z.H.; Xi, H.; Kou, C.L.; He, Z.C.; Shen, A.L. Long-term Decomposed Straw Return Positively Affects the Soil Microbial Community. *J. Appl. Microbiol.* **2020**, *128*, 138–150. [CrossRef]
79. Cai, G.; Zhang, T.; Liu, S.; Li, J.; Jie, D. Stabilization Mechanism and Effect Evaluation of Stabilized Silt with Lignin Based on Laboratory Data. *Mar. Georesour. Geotechnol.* **2016**, *34*, 331–340. [CrossRef]
80. Zhang, T.; Cai, G.; Liu, S. Application of Lignin-Based by-Product Stabilized Silty Soil in Highway Subgrade: A Field Investigation. *J. Clean. Prod.* **2017**, *142*, 4243–4257. [CrossRef]
81. Zhang, T.; Liu, S.; Zhan, H.; Ma, C.; Cai, G. Durability of Silty Soil Stabilized with Recycled Lignin for Sustainable Engineering Materials. *J. Clean. Prod.* **2020**, *248*, 119293. [CrossRef]
82. He, L.; Dai, Z.; Liu, X.; Tang, C.; Xu, J. Effect of Alkaline Lignin on Immobilization of Cadmium and Lead in Soils and the Associated Mechanisms. *Chemosphere* **2021**, *281*, 130969. [CrossRef]
83. Haddad, S.A.; Lemanowicz, J.; Abd El-Azeim, M.M. Cellulose Decomposition in Clay and Sandy Soils Contaminated with Heavy Metals. *Int. J. Environ. Sci. Technol.* **2019**, *16*, 3275–3290. [CrossRef]
84. Jamshaid, A.; Hamid, A.; Muhammad, N.; Naseer, A.; Ghauri, M.; Iqbal, J.; Rafiq, S.; Shah, N.S. Cellulose-Based Materials for the Removal of Heavy Metals from Wastewater—An Overview. *ChemBioEng Rev.* **2017**, *4*, 240–256. [CrossRef]
85. Guleria, A.; Kumari, G.; Lima, E.C. Cellulose-*g*-poly-(acrylamide-*co*-acrylic acid) Polymeric Bioadsorbent for the Removal of Toxic Inorganic Pollutants from Wastewaters. *Carbohydr. Polym.* **2020**, *228*, 115396. [CrossRef] [PubMed]
86. Zhou, D.; Wang, H.; Guo, S. Preparation of Cellulose/Chitin Blend Materials and Influence of Their Properties on Sorption of Heavy Metals. *Sustainability* **2021**, *13*, 6460. [CrossRef]
87. Cretoiu, M.S.; Korthals, G.W.; Visser, J.H.M.; van Elsas, J.D. Chitin Amendment Increases Soil Suppressiveness toward Plant Pathogens and Modulates the Actinobacterial and Oxalobacteraceal Communities in an Experimental Agricultural Field. *Appl. Environ. Microbiol.* **2013**, *79*, 5291–5301. [CrossRef]
88. Hui, C.; Jiang, H.; Liu, B.; Wei, R.; Zhang, Y.; Zhang, Q.; Liang, Y.; Zhao, Y. Chitin Degradation and the Temporary Response of Bacterial Chitinolytic Communities to Chitin Amendment in Soil under Different Fertilization Regimes. *Sci. Total Environ.* **2020**, *705*, 136003. [CrossRef]
89. Duan, Y.; Freyburger, A.; Kunz, W.; Zollfrank, C. Lignin/Chitin Films and Their Adsorption Characteristics for Heavy Metal Ions. *ACS Sustain. Chem. Eng.* **2018**, *6*, 6965–6973. [CrossRef]
90. Beesley, L.; Moreno-Jiménez, E.; Gomez-Eyles, J.L.; Harris, E.; Robinson, B.; Sizmur, T. A Review of Biochars' Potential Role in the Remediation, Revegetation and Restoration of Contaminated Soils. *Environ. Pollut.* **2011**, *159*, 3269–3282. [CrossRef]
91. Zhang, X.; Wang, H.; He, L.; Lu, K.; Sarmah, A.; Li, J.; Bolan, N.S.; Pei, J.; Huang, H. Using Biochar for Remediation of Soils Contaminated with Heavy Metals and Organic Pollutants. *Environ. Sci. Pollut. Res.* **2013**, *20*, 8472–8483. [CrossRef]
92. Liang, M.; Lu, L.; He, H.; Li, J.; Zhu, Z.; Zhu, Y. Applications of Biochar and Modified Biochar in Heavy Metal Contaminated Soil: A Descriptive Review. *Sustainability* **2021**, *13*, 14041. [CrossRef]
93. Taraqqi-A-Kamal, A.; Atkinson, C.J.; Khan, A.; Zhang, K.; Sun, P.; Akther, S.; Zhang, Y. Biochar Remediation of Soil: Linking Biochar Production with Function in Heavy Metal Contaminated Soils. *Plant Soil Environ.* **2021**, *67*, 183–201. [CrossRef]
94. Mandal, S.; Pu, S.; Adhikari, S.; Ma, H.; Kim, D.-H.; Bai, Y.; Hou, D. Progress and Future Prospects in Biochar Composites: Application and Reflection in the Soil Environment. *Crit. Rev. Environ. Sci. Technol.* **2021**, *51*, 219–271. [CrossRef]
95. Liu, Z.; Xu, J.; Li, X.; Wang, J. Mechanisms of Biochar Effects on Thermal Properties of Red Soil in South China. *Geoderma* **2018**, *323*, 41–51. [CrossRef]
96. Lu, H.; Zhang, W.; Yang, Y.; Huang, X.; Wang, S.; Qiu, R. Relative Distribution of Pb^{2+} Sorption Mechanisms by Sludge-Derived Biochar. *Water Res.* **2012**, *46*, 854–862. [CrossRef] [PubMed]
97. Wang, Y.; Zhong, B.; Shafi, M.; Ma, J.; Guo, J.; Wu, J.; Ye, Z.; Liu, D.; Jin, H. Effects of Biochar on Growth, and Heavy Metals Accumulation of Moso Bamboo (*Phyllostachy pubescens*), Soil Physical Properties, and Heavy Metals Solubility in Soil. *Chemosphere* **2019**, *219*, 510–516. [CrossRef]
98. Githinji, L. Effect of Biochar Application Rate on Soil Physical and Hydraulic Properties of a Sandy Loam. *Arch. Agron. Soil Sci.* **2014**, *60*, 457–470. [CrossRef]
99. Berihun, T.; Tadele, M.; Kebede, F. The Application of Biochar on Soil Acidity and Other Physico-chemical Properties of Soils in Southern Ethiopia. *J. Plant Nutr. Soil Sci.* **2017**, *180*, 381–388. [CrossRef]
100. Kim, H.-S.; Kim, K.-R.; Kim, H.-J.; Yoon, J.-H.; Yang, J.E.; Ok, Y.S.; Owens, G.; Kim, K.-H. Effect of Biochar on Heavy Metal Immobilization and Uptake by Lettuce (*Lactuca sativa* L.) in Agricultural Soil. *Environ. Earth Sci.* **2015**, *74*, 1249–1259. [CrossRef]
101. Park, J.H.; Lee, S.-J.; Lee, M.-E.; Chung, J.W. Comparison of Heavy Metal Immobilization in Contaminated Soils Amended with Peat Moss and Peat Moss-Derived Biochar. *Environ. Sci. Process. Impacts* **2016**, *18*, 514–520. [CrossRef]

102. Egene, C.E.; Van Poucke, R.; Ok, Y.S.; Meers, E.; Tack, F.M.G. Impact of Organic Amendments (Biochar, Compost and Peat) on Cd and Zn Mobility and Solubility in Contaminated Soil of the Campine Region after Three Years. *Sci. Total Environ.* **2018**, *626*, 195–202. [CrossRef]
103. Huang, D.; Liu, L.; Zeng, G.; Xu, P.; Huang, C.; Deng, L.; Wang, R.; Wan, J. The Effects of Rice Straw Biochar on Indigenous Microbial Community and Enzymes Activity in Heavy Metal-Contaminated Sediment. *Chemosphere* **2017**, *174*, 545–553. [CrossRef] [PubMed]
104. Igalavithana, A.D.; Lee, S.-E.; Lee, Y.H.; Tsang, D.C.W.; Rinklebe, J.; Kwon, E.E.; Ok, Y.S. Heavy Metal Immobilization and Microbial Community Abundance by Vegetable Waste and Pine Cone Biochar of Agricultural Soils. *Chemosphere* **2017**, *174*, 593–603. [CrossRef] [PubMed]
105. Abulaiti, A.; She, D.; Liu, Z.; Sun, X.; Wang, H. Application of Biochar and Polyacrylamide to Revitalize Coastal Saline Soil Quality to Improve Rice Growth. *Environ. Sci. Pollut. Res.* **2022**, *30*, 18731–18747. [CrossRef] [PubMed]
106. Wiśniewska, M.; Fijałkowska, G.; Szewczuk-Karpisz, K.; Herda, K.; Chibowski, S. Ionic Polyacrylamides as Stability-Modifying Substances of Soil Mineral Suspensions Containing Heavy Metal Impurities. *Processes* **2022**, *10*, 1473. [CrossRef]
107. Rabiee, A.; Gilani, M.; Jamshidi, H.; Baharvand, H. Synthesis and Characterization of a Calcium- and Sodium-Containing Acrylamide-Based Polymer and Its Effect on Soil Strength. *J. Vinyl Addit. Technol.* **2013**, *19*, 140–146. [CrossRef]
108. Asghari, S.; Abbasi, F.; Neyshabouri, M.R. Effects of Soil Conditioners on Physical Quality and Bromide Transport Properties in a Sandy Loam Soil. *Biosyst. Eng.* **2011**, *109*, 90–97. [CrossRef]
109. Dhiman, J.; Prasher, S.O.; ElSayed, E.; Patel, R.M.; Nzediegwu, C.; Mawof, A. Effect of Hydrogel Based Soil Amendments on Heavy Metal Uptake by Spinach Grown with Wastewater Irrigation. *J. Clean. Prod.* **2021**, *311*, 127644. [CrossRef]
110. Dhiman, J.; Prasher, S.O.; ElSayed, E.; Patel, R.; Nzediegwu, C.; Mawof, A. Use of Polyacrylamide Superabsorbent Polymers and Plantain Peel Biochar to Reduce Heavy Metal Mobility and Uptake by Wastewater-Irrigated Potato Plants. *Trans. ASABE* **2020**, *63*, 11–28. [CrossRef]
111. Li, D.; Zhou, J.; Zhang, Y.; Sun, T.; An, S.; Jia, H. Effects of Amendments on Physicochemical Properties and Respiration Rate of Soil from the Arid Region of Northwest China. *Sustainability* **2021**, *13*, 5332. [CrossRef]
112. Zhao, T.; Zhang, K.; Chen, J.; Shi, X.; Li, X.; Ma, Y.; Fang, G.; Xu, S. Changes in Heavy Metal Mobility and Availability in Contaminated Wet-Land Soil Remediated Using Lignin-Based Poly(acrylic acid). *J. Hazard. Mater.* **2019**, *368*, 459–467. [CrossRef]
113. Zhou, S.; Xue, A.; Zhao, Y.; Wang, Q.; Chen, Y.; Li, M.; Xing, W. Competitive Adsorption of Hg^{2+}, Pb^{2+} and Co^{2+} Ions on Polyacrylamide/Attapulgite. *Desalination* **2011**, *270*, 269–274. [CrossRef]
114. Basuki, K.T.; Swantomo, D.; Sigit; Sanyoto, N.T. Characterization of Chitosan-Acrylamide Hydrogels as Soil Conditioner. *AMR* **2015**, *1112*, 414–417. [CrossRef]
115. Niu, Y.; Li, H. Controlled Release of Urea Encapsulated by Starch-*g*-poly(vinyl acetate). *Ind. Eng. Chem. Res.* **2012**, *51*, 12173–12177. [CrossRef]
116. Mahdavi, A.E.B.; Panahpour, E.; Kalbasi, R.J.; Gholami, A. Preparation, Characterization, and Application of Polyacrylamide-Polystyrene/Bentonite Nanocomposite as an Effective Immobilizing Adsorbent for Remediation of Soil. *ChemistrySelect* **2020**, *5*, 4538–4547. [CrossRef]
117. Song, P.; Xu, D.; Yue, J.; Ma, Y.; Dong, S.; Feng, J. Recent Advances in Soil Remediation Technology for Heavy Metal Contaminated Sites: A Critical Review. *Sci. Total Environ.* **2022**, *838*, 156417. [CrossRef]
118. Verma, S.; Kuila, A. Bioremediation of Heavy Metals by Microbial Process. *Environ. Technol. Innov.* **2019**, *14*, 100369. [CrossRef]
119. Liu, W.; Luo, Y.; Teng, Y.; Li, Z.; Ma, L.Q. Bioremediation of Oily Sludge-Contaminated Soil by Stimulating Indigenous Microbes. *Environ. Geochem. Health* **2010**, *32*, 23–29. [CrossRef]
120. Ren, W.-X.; Li, P.-J.; Geng, Y.; Li, X.-J. Biological Leaching of Heavy Metals from a Contaminated Soil by *Aspergillus niger*. *J. Hazard. Mater.* **2009**, *167*, 164–169. [CrossRef]
121. Basyal, B.; Emery, S.M. An Arbuscular Mycorrhizal Fungus Alters Switchgrass Growth, Root Architecture, and Cell Wall Chemistry across a Soil Moisture Gradient. *Mycorrhiza* **2021**, *31*, 251–258. [CrossRef]

toxics

MDPI

Review

Biochar-Derived Persistent Free Radicals: A Plethora of Environmental Applications in a Light and Shadows Scenario

Silvana Alfei [1,*] and Omar Ginoble Pandoli [1,2]

[1] Department of Pharmacy (DIFAR), University of Genoa, Viale Cembrano 4, 16148 Genoa, Italy; omar.ginoblepandoli@unige.it or omarpandoli@puc-rio.br

[2] Department of Chemistry, Pontifical Catholic University, Rua Marquês de São Vincente 225, Rio de Janeiro 22451-900, Brazil

* Correspondence: alfei@difar.unige.it

Abstract: Biochar (BC) is a carbonaceous material obtained by pyrolysis at 200–1000 °C in the limited presence of O_2 from different vegetable and animal biomass feedstocks. BC has demonstrated great potential, mainly in environmental applications, due to its high sorption ability and persistent free radicals (PFRs) content. These characteristics enable BC to carry out the direct and PFRs-mediated removal/degradation of environmental organic and inorganic contaminants. The types of PFRs that are possibly present in BC depend mainly on the pyrolysis temperature and the kind of pristine biomass. Since they can also cause ecological and human damage, a systematic evaluation of the environmental behavior, risks, or management techniques of BC-derived PFRs is urgent. PFRs generally consist of a mixture of carbon- and oxygen-centered radicals and of oxygenated carbon-centered radicals, depending on the pyrolytic conditions. Here, to promote the more productive and beneficial use of BC and the related PFRs and to stimulate further studies to make them environmentally safer and less hazardous to humans, we have first reviewed the most common methods used to produce BC, its main environmental applications, and the primary mechanisms by which BC remove xenobiotics, as well as the reported mechanisms for PFR formation in BC. Secondly, we have discussed the environmental migration and transformation of PFRs; we have reported the main PFR-mediated application of BC to degrade inorganic and organic pollutants, the potential correlated environmental risks, and the possible strategies to limit them.

Keywords: biochar (BC); pyrolysis; biochar-derived permanent free radicals (PFRs); reactive oxygen species (ROS); PFR-mediated BC applications; environmental risk

1. Introduction

Biochar (BC) is a stable carbon-rich black solid substance produced from vegetable or animal biomass feedstocks when pyrolyzed. Pyrolysis is a procedure that involves the heating of substrates at 200–1000 °C under oxygen-limited conditions [1]. The term "biochar" derives from the combination of "bio-," which stands for "biomass," and "char," meaning "charcoal." In recent years, BC has received widespread attention due to its potential application in carbon sequestration, soil amendment/remediation, wastewater treatment, and catalysis [2–11]. In particular, several ground-breaking studies have been carried out to investigate the potential of BC in alternative energy production and in the recovery of value-added chemicals/by-products [3]. In this regard, Lee et al. have used BC as briquettes and electrodes for microbial fuel cells (MFCs) finalized for alternative energy production [5]. Zhang et al. have employed BC as an additive/catalyst in anaerobic digestion and transesterification reactions for biogas [6], while Behera et al. employed BC to produce biodiesel [4]. Environmental applications of BC for reducing gaseous emissions, mainly by carbon and nitrogen sequestration, are also gaining attention [8].

BC has been applied to prevent eutrophication by recovering excessive nutrients, including nitrogen and phosphorus, from wastewater [11], while the agronomic application of BC, due to its characteristics of high cation exchange capacity (CEC) and specific surface area (SSA), have recently been reported by Zhao et al. [10]. Batch and column sorption experiments have shown that certain types of BC have good adsorption performance for heavy metals, dyes, or phosphate from aqueous solutions and are being investigated as cost-effective, promising, and eco-friendly alternative adsorbent materials [12]. Additionally, BC has demonstrated high efficiency in removing pharmaceuticals [13], pesticides [14], polycyclic aromatic hydrocarbons (PAHs) [15], and petroleum derivatives [16]. Also, the metal ion absorbent capacity of BC has been extensively reported in both the absence and presence of fulvic acid and humic acid [17]. As a soil improver, BC can reduce soil acidity and help maintain soil moisture and nutrient levels. Through its carbon sequestration action, BC performs climate restoration. Moreover, due to its strong adsorption capacity, BC can remove environmental xenobiotics, thus preventing their uptake in plants, animals, and humans [18–21]. Additionally, BC derived from the thermal treatment of organic material generally contains persistent free radicals (PFRs) bound to the external or internal surfaces of its solid particles [22,23]. Such BC-bound PFRs, which are reactive due to unpaired electrons, can persist for minutes and up to several months, in contrast to traditional transient radicals [24], thus conferring on BC the capacity to degrade organic pollutants through the generation of other reactive oxygen species (ROS) and sulfate radicals [25–27]. In this context, BC-bound PFRs have been investigated to activate persulfate ($S_2O_8^{2-}$) to obtain sulfate radicals, which have efficiently degraded phenolic compounds and polychlorinated biphenyls [28], acid orange 7 [29], and sulfamethoxazole [30,31]. In the presence of PFRs, hydrogen peroxide (H_2O_2) or oxygen (O_2) have been activated to produce hydroxyl radicals (OH•) and superoxide radicals ($O_2•^-$), which succeeded in efficiently degrading chloro-biphenyl [32], diethyl phthalate [33,34], and ciprofloxacin [35]. By the PFR-mediated activation of peroxyl mono sulfate (PMS), radical species such as $SO_4•^-$, •OH, and $O_2•^-$, as well as non-radical species such as $^1O_2$ formed, which were the main contributors in the antibiotics' degradation [36]. On the other hand, by stimulating the production of ROS, PFRs can inhibit seed germination and retard the growth of roots and shoots [37]. Additionally, BC production itself may cause the release of xenobiotics such as polycyclic aromatic hydrocarbons (PAHs), toxic inorganic elements, and dioxins, thus posing potential risks to human health and the environment [1]. The scientific community should evaluate BC and BC-bound PFRs' positive and negative impacts before their extensive ecological application. Although the environmental behavior and risks of BC and BC-associated PRFs are increasingly attracting research attention, in the last ten years (the year 2024 excluded), studies into their toxicity remain very limited (96 publications), compared to those concerning PFRs in general (542 publications) and their degradation capability (402 publications) (Figure 1).

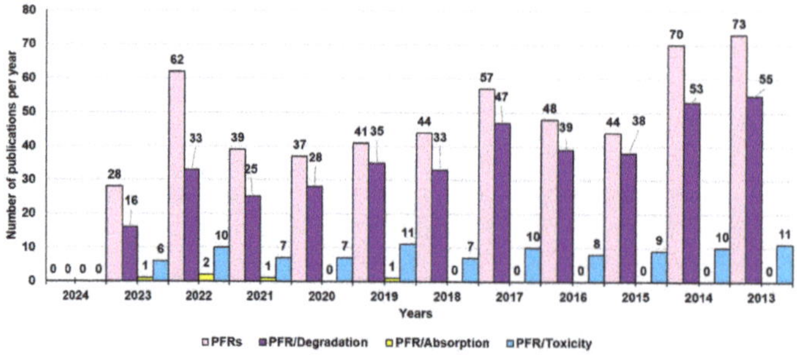

Figure 1. Number of publications on PFRs and their degradation, absorption, and possible toxic actions published over the last ten years (current year excluded) according to the PubMed dataset. The survey was carried out using the following keywords: permanent free radicals (pink bars); permanent free radicals AND degradation (purple bars); permanent free radicals AND absorption (yellow bars); permanent free radicals AND toxicity (light blue bars).

However, to safeguard the environment from BC-related PFRs' potential adverse effects, it is necessary to comprehensively and systematically consider their environmental risks, formation mechanisms, and controlling factors [27], as well as the corresponding possible mitigation actions. Studies have shown that the type of biomass feedstock used to produce BC is pivotal in determining the physicochemical properties of the resulting BC, which also strongly affect the formation and characteristics of PFRs. To date, the types of biomasses used to prepare BC and involved in the investigation of BC-bound PFRs mainly include lignocellulosic biomasses (hemicellulose, cellulose, and lignin) from sources such as pine needles, wheat straw, lignin, cow manure, rice husk, and maize straw [38–40]. Additionally, cow dung (CD), sheep manure (SM), lotus stem (LS), and eggshell (ES) biomasses, representative of farm wastes, have been reported to provide BC containing PFRs [41]. Bamboo is an emerging starting material that is perfect for synthesizing BC and activated carbon (AC) due to its inexpensive cost, high biomass yield, and accelerated growth rate [42,43]. However, only a few researchers and scientists have used bamboo as a unique source for developing BC so far, as established by the number of publications on bamboo-derived BC developed in the last ten years (current year excluded) (331) vs. those on BC derived from different sources (13,630) (Figure 2).

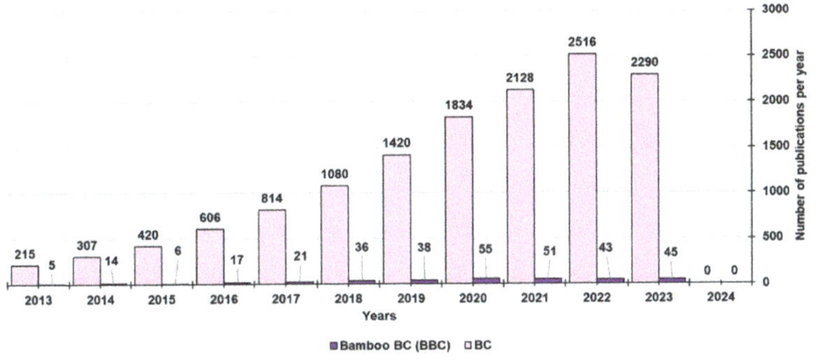

Figure 2. Number of publications on BCs and on bamboo-derived BCs published in the last ten years (current year excluded) according to the PubMed dataset. The survey used the following keywords: biochar (pink bars) and bamboo biochar (purple bars).

Among the 331 publications on bamboo-derived BC, most were about their adsorption activity (119), followed by those on their degradation capacity (87). At the same time, few studies were conducted on their possible toxic action (Figure 3).

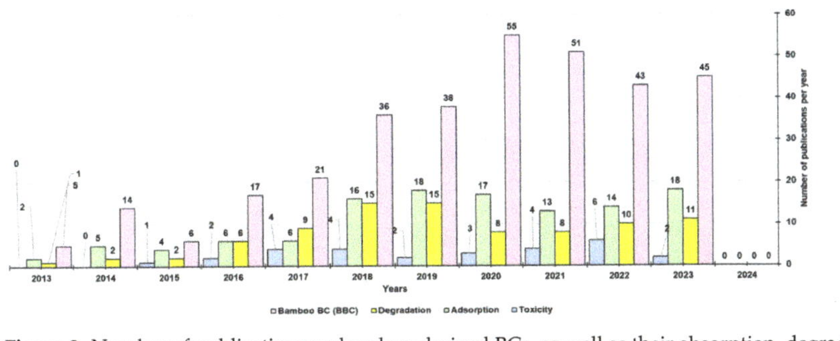

Figure 3. Number of publications on bamboo-derived BCs, as well as their absorption, degradation, and toxic properties, published in the last ten years (current year excluded) according to the PubMed dataset. The survey was carried out using the following keywords: bamboo biochar (pink bars); bamboo biochar AND adsorption (green bars); bamboo biochar AND degradation (yellow bars); bamboo biochar AND toxicity (light blue bars).

In this scenario, to promote the more productive and beneficial use of BC and the related PFRs and stimulate further studies to make them environmentally safer and less hazardous to humans, we have first reviewed the most common methods used to produce BC and its main environmental applications, as well as the reported mechanisms for PFR formation in BC. The main factors influencing the physicochemical properties of BC have also been reported. Secondly, we have discussed the environmental migration and transformation of PFRs, the main PFR-mediated applications of BC to remediate inorganic and organic pollutants described in the last five years, the correlated potential risks to the environment and humans, and the possible strategies to limit them. To confirm the relevance and essentiality of the present paper, a recent survey of the PubMed dataset has evidenced that, although the number of studies on BC-related PFRs has increased in the last few years, they are still very limited compared to those on BC in general (134 vs. 11646 from 2014 to the present). Additionally, the review articles on BC-associated PFRs that, by gathering information on the topic, could stimulate more research on it are indeed limited (16). Some recent examples could be considered works by Zhang et al., Liu et al., Luo et al., and Yi et al. [1,44–46]. In this scenario, this review can be considered essential because it offers an all-round and complete overview of both BC and BC-related PFRs, via an extensive discussion on both their beneficial impact and the possible risks to humans and the environment that could derive from their widespread and indiscriminate application. In the case of the present paper, we have provided a reader-friendly work where the information has mostly been organized into easy-to-read tables, schemes, and statistical graphs that could have a profound impact on its readers' understanding.

2. Biochar (BC)

The constant growth in world population translates into a continued increase in the global energy requirement by all sectors and a dramatic decrease in fossil fuels, the primary energy source [47–49]. Furthermore, the effect of the resulting CO_2 emissions on the environment determines additional global energy issues, which make the replacement of fossil fuels necessary and urgent [50]. In this regard, biochar (BC), mainly obtained from organic waste and possessing the capability of sequestering carbon, represents a rich carbon source and an alternative to fossil fuels [51,52]. Table S1 in the Supplementary Materials reports the biomasses commonly used for BC production [53–66].

BC obtained by the combustion of various biomasses, as reported in Table S1, has been demonstrated to possess unparalleled physicochemical properties such as a large surface area, high porosity, the presence of several functional groups, high cation exchange capacity (CEC), long-term stability, etc. (Figure 4). Such properties make BC suitable for various applications, including, but not limited to, carbon sequestration, soil amendment, energy storage, and catalysis [67–75] (Table S1). Additionally, BC is cost-effective, has an eco-friendly nature, and is endowed with reusability (Figure 4) [76,77]. Mainly, BC is increasingly gaining attention from many researchers as a material to efficiently remove various environmental contaminants, including antibiotics, thus reducing the emergence of microbial resistance [72,74].

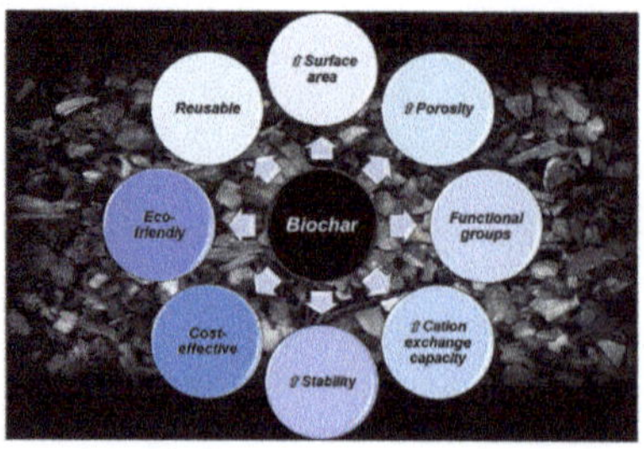

Figure 4. Some of the advantages and physicochemical properties of BCs. ⇑ = high.

Among the biomass waste materials appropriate for BC production, crop residues from agriculture, forestry, municipal solid waste, food, and animal manure have high potential [78–83].

2.1. Main Methods of Producing Biochar

As reported in the following Table 1, BC can be prepared rapidly using thermochemical conversion techniques such as pyrolysis, hydrothermal carbonization, gasification, flash carbonization, and torrefaction [84,85], with pyrolysis the most widely adopted (Section 2.1.1).

Table 1. Main BC production methods, temperature conditions, and yields.

	Temperature (°C)	Residence Time	Biochar (%)	Bio-Oil (%)	Syngas (%)	Refs
Pyrolysis	200–700	0.5–2 s	35	30	35	[86]
	500–1000	Hours/day	12	75	13	
HC	180–300	1–16 h	50–80	5–20	2–5	[87]
Gasification	750–900	10–20 s	10	5	85	[88]
Torrefaction	290	10–60 min	80	0	20	[89]
Flash carbonization	300–600	<30 min	37	--	--	[90]

HC = Hydrothermal carbonization.

2.1.1. Pyrolysis

Pyrolysis is a thermochemical process wherein the organic compounds present in the biomass are decomposed at a specific temperature [91]. Mainly, during pyrolysis, the thermal decomposition of organic materials occurs in an oxygen-free or oxygen-limited environment within a temperature range of 250–1000 °C [92]. In these conditions, the lignocellulosic components of biomass, such as cellulose, hemicellulose, and lignin, go through chemical reactions like depolymerization, fragmentation, and cross-linking, depending on the adopted temperatures. There are three principal possible products, including solid, liquid, and gas physical state materials. The solid products comprise BC and ash, while the liquid ones encompass bio-oils and tar, and the gaseous products (syngas) comprise carbon dioxide, carbon monoxide, hydrogen, and C1-C2 hydrocarbons [86]. As shown in Figure 5, during pyrolysis, the process parameters, including temperature, the type and nature of the biomass, residence time, heating rate, pressure, etc., could strongly affect BC yield and its physicochemical characteristics [93,94]. Moreover, although BC samples derived from different biomasses are all entirely made of carbon content and ash, their elemental composition, as well as their physical characteristics and properties, could differ enormously based on the type of biomass, reaction conditions, and type of reactor used during the carbonization process [95] (Figure 5). Consequently, every experimental condition and the starting raw material should be considered as a proof-of-concept of the future industrial application of BC.

The most widely used reactors for the chemical transformation of different biomasses include paddle kilns, bubbling fluidized beds, wagon reactors, tubular ovens, and agitated sand rotary kilns. However, temperature remains the primary operating process condition that governs the yield in BC, vs. those of the oily and gaseous products. Usually, BC yield decreases and syngas production increases when the pyrolysis temperature is improved [96]. Based on the heating rate, temperature, residence time, and pressure, pyrolysis can be categorized as fast or slow, as summarized in Table S2 in the Supplementary Materials [97]. Generally, fast pyrolysis is employed to maximize the liquid product yield, while slow pyrolysis is employed to maximize the solid product yield [98].

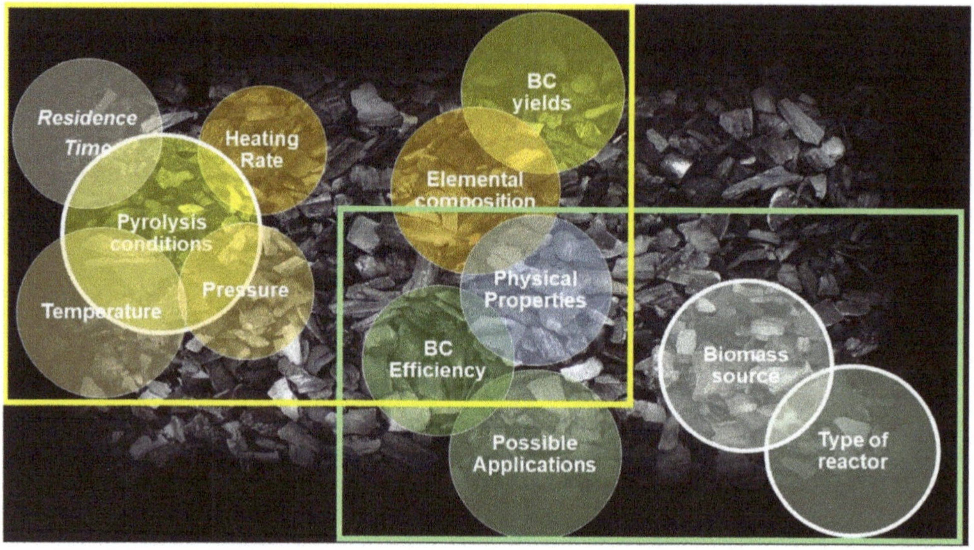

Figure 5. Parameters that mainly influence the pyrolysis process outcomes and the prepared BC elemental composition, physicochemical properties, and possible applications.

2.2. Biochar Characterization and Main Properties

The characterization of BC to determine its elemental composition is carried out by performing elemental analyses. Otherwise, its physicochemical, surface, and structural characterization is carried out by determining its surface functional groups, stability, and structure by employing various modern techniques reported in Table S3 in the Supplementary Materials [57].

As mentioned, the source of feedstock and the heat treatment temperatures during preparation are two significant factors that determine the physiochemical properties of BC.

The properties of pristine biomass that mainly influence the related BC include moisture content, ash content, calorific value, the percentage of lignin, cellulose, hemicellulose, fractions of fixed carbon, and volatile components [98]. High-yield BC with high porosity is achievable using biomasses possessing more lignin and less cellulose. Additionally, the volatile component, water content, particle size, and shape of the original biomass can also affect the properties of BC [98]. Table 2 reports the general chemical and physical features of BC, while Table S4 in the Supplementary Materials reports some characteristics of BCs produced from specific feedstocks at various production temperatures [99].

Table 2. Main chemical and physical features of BCs.

	Properties	Discussion
Chemical properties	Atomic ratio	⇓ O/C and H/C ratio for untreated biomass
	Elemental composition	⇑ Carbon content (> 95%) * ⇓ Hydrogen content (< 5%) * ⇓ Oxygen content (< 2%) *
	Energy content	⇑ Energy content with temperature (From 15–20 MJ/kg [a] to 30–35 MJ/kg [b] at 700 °C
	Fixed carbon (FC) ** Volatile matter (VM)	⇑ in FC (from 10% [a] to 90% [b] at 700 °C) ⇓ in VM (from 90% [a] to 10% [b] at 700 °C)

Table 2. *Cont.*

	Properties	Discussion
	Structural composition	Partially decomposed cellulose [c] Near totally decomposed hemicellulose [c] Partially decomposed lignin [c]
		Release of O_2 and H_2 ⇓ Oxygenated functional groups in BC (OH and C=O groups) * ⇑ Highly stable aromatic structures in BC * (with maximum aromaticity at 500–800 °C) ⇑ Alkalinity and ability to neutralize acids in soils * ⇑ Unpaired negative charges that enable BC to accept protons
	pH value	⇑ pH value (from 5–7.5 [a] to 10–12 [b] at > 500 °C) ⇑ Ash
	Cation exchange capacity (CEC)	⇑ CEC for BCs produced at relatively ⇓ low temperatures
	Ash content (SiO_2, CaO, K_2O, P_2O_5, Al_2O_3, MgO)	⇑ With temperature
	Self-heating degradation during storage	⇓ Highly volatile content in BC ⇓ Risk of self-heating ⇑ Thermal stability ⇓ Risk of spontaneous combustion ⇓ Water content and microbial
Physical properties	Density and porosity	⇑ Weight-based energy density * at ⇑ temperature ⇓ Bulk density * (the volume-specific weight of a bulk material in a heap or pile) ⇑ Porosities at ⇑ temperature
	Surface area	⇑ Total surface area * (<800 °C) ⇓ Total surface area * (>800–1000 °C)
	Pore volume distribution Pore size distribution	⇑ Total pore volume * with ⇑ temperature Macropores (1000–0.05 μm) Mesopores (0.05–0.002 μm) Micropores (0.05–0.0001 μm (more than 80% of the total pore volume)
	Hydrophobicity Water-holding capacity (WHC)	⇑ Hydrophobicity ⇓ Affinity to water ⇑ Porosity and amount of water that can be absorbed
	Mechanical stability	⇓ Mechanical stability during carbonatization ⇓ Structural complexity during carbonization
	Grindability	⇑ Grindability compared to the parent material
	Thermal conductivity Heat capacity	⇓ Thermal conductivity in BC (from 1300 J/(kgK) [a] to 1000 J/(kgK) [b] at 500 °C)
	Electromagnetic properties	⇑ Conductivity ⇑ Electromagnetic shielding efficiency

* Depending on pyrolysis temperature: a higher degree of carbonatation at higher temperatures; ** after removing the volatile components, the carbon content that remains in the solid structure is called fixed carbon; [a] row biomass; [b] BC; [c] depending on biomass and pyrolysis temperatures involved (<650 °C decomposes almost all of the holocellulose (cellulose and hemicellulose); the temperatures required for decomposing lignin are several hundred degrees higher than those for holocellulose; CEC is defined as the number of exchangeable cations (e.g., Ca^{2+}, Mg^{2+}, K^+, Na^+, NH_4^+) that a material can capture, which directly depends on the surface structure and the presence of functional groups providing surface negative charges; ⇑ = high, higher, or an increase; ⇓ = low, lower or a decrease.

The Question of Temperature

As already reported, the pyrolysis temperature and feedstock greatly influence the physicochemical properties of BC, including pH, specific surface area, pore size, CEC, volatile matter, ash, and carbon content. CEC and volatile matter decrease with increasing pyrolysis temperature, whereas pH, specific surface area, ash, carbon content, and pore volume increase with an increase in pyrolysis temperature [100]. Increasing temperature also causes a decrease in the number of acidic functional groups, especially carboxylic functional groups, and causes the appearance of carbonylic functional groups and alkalinity [101]. In particular, unpaired negative charges forming during pyrolysis at higher temperatures enable BC to accept protons [101]. Although BC's alkalinity increases with higher pyrolysis temperatures, thus improving its capacity to neutralize acids in soils, lower temperatures are necessary to preserve functional groups and obtain BC with higher CEC [102]. Low water content in BC, which reduces the possible microbial activity, promoting self-healing and degradation, is achievable at a higher temperature. However, the highly porous structure of BC obtained in such conditions causes the ready adsorption of moisture from the surroundings, thus increasing water content, re-enabling microbial activity, and contributing to self-heating and degradation [100].

During biomass decomposition to BC, the total surface area changes like the porosity due to the escaping of volatile gases and increases with increasing temperature [103]. In this regard, a large surface area affects CEC and water-holding capacity (WHC). Curiously, during pyrolysis, the hydro-properties of the initial biomass undergo several modifications depending on the pyrolysis temperature, which can translate into contradictory findings. Notably, with increasing temperature, due to a decrease in functional oxygenated groups and an increase in aromatic structure, the material's affinity to water is altered, the hydrophobicity of BC becomes higher than that of pristine biomass, and its capacity to retain water will be lower. Conversely, thanks to increased porosity, which changes the amount of water that can be adsorbed, BC produced at high temperatures can hold more water in its porous structure than BC prepared at lower ones [104].

The mechanical stability of biomass usually decreases during pyrolysis and correlates inversely with the porosity and directly to the density of the BC and temperature. The electric conductivity increases with higher thermal treatment, improving the graphitic carbons' crystallinity and the carbon-packed domains' density [105]. BC with high mechanical stability can be produced from feedstocks with high density and lignin content, making lignin, the constituent, more resilient to decomposition and the loss of structural complexity. Conversely, BC with higher grindability can be obtained by the torrefaction of biomass with a larger amount of hemicellulose (e.g., agricultural residues) compared to woody biomass. The decomposition of biomass to BC causes a reduction in its bulk density and an increase in its porosity and, therefore, a decrease in its thermal conductivity, depending on the pyrolysis temperature. Concerning the electric properties of BC, the reduction in oxygenated functional groups and the appearance of conjugated double bonds cause an increase in conductivity and electromagnetic shielding efficiency, which make BC suitable as an additive in various composite materials (e.g., building materials such as cement). Furthermore, the effectiveness of shielding against electromagnetic interference is enhanced concerning the pristine biomass.

2.3. Possible Biochar Applications

The various properties of BC reported above, including its high carbon content, larger surface area, well-developed porous structure, and a surface sufficiently enriched with functional groups, render it potentially pertinent for various applications. In Table 3, we have reported the current possible environmental BC applications.

Table 3. Main possible applications of BC.

Application	Mechanisms	Refs.
Climate change mitigation	Sequestering carbon in soil ⇓ CO_2 emissions into the atmosphere ⇓ NO_2 emissions ⇓ CH_4 emissions Tackling 12% of current anthropogenic carbon emissions	[73]
Soil improvement	⇑ Physicochemical and biological properties of soils ⇑ Water retention capacity of soil ⇓ Nutrient leaching ⇓ Acids in soils ⇑ Microbial population and microbial activity in soils Positive impacts on the earthworm population Preventing desiccation	[68]
Waste management	By pyrolyzing waste biomass *	[106]
Energy production	By conversion of waste biomass to BC by fast pyrolysis, thus providing liquid fuel (bio-oil)	[71]
Capturing contaminants	By adsorption of both organic pollutants and/or metal ions from soil and water	[72,74]
Composting	⇑ Physicochemical properties of composting ⇑ Composting microbial activities ⇑ Organic matter decomposition	[69]

* Including crop residues, forestry waste, animal manure, food processing waste, paper mill waste, municipal solid waste, and sewage sludge; ⇑ = high, higher, improved, or enhanced; ⇓ = low, lower, reduced, or decreased.

BC production could be an alternative to mitigate climate change by carbon sequestration in soil, thus retaining half of the carbon fixed in biomass during photosynthesis and reducing CO_2, NO_2, and CH_4 emissions [73]. Mainly, BC shows long-term stability in soil. The mean carbon residence time in BC has been estimated to be around 90–1600 years, depending upon the labile and intermediate stable carbon components [73]. Due to these characteristics, BC can sequester carbon in soil, thus decreasing carbon dioxide emissions into the atmosphere and those of nitrous oxide and methane by biotic and abiotic mechanisms [73]. Experiments have demonstrated that the emission of greenhouse gases (including CH_4 and N_2O) can be avoided by pyrolyzing waste biomasses [107]. Concurrently, the pyrolysis process balances fossil fuel consumption by producing bioenergy.

Interestingly, BC has been estimated to be capable of tackling 12% of the current anthropogenic carbon emissions. Furthermore, thanks to its high carbon content, BC can work as a soil conditioner, mainly by improving the soil's physicochemical and biological properties. BC increases soil water retention capacity by ~18%, reduces nutrient leaching [68], and neutralizes acidic soils, thereby enhancing plant productivity, seed germination, plant growth, and crop yields. Additionally, wet BC prevents soil desiccation [68]. While it has been reported that soils treated with BC demonstrated improved microbial population and activity [108], null or positive effects were observed in the earthworm population in soils amended with wood-based BC [109].

The production of BC itself is an economical and mutually beneficial strategy to manage and eliminate waste from animals and plants and reduce the pollution associated with it [108]. Furthermore, when waste biomass that is derived mainly from animal manure and sewage sludge is pyrolyzed, the hazardous microbial population that is possibly present is killed, thus reducing its possible negative impact on the environment and humans. Unfortunately, toxic heavy metals from sewage sludge and municipal solid waste could persist in BC, which must be carefully checked and handled correctly before long-term soil application [110].

A remarkable potential use of BC, one that is still too little investigated and controversial, is the production of bioenergy as an alternative to fossil fuel that could lower carbon emissions. In this regard, while slow pyrolysis allows a lower yield of liquid fuel and more BC, fast pyrolysis provides more liquid fuel (bio-oil) and less BC [111]. Evidence has demonstrated that BC can be successfully applied in environmental remediation because it is capable of adsorbing both organic and inorganic contaminants, such as pesticides, herbicides, PAH, dyes, and antibiotics, as well as non-biodegradable metal ions, which are highly toxic to all living organisms [72,74]. BC can enhance the composting process by improving its physicochemical properties and microbial activities and promoting the decomposition of organic matter. Also, more investigations are needed to evaluate BC compost's agricultural/environmental performance [69]. Table 4 summarizes some of the advantages and disadvantages associated with the production and use of BC.

Table 4. Advantages and disadvantages associated with the production and use of BC.

Advantages	Disadvantages
Obviate to significant modification on Earth	Gaseous aerosol emissions during improper pyrolysis
Enhanced soil productivity	Environmental pollution from dust; erosion and leaching of BC particles
Higher food security	Ash could be at risk for respiratory diseases.
Solution of xenobiotics danger	BC can sequester water and nutrients not further available for crops
Addressing waste management	Not desired sorption of residual herbicides and pesticides
Reduced utilization of fossil fuels	Long-term removal of crop residues for producing BCs can reduce overall soil health by diminishing the number of soil microorganisms and disrupting internal nutrient cycling
Less expensive than activated carbon (AC)	Possible negative impact on soil biota
Improvement of living microbiology in soil	Short-term adverse effects on earthworm population density
Greater WHC than AC Enhanced food web in soil Improved aeration in the soil Reduced loss of nutrients through leaching	No universal reduction in nitrous oxide emissions

AC = Activated carbon; WHC = water-holding capacity.

As evidenced in Table 4, we benefit from additional advantages by producing BC from biomass, including waste biomass. The cost necessary to produce BC is six-fold lower than that of commercially available activated carbon (AC), which, unlike BC, is deprived of some properties of BC, such as its ion exchange capacity [112]. Generally, BC does not require further processing to be activated, and, thanks to its non-carbonized fraction and maintained oxygen-containing groups such as carboxyl, hydroxyl, and phenolic surface functional groups, BC is capable of adsorbing both organic as well as inorganic contaminants and of interacting with soil contaminants [72,74]. BCs produced from sewage sludge and manure have a high nutrient content for soils, thus enriching their quality [68]. However, apart from the advantages of using BC, there is a series of possible fallouts, as reported, that need consideration. Among these, the long-term removal of crop residues, like stems, leaves, and seed pods, for producing BC could reduce the overall soil health by diminishing the number of soil microorganisms and disrupting internal nutrient cycling, with a possible negative impact on soil biota, including short-term adverse effects on earthworm population density. In this scenario, there is a dire need for further extensive research so that any possible issues associated with its usage can be effectively resolved.

2.3.1. Xenobiotics Removal by Biochar (BC)

As reported in the previous section, BC is a porous material, and its porosity, depending on the production temperature, allows it to interact with water nutrients and other materials, including inorganic metal cations and organic pollutants. Due to its enhanced porous structure, surface area, functional groups, and mineral components, BC is an optimal absorbent material for solutions. Although BC that is produced through pyrolysis has a relatively moderate adsorption capacity (3.6–6.3 g/g for BC prepared at a temperature range of 300–700 °C) [113], this can be enhanced by modifying its physicochemical properties through acid, alkali, or oxidizing treatments, while the surface area can be altered mainly using acid treatments [114–116]. As an adsorbent, BC can absorb organic and inorganic contaminants, such as PAH and phthalate acid esters, and its help in improving the treatment of sewage wastewater containing organic xenobiotics has been widely reported [117]. In this context, there are several main mechanisms used by BC for capturing inorganic or organic pollutants, which have been included and are discussed in Table 5.

Table 5. Main mechanisms by which BC can capture inorganic or organic contaminants.

Capturing Mechanism	Influencing Factors #, Details °, Examples §	Ref.
Sorption *	⇑ Surface area # Microporosity of BC # pH #	[117]
Hydrogen bond formation **	For polar compounds °,**	
Electrostatic attraction/repulsion	For cationic compounds ° Interaction between positively charged cationic organic contaminants and negatively charged BC surfaces °,**	
Electrostatic outer sphere complexation	Due to metallic exchange with K^+ and Na^+ available in BC °,**	
Hydrophobic interactions ***	For non-polar compounds °	
Diffusion	Non-ionic compounds can diffuse into the non-carbonized and carbonized fractions of BC °	
Formation of surface complexes **	pH # Ionic radius # Between metal cations and -OH, -COOH on BCs °	
Precipitation	Lead precipitates as lead-phosphate-silicate in BC § Co-precipitates and inner-sphere complexes can form between metals and organic matter/mineral oxides of BC §	

* From water/soil onto biochar; ** for BCs produced at relatively lower temperatures; *** for BCs produced at higher temperatures; ⇑ = high, higher, improved, enhanced; In the table, the voices with # are influencing factors, those with ° are details, and those with § are examples.

Interestingly, BCs produced at higher temperatures exhibited higher sorption efficiency for the remediation of organic and metallic contaminants in soil and water. Additionally, it is worth mentioning that the sorption of organic xenobiotics by BC is more favorable than that of inorganic ones. Concerning complexation with metal cations, the smaller the ionic radius of metals, the greater the adsorption capacity by BC.

2.3.2. Not Only Adsorption

It is commonly reported that the principal mechanism by which BC removes toxic heavy metals and other contaminants, including organic pollutants, is adsorption. Its adsorptive efficiency mainly depends on the type and number of functional groups, surface area, CEC, etc. However, previous research studies and reviews on BC have evidenced the presence, either on the surface or inside its particles, of free radicals known as persistent free radicals (PFRs), the nature of which depends strongly on the pyrolysis conditions and the formation and characteristics of which mainly differ based on the feedstock types. In this regard, several recent studies have mainly focused on the role of BC-related PFRs in the degradation of organic xenobiotics, in addition to their adsorptive capacity. Odinga et al., in their recent work, reviewed the application of BC-derived PFRs in environmental pollution remediation [27], while Fang et al. investigated the reactivity of PFRs in BC and their catalytic ability to activate persulfate to degrade pollutants [28]. However, Odinga et al. also considered and commented on the possible environmental risks of PFRs from BCs, which represent the shadows associated with these chemicals and need further study, knowledge, and regulation before their extensive application [27].

3. Biochar-Derived Free Radicals

As previously mentioned, BC has a broad-prospective use in the treatment of environmental xenobiotics, in soil amendment, in photocatalytic and photothermal systems, for photothermal conversion, as electrical and thermal devices, and as 3D solar vapor-generation devices for water desalination [118–121]. All these potential uses are due to its high surface area and rich pore structure, which provide great physical absorptivity. They also depend on the chemical characteristics of BC, including the presence of PFRs [122,123]. In this regard, it is of paramount importance to clarify the formation mechanism of free radicals in BC for the optimal management of their properties and their more efficient and safer utilization [124].

3.1. Persistent Free Radicals (PFRs)

An atom or molecule with at least a lone pair of electrons is a chemical species characterized by significant instability and high chemical activity and is referred to as a free radical species [107]. Usually, free radicals are highly unstable and rapidly react with each other, thus being destroyed as soon as they form, with a consequent very short half-life. However, it has been found that in BC, some free radicals, named persistent free radicals (PFRs), like the radicals that naturally occur in the environment, known as environmental persistent free radicals (EPFRs), can remain stable for months and play a crucial role in the subsequent reactions of oxidative degradation carried out by BC containing them [25,107,125,126] (Figure 6).

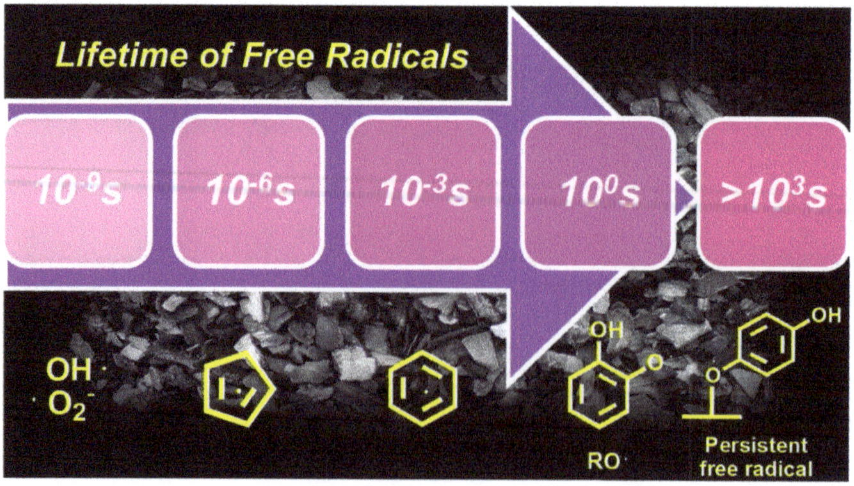

Figure 6. The lifetimes of different free radicals.

Unlike other free radicals, PFRs are resonance-stabilized since they are bound to the external or internal surface of solid particles of BC. They can be analyzed by electron paramagnetic resonance spectroscopy (EPR) [25]. Figure 7a provides an example of an EPR analysis of the PFRs present on a solid N-doped hydro char prepared in a tube furnace at a temperature of up to 600 °C for 1 h under a N_2 atmosphere [127].

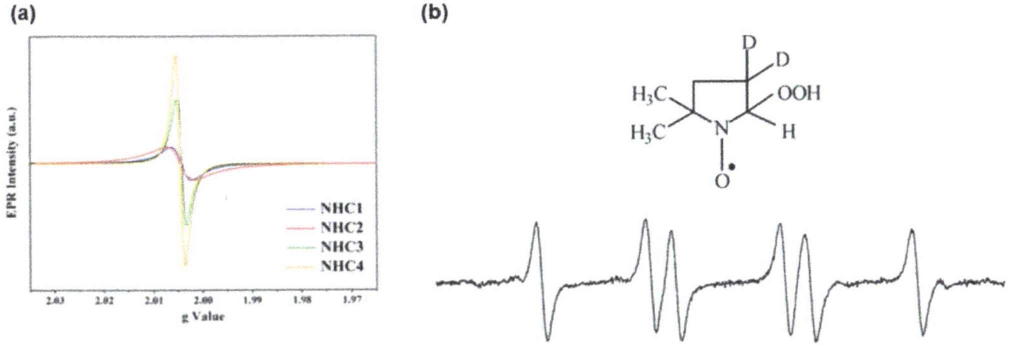

Figure 7. EPR analysis of PFRs generated on a N-doped hydro char prepared in a tube furnace heated to 600 °C for 1 h under N_2 atmosphere (**a**) [127]. The image has been reproduced with permission from Elsevier under license number 5754160784661, received on 22 March 2024; EPR spectrum of the unstable free radical superoxide ($O_2{\bullet}^-$) when trapped by DMPO to form a long-lived nitroxide (DMPO-OOH) (**b**).

Their lifetime under a vacuum appears infinite, while they react with molecular oxygen in the air, resulting in decay with time and the simultaneous production of reactive oxygen species (ROS). In this regard, PFRs act as transition metals like Fe^{2+}, stimulating ROS production in aqueous systems. Unlike PFRs, ROS are detectable by EPR only when captured by a proper radical scavenger, such as 5,5-dimethyl-1-pyrroline-N-oxide (DMPO). Figure 7b provides an example of the EPR spectrum of the unstable free radical superoxide ($O_2{\bullet}^-$) when trapped by DMPO to form a long-lived nitroxide radical (DMPO-OOH). In BC analyzed using the EPR technique, PFRs were previously detected in combustion-generated particulate matter (PM), sediments, and soils. PFRs can be categorized into three classes, i.e., oxygen-centered PFRs (OCPFRs), carbon-centered PFRs (CCPFRs), and oxygenated carbon-centered radicals (CCPFRs-O). The EPR analyses provide three parameters: the PFR concentration, the g-value, and the line width [107] (Figure 8).

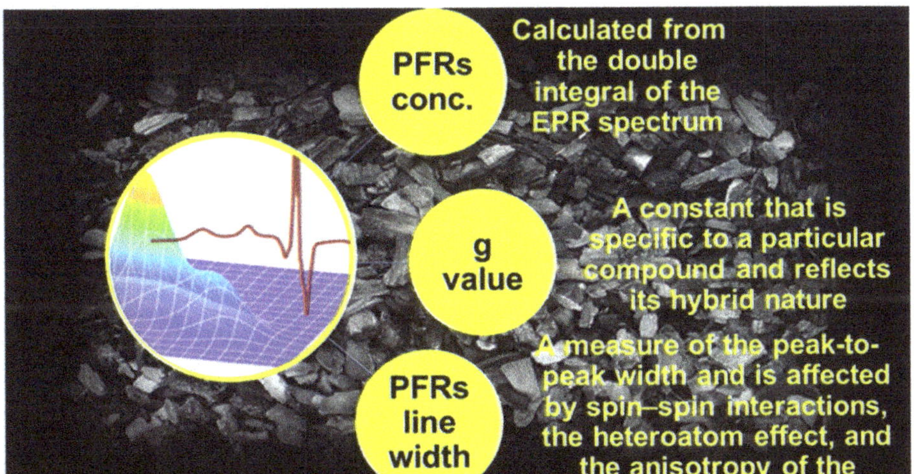

Figure 8. Information derived from EPR analyses.

PFR concentration is calculated from the double integral of the EPR spectrum and can reflect the content of PFRs in BC [128]. The g-value of PFRs is a constant specific to a particular compound, reflects its hybrid nature, and provides information about the type of radical [129]. The PFR line width in the EPR spectrum measures the peak-to-peak width. This is affected by spin–spin interactions (including electron–proton interaction and electron–electron interaction), the heteroatom effect, and the anisotropy of the spectrum [107].

The line width reflects the relaxation time of spinning electrons [130]. It has been reported that the oxidation processes that can occur using BCs mainly depend on PFRs and these parameters [125,131,132]. These parameters are, in turn, affected significantly by pyrolysis conditions, biomass types, the elemental composition of pristine biomass, and the presence of external transition metals (Table 6).

Table 6. Factors influencing PFR formation in BC.

Parameter	Influencing Factors	Specifications	Observations	Ref.
PFRs concentration	Biomass type	Cow manure, rice husk, others (< 500 °C)	≠ Concentrations	[132,133]
		Non-lignocellulosic biomass with ⇓ H/C and O/C	⇓ Concentration	[134]
		Lignocellulosic biomass	⇑ Concentration	
	Temperature	300 °C, 700 °C	≠ Concentrations	[132]
		Maximum concentration at 600 °C		[135]
		Maximum of concentration at 500–600 °C		[24,136]
	Transition metals	Adsorb onto biomass and transfers electrons from polymer to metal center during pyrolysis	⇑ Concentration	[32]
Type of PFRs	Temperature	200–300 °C	Oxygen centered radicals	[24]
		400 °C	A mixture of oxygen and carbon-centered radicals	
		500–700 °C	Exclusively carbon centered radicals	

≠ = Different; ⇓ = low, lower, reduced, decreased; ⇑ = high, higher, increased, improved.

Qin et al. [132] found that the PFR concentrations in the same BC that were obtained at different temperatures and those in different kinds of BC obtained at the same temperature were significantly different. Tao et al. [135], as well as Xiang [136] and Huang et al. [24], found that in BCs from different feedstocks, the PFR concentrations first increased with increasing temperature, reaching a maximum around 500–600 °C, and then decreased with a further increase in temperature. The relations between the feedstocks' properties or the BCs' composition with the PFR concentrations were also demonstrated [133], and non-lignocellulosic-biomass produced lower PFR concentrations than lignocellulosic-biomass under the same pyrolytic conditions, perhaps due to their lower H/C and O/C atomic ratios [134]. The types of PFRs that can be produced during pyrolysis change in the pyrolysis process and change along with a temperature rise, as reported in Table 6. One study speculated that the reduction in oxygen content during biomass pyrolysis may account for the progressive conversion of oxygen-centered radicals to carbon-centered radicals [134].

Mechanism Proposed for PFR Formation during Biomass Pyrolysis

The several environmental sources of PFRs include atmospheric particulate matter (PM), contaminated soil, materials from thermal treatments of plastic and hazardous waste, tar balls, and the products deriving from the pyrolysis of biodiesel and biomass waste feedstocks at high temperatures [25]. Concerning BC-derived PFRs, it was observed that they mainly form in the post-flame and cool-zone regions of combustion systems and other thermal conversion processes. Although the actual mechanism by which PFRs form during pyrolysis remains not fully clarified, transition metals capable of electron transfer and the substituted aromatics molecules present in lignin have been recognized as critical factors of PFR formation. However, high concentrations of PFRs have also been detected in the product of combustion of non-aromatic cellulose in the absence of transition metals [135]. Based on the temperature of pyrolysis processes, during the production of BC, highly heterogeneous composite structures occur, comprising both labile and recalcitrant organic molecules, such as PAH, furans, and dioxins, as well as inorganic fractions including oxides, cations, anions, and free radicals [137]. These fractions, products of the incomplete combustion of biomass, may gradually form PFRs by different pathways, including or not including transition metals. The formed PFRs could be either only surface-stabilized or be surface-stabilized in metal-radical complexes [27]. Generally, the breaking of covalent bonds by heat, light, electricity, and chemical energy, is essential to form free radicals; during the pyrolysis process of lignocellulosic biomasses. Their main constituents, namely, cellulose, hemicellulose, and lignin, undergo different reaction pathways at various destructive pyrolysis temperatures of 300 °C, 300–400 °C, and 350–450 °C. Anyway, the presence of transition metals can strongly affect the possible formation of PFRs during pyrolysis. Figure 9 attempts to describe the possible series of events occurring during biomass pyrolysis that could lead to the formation of PFRs, which are also chemically described in Scheme 1 (concerning lignin) and Scheme 2 (concerning cellulose and hemicellulose).

First, the C-O and C-C covalent bonds of constituents of lignin are broken under heat, either via electron transfer by transition metals or not, to form free radical fragments, phenols, chinones, and other products of incomplete combustion. Simultaneously, the cleavage of the glucoside bonds of cellulose and hemicellulose that are present in biomass feedstocks occurs, causing depolymerization and the formation of other radicals. These first radicals can couple to form bio-oil, pyrolysis syngas (CO_2, CO, CH_3CH_3, and CH_4), and BC simultaneously or may abstract hydrogen from other molecules, forming further radicals [133,138]. Several chemical reactions can occur, including dehydration, decarboxylation with further emissions of CO_2, CO, and H_2O, aromatization, and intra-molecular condensation leading to the formation of the crystalline graphene structure and graphitic radicals. During pyrolysis, the elemental composition of biomass undergoes changes that cause mutations in the types of radicals that, upon entrapment onto the BCs' surface and/or the formation of metal–radical complexes, form stable PFRs [24].

According to findings reported in the literature, the possible types of PFRs comprise (i) transition metal-mediated PFRs or (ii) PFRs forming inside organic matrices during biomass pyrolysis to give BC [139]. The transition metal-mediated PFR formation starts with the initial physisorption of an aromatic substituted molecule or of its degradation intermediate radicals, generated at 150–400 °C or under UV irradiation onto transition metal oxides such as ZnO, NiO, CuO, Fe_2O_3, and TiO_2 or transition metal ions [140]. Then, chemisorption occurs by forming a chemical bond eliminating water or hydrogen chloride. Finally, a single electron is transferred from the substituted aromatics to the center of the transition metals, leading to the simultaneous reduction of metal and the formation of PFRs [140], the stability of which can be attributed to the synergy of metals and aromatic compounds [140]. A transition metal accepts an electron, and its valence changes from high to low during this process.

Unlike the PFRs discussed previously, PFRs formed inside the matrix of organic moieties are not related to the presence of transition metals [139]. Still, they are highly dependent on the relevant organic matter, while their concentration is significantly and positively correlated with the elemental carbon content [139]. In this case, PFRs are compared in terms of thermally treated particles, for which the breaking of chemical bonds in the precursor molecules during pyrolysis is the primary reason. At the initial pyrolysis stage, the homolytic cleavage of weak linkage bonds like the α- and β-alkyl aryl ether bonds, C-C, and C-O linkage resulted in the forming of free radicals in BC. The outer-surface free radicals would rapidly react and dissipate, resulting in a decrease in EPR signals. The free radical concentrations then increased with extended pyrolysis and during the cooling stage, thus accumulating many free radicals on the BC's surface [139] and dramatically boosting the EPR signals. The free radicals formed in the matrix of the produced BC are probably protected from reacting with each other or other chemicals and are thus stabilized.

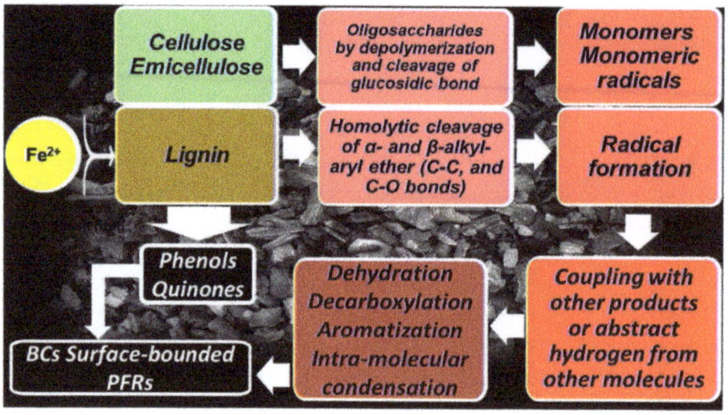

Figure 9. The PFR formation process.

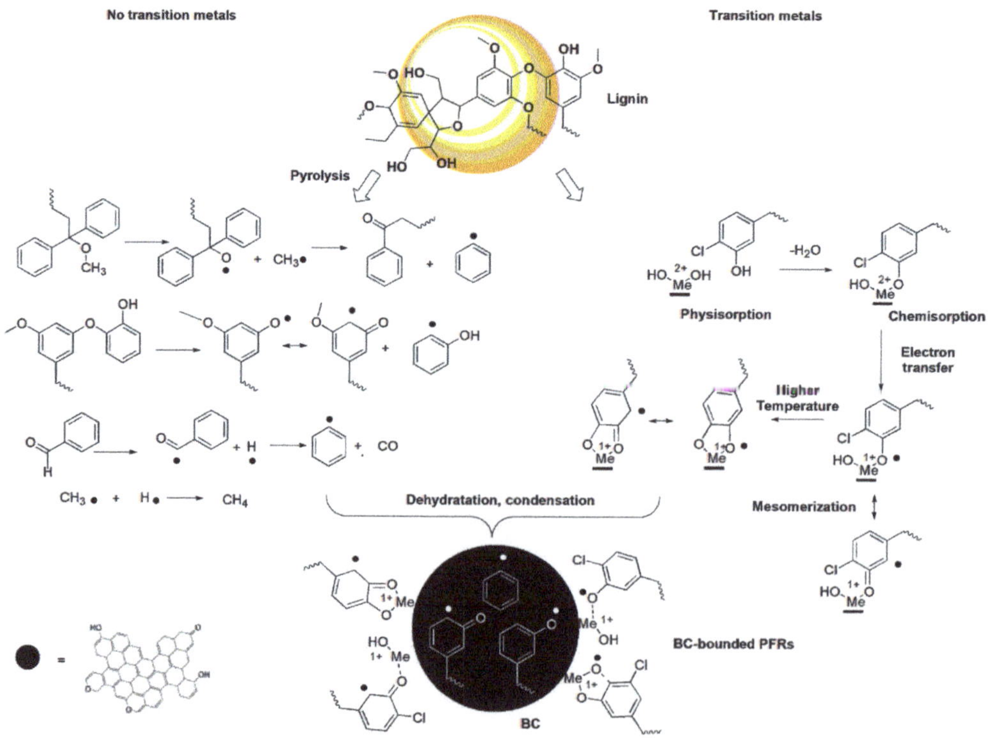

Scheme 1. Possible mechanisms leading to the formation of BC-bounded PFRs from lignin. The orange sphere represents biomass, while the black sphere represents BC, the hypothetical structures of which, depending on pyrolysis condition, are shown at the bottom of the scheme.

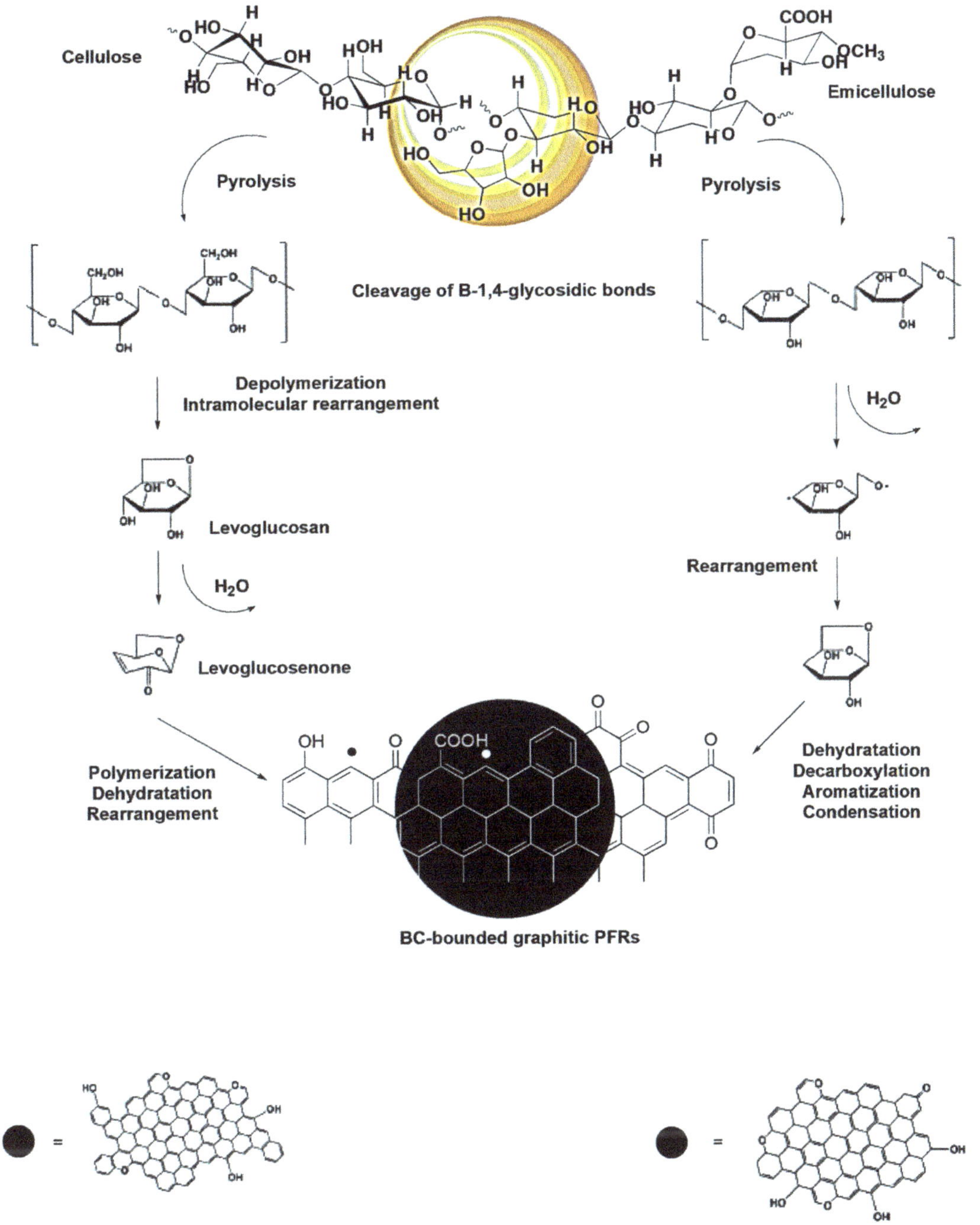

Scheme 2. Possible mechanisms leading to the formation of BC-bounded graphitic PFRs from cellulose (**left side**) and hemicellulose (**right side**). The orange sphere represents biomass, while the black sphere represents BC, the hypothetic structures of which, depending on pyrolysis condition, are shown at the bottom of the scheme.

As mentioned earlier, the type of biomass and its elemental composition, the presence of oxygenated functional groups, the pyrolysis conditions (temperature, heating time, and heating rate), and the presence of external transition metal as well as phenolic compounds strongly affect both the concentration, structure, and type of PFRs present in BC. Notably, no radical is produced during the first stage of pyrolysis, providing the transition char (< 300 °C). Subsequently, in the second stage of pyrolysis (300–500 °C), amorphous char is produced, and oxygen-centered radicals and oxygenated carbon-centered radicals appear. In the third stage of pyrolysis at 500–700 °C, composite char is created, wherein the concentrations of PFRs, including carbon-centered and oxygenated carbon-centered radicals, drastically decrease. Finally, when turbostratic char is produced (> 700 °C), little or no PFRs are subsequently produced [27]. In the EPR, the g-factor values, even if they could change due to the presence of metal ions and temperature changes, are specific for a type of radical. Table 7 reports the main types of radicals recognizable in BC and their specific g-values.

Table 7. Features and g-values of the main PFRs forming in BC.

Radicals	g-Value	Features
Carbon-centered radicals	<2.003	Susceptible to oxidation in air
Carbon-centered radicals adjacent to an oxygen atom (oxygenated carbon-centered radicals)	2.003–2.004	Susceptible to oxidation in air
Oxygen-centered radicals	>2.004	More stable in an atmospheric environment
Semiquinone radicals (oxygen-centered)	>2.0045	More resistant to reacting with molecular oxygen in the ambient environment
Phenoxy radicals (oxygenated carbon-centered radicals)	2.0030–2.0040	Susceptible to oxidation in air
Cyclopentadienyls (carbon-centered radicals)	<2.003	Susceptible to reacting with molecular oxygen in the ambient environment

3.2. *PFRs: Light and Shadows*

3.2.1. PFR Light

It has been demonstrated that PFRs originating in BC by combustion in the presence or absence of external transition metals could play a vital role in several beneficial reactions, such as the PFR-mediated remediation and degradation of organic and inorganic pollutants by different actions and mechanisms, including oxidative and reductive processes (Table 8).

Table 8. Main actions and mechanisms by which PFRs remediate/degrade environmental xenobiotics.

EPFRs Actions	Degraded Substances *	Mechanism	Refs.
Activation of H_2O_2 by single electron transferring	SMX, CIP, SMT, TC, OG, MNZ, ERF benzene	Oxidation by the production of ROS (OH• #, HO_2•, O_2•$^-$)	[32,141]
Activation of O_2 by single electron transferring	Degradation of organic compounds Chloro-biphenyl Phenolic compounds Polychlorinated biphenyls Diethyl phthalate Thiacloprid Bisphenol A	Oxidation by the production of radical superoxide (O_2•$^-$)	[21,25,33,123]
Activation of persulfate ($S_2O_8^{2-}$)	X-3B, SMT, CTC, SMX, TC, MB, SDZ, OG	Oxidation by the production of sulfate radicals (SO_4•$^-$)	[141]

Table 8. *Cont.*

EPFRs Actions	Degraded Substances *	Mechanism	Refs.
Direct activity of macromolecular radicals on the BC surface	Direct degradation of organic chemicals	Oxidation	[74]
Direct activity of semiquinone-type radicals	As (III) removal	Oxidation	[142]
Direct activity of PFRs	Removal of Cr (VI)	Reduction to Cr (III)	[143–147]
Catalytic effects	Detoxification of environmental xenobiotics	Generation of activated species Stimulation of the microbial biotransformation	[74]
Ions' exchange	Enhancement of agricultural soil performance	Maintenance of CEC in soils	[148]
Electron-hole pair formation	Photocatalytic degradation of contaminants under Vis irradiation	Electrons in free radicals can be transformed from the valence band to the conduction band under irradiation	[45]

* Degraded or removed; SMX = sulfamethoxazole; CIP = ciprofloxacin; SMT = sulfamethazine; TC = tetracycline; OG = orange; MNZ = metronidazole; # free or surface bond; KET = ketoprofen; CTC = chloro-tetracycline; SDZ = sulfadiazine; MB = methylene blue; ERF = enrofloxacin (photocatalytic degradation); X-3B = reactive brilliant red X-3B.

For instance, PFRs on BC can activate hydrogen peroxide (H_2O_2) or oxygen (O_2), as well as persulfate ($S_2O_8^{2-}$), to produce different free oxygenated radicals (ROS) that are capable of efficiently degrading organic contaminants such as chloro-biphenyl [32], phenolic compounds and polychlorinated biphenyls [21], diethyl phthalate [33], thiacloprid [123], and bisphenol A [25]. Moreover, organic chemicals can also be directly degraded on the BC surface by macromolecular free radicals without adding any radical activators [74]. The semiquinone-type radicals present in BC can oxidize As (III) [142]. At the same time, BCs can also exhibit the highly effective removal of Cr (VI) by reduction to Cr (III) using PFRs for industrial wastewater remediation [143–147].

Unfortunately, PFRs, by generating surface-bound hydroxyl radicals and free hydroxyl radicals in aqueous solution and also in the absence of H_2O_2, can induce various types of cardiovascular and pulmonary disease through ROS-induced oxidative stress (OS) [25]. PFRs and OH radicals that were detected in biological fluids generated ROS that induced an oxidant injury and modulated toxic responses in biological tissues [149]. Moreover, quinoid redox cycling is another possible path causing the formation of ROS from material containing semiquinone-type radicals, which could exert toxicity like that exercised by the combustion products present in cigarette smoke [150]. Although BC has beneficial effects on agricultural soil, the PFRs in BC could inhibit plant germination and growth when used in soil remediation. BC addition as a soil amendment has been reported to positively affect plant germination, growth, and yield [151,152]. In contrast, a negative impact has also been documented when BC-bounded PFRs induce ROS, which can inhibit seed germination and retard the growth of roots and shoots [32,35]. As shown in Figure 10, the formation and presence of PFRs in the BC produced by several biomasses have been widely documented and studied since 2014.

In this regard, in Table 9, we have reported a random selection of the main experimental works regarding the PFRs found in BCs, obtained by different biomasses conveyed in recent years (2019–2024). Table 9 also summarizes their reported applications, including mainly the oxidative degradation of organic environmental pollutants (51 papers), the removal of hazardous inorganic compounds from wastewater such as As (III) and Cr (VI) (12 papers), the degradation of biological samples, including bacteria (3 papers), hormones (4 works), genes of bacterial resistance (1 paper), and their use as electrical devices due to their electron and electron donor capacities (EAC and EDC).

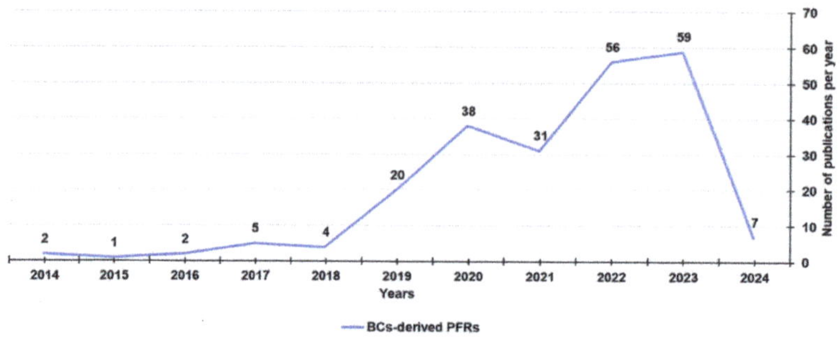

Figure 10. Number of publications on BC-derived PFRs from the year 2014 to the present, according to the Scopus dataset (reviews and chapters in the books included). The survey used the following keywords: persistent AND free AND radicals AND biochar.

Table 9. BC-derived PFRs and their applications as described previously, reported in the years 2019–2024.

Biomass	Pyrolysis °C/Time	BC-Name	Active Radicals	Radical Mechanisms	Application [1] Degraded Compound [2]	Refs.
Sawdust	700 °C/1 h	Fe^0-BC-700	$SO_4{\bullet}^-$ PFRs OH•	Activation of PMS by Fe^0 Activation of PMS by PFRs	BPA [2]	[153]
Waste wood	500 °C, 700 °C	Fe^0-BC	$SO_4{\bullet}^-$ PFRs OH•	Production of PFRs by Fe^0 Activation of PS by Fe^0 Activation of PS by PFRs	TDWW [2]	[154]
Camellia seed husks	400 °C/2 h	OBC-Fe_3O_4	$SO_4{\bullet}^-$ PFRs OH•	Activation of PS	TC [2]	[155]
Maize straw	900 °C/2 h	NBC1	$\bullet O_2^-$ $SO_4{\bullet}^-$ PFRs OH•	Activation of PS	(86.6%) TC [2]	[156]
Sawdust	300 °C, 700 °C	SBC	$SO_4{\bullet}^-$ PFRs OH•	Activation of PS	AO-7 [2]	[157]
N.R.	200 °C, 500 °C	N.R.	PFRs $\bullet O_2^-$	UV-induced interaction PFRs/DOM and $\bullet O_2^-$ production	RhB [2]	[158]
Sewage sludge	500 °C/4 h	HNO_3-BC	PFRs $\bullet O_2^-$ •OH $\bullet O_2H$	Activation of H_2O_2	CIP [2]	[35]
Wheat straw	500 °C/2 h	BC/Fe (III)	$SO_4{\bullet}^-$ PFRs OH•	Activation of PS by PFRs	SMX [2]	[159]
Sawdust	700 °C	BC700	$SO_4{\bullet}^-$ PFRs OH•	Activation of PDS by PFRs	CA [2]	[160]
Pine needle	500 °C/2 h	Fe/Mn/BC	•OH	Activation of H_2O_2 by Fe (II), Mn (II) and PFRs (FeMn/BC/H_2O_2 photo-Fenton system)	Naphthalene [2]	[161]

Table 9. *Cont.*

Biomass	Pyrolysis °C/Time	BC-Name	Active Radicals	Radical Mechanisms	Application [1] Degraded Compound [2]	Refs.
Sewage sludge	500 °C/4 h	SS-BC	PFRs • O_2^- •OH •O_2H	Activation of O_2 and H_2O_2 by PFRs Degradation of PNP by PFRs	CIP [2]	[162]
Swine manure	600 °C	SBC	OCPFRs CCPFRs-O • OH •O_2H	Activation of oxygenated species by OCPFRs and CCPFRs-O (heterogeneous Fenton-like systems SBC/H_2O_2)	SMT [2]	[163]
Wheat straw	300 °C, 600 °C	BC300 BC600	•OH •O_2H	Goethite (Gt)-mediated activation of H_2O_2 (Fenton-like system)	OFX [2]	[164]
Wheat straw	500 °C/2 h, 800 °C/2 h	CoBCX	SO_4•$^-$ PFRs OH•	Cobalt and PFR-mediated activation of PMS via O_2	ATZ [2]	[165]
Various crop straws	450 °C, 550 °C 650 °C	BC450,550 BC650	SO_4•$^-$ • O_2^- OH•	BC-mediated activation of PS by electron transfer	SDZ [2]	[166]
Tobacco steam	300°C, 500°C 700°C	T-BC	ROS	OCPFR-mediated activation of O_2 in the water	PNP [2]	[167]
Pruning wastes of apple trees	400 °C, 550 °C 700 °C	BC400, BC550 BC700	SO_4•$^-$ PFRs	BC and PFR-mediated activation of PS	ACT [2]	[168]
Camphor leaves	400 °C/6 h	Fe (TPFPP)/BC	SO_4•$^-$ PFRs OH•	PFRs-mediated electrons transferring to iron porphyrin-loaded BC [3]	PFOA [2]	[169]
Corn stalks	240 °C/4 h	hydrochar	•OH	Electrode and PFR-mediated generation of ROS	2,4-DCP [2]	[170]
Wheat straw	450 °C/4 h	Co3O4-BC	SO_4•$^-$ PFRs OH•	Co_3O_4-BCmediated activation of PMS	CAP [2] FF [2] TAP [2]	[171]
Wheat straw Urea Iron salts	800 °C/1 h	Fe-N-BC	SO_4•$^-$ PFRs •OH • O_2^-	Fe, N co-doped BC and PFR-mediated activation of O_2 and PS	AO7 [2]	[172]
Candida utilis	700 °C/2 h	NCS-x	SO_4•$^-$ PFRs OH•	Activation of PMS by nitrogen-doped biochar nanosheets (NCS-x) using molten salt (NaCl and KCl) in the pyrolysis process	BPA [2] BPF [2] BPS [2] BPAF [2]	[173]
Pine needles	500 °C	nFe3O4/BC	PFRs •O_2H •OH • O_2^-	Activation of H_2O_2 by nano-magnetite supported biochar via Fe (III)/Fe (II) cycling and electron transfer with the PFRs	Ethylbenzene [2]	[174]
Sewage sludge	800 °C/3 h	SM-(0.5:1)	SO_4•$^-$ PFRs OH•	Activation of PMS by nitrogen-doped sludge biochar with different ratios of melamine in acidic	Cationic/anionic dyes [2]	[175]

Table 9. *Cont.*

Biomass	Pyrolysis °C/Time	BC-Name	Active Radicals	Radical Mechanisms	Application [1] Degraded Compound [2]	Refs.
Elephant grass	350 °C, 600 °C 900 °C 30–120 min	EG	OCPFRs	OCPFR-mediated oxidation	CV [2]	[176]
Sunflower-straw	N.R.	SSBC	SO_4•$^-$ PFRs OH•	Enhanced Fe (II) activation of PS via BC and PFRs	Benzoic acid [2]	[177]
Pine chips	500 °C	OP5 RP5	SO_4•$^-$ PFRs •OH • O_2^- •O_2H	EDC-involved structures, Fe^{3+} and BC (PFR)-mediated activation of PS in a Fenton-like reaction system using H_2O_2 and NaBH4	2,4-DCP [2]	[178]
Rice straw	350 °C, 500 °C 700 °C	BCs MBCs BDOMs	PFRs •OH	Direct photocatalytic degradation in BCs and MBCs solutions by Xenon-lamp Oxygen reduction by FPRs of BCs and MBCs BDOM-mediated generation of ROS	SMX [2] CAP [2]	[179]
Pomelo peels	600 °C	Fe@PP-Hy-Py	PFRs •OH • O_2^-	Amorphous Fe (0)-mediated formation of PFRs Fe (0)-mediated reduction of PNP EPFR-mediated oxidation of PNP via ROS (O_2 and H_2O_2) activation	PNP [2]	[180]
Softwood pine	823–873 K	US-BC BC-P BC-P-DEA US-BC-P-DEA US-BC-P-DEA	PFRs •OH • O_2^- •O_2H	Reinforcement of PFRs concentration doping BCs with Ni and Pb Activation of H_2O_2 by PFRs	Phenol [2]	[181]
Camphor leaves	500 °C/1 h	Fe (VI)/BC-2	Fe(V)/Fe(IV) PFRs •OH	Fe (VI)-BC (PFRs)-mediated electron transferring and generation of ROS	AZT [2]	[182]
Bagasse powder	800 °C	DBC800 PBC800-A	SO_4•$^-$ PFRs •OH• O_2^- •O_2H	Enhanced BC-mediated activation of PS Improved PFR generation by natural endogenous minerals	TC [2]	[183]
Eichhornia crassipes Iron salts	400 °C/2 h	MBC	PFRs •OH• O_2^-•O_2H	Fe (II)-BC-mediated activation of H_2O_2 (Fenton-like system)	MNZ [2]	[184]
Poplar and pine sawdust	300–500 °C	PO xxx PI xxx	SO_4•$^-$ PFRs •OH• O_2^-	Activation of PMS by CCEPFRs-O and CCEPFRs in BC	TC [2] CTC [2] DOX [2]	[185]

Table 9. *Cont.*

Biomass	Pyrolysis °C/Time	BC-Name	Active Radicals	Radical Mechanisms	Application [1] Degraded Compound [2]	Refs.
S. alfredii	Air-dried	Metal@P	$\bullet O_2H$	PFR generation by the thermochemical behaviour of Mn and Zn Electron transfer Activation of PDS by PFRs in Fe/Zn@PB9/PDS system AOPs	Imidacloprid [2]	[186]
Sludge	N.R.	N.R.	$SO_4\bullet^-$ PFRs $\bullet OH\bullet$ O_2^-	Production of ROS via PFRs Mn-mediated electron transfer through Mn-doped sludge-based biochar (BC) after mediation	CIP [2]	[187]
Cellulose Lignin	200–1000 °C	C200, C500 C1000 L200, L500 L1000	$SO_4\bullet^-$ PFRs $\bullet$ O_2^-	Activation of PS adsorbed onto BCs via PFRs, oxygen-containing functional groups, and defective structures of BCs	OFX [2]	[188]
Chestnut shell $KMnO_4$	700 °C/1 h 400 °C/1 h	Mn-BC	PFRs	Mn-improved electron-transfer	OTC [2]	[189]
Spent coffee TiO_2	300 °C 500 °C 600 °C	SBC500	PFRs $\bullet OH\bullet$ $O_2^-\bullet O_2H$	Activation of H_2O_2 by Ti-doped H_2SO_4-modified biochar (SBCs) (Photo-Fenton-like system)	MO [2]	[190]
RS	550 °C/2 h	BC-α-Fe2O3/MgO	PFRs $\bullet OH\bullet$ $O_2^-\bullet O_2H$	UV light activation of PFRs Production of O_2 upon NPA degradation O_2 activation by PFRs	NPA [2]	[191]
Sewage sludge	400 °C/2 h	SDBC	PFRs $\bullet OH\bullet$ $O_2^-\bullet O_2H$	O_2 activation by PFRs promoted by HNO_3 or NaOH environmental	*p*-Chlorophenol [2]	[192]
Peanut hull	700 °C/2 h	BC-Fe-1-Zn	$SO_4\bullet^-$ PFRs $\bullet OH$	Activation of PS by bimetal-modified peanut hull-derived biochar via Fe and Zn oxides and oxygen-containing functional groups active sites	TC [2]	[193]
Blue algae	700 °C	Z-700 FeOX@BC	$SO_4\bullet^-$ PFRs $\bullet OH$ $\bullet O_2^-\bullet O_2H$	$\bullet$ O_2^- production by FeOX (zero-valent iron and iron oxide) C=C, C=O, O-C=O, Fe-O functional groups and PFRs promoted the activation of PDS	TC [2]	[194]

Table 9. *Cont.*

Biomass	Pyrolysis °C/Time	BC-Name	Active Radicals	Radical Mechanisms	Application [1] Degraded Compound [2]	Refs.
Biomass	300–1000 °C	N.R.	$SO_4•^-$ PFRs •OH • O_2^-	Activation of PS and PMS by physically and chemically modified BCs using acid/alkali treatment and metal doping via PFRs	PPCPs [2]	[195]
Chicken feathers	350 °C/4 h 800 °C/4 h	MBC35@FH MBC80@FH	$SO_4•^-$ PFRs •OH • O_2^-	Activation of PDS by the transformation of Fe species, oxygen-containing functional groups, pyrrolic nitrogen, and PFRs to produce ROS	TPhP [2]	[196]
Pine needles	300–900 °C	BC300-900	$SO_4•^-$ PFRs •OH • O_2^-	Activation of PMS by BC via ROS production or electron transfer capability	OFX [2] ENR [2] FLE [2]	[197]
PolyS	220 °C/2 h [$] 500 °C /2 h [#] 900 °C /2 h [#]	BC500 + PS BC900 + PS BC500 BC900	$SO_4•^-$ PFRs •OH • O_2^-	Activation of PMS using CCEPFRs on BC aged by microbial fermentation for ROS production	SDZ [2] OFX [2] DOX [2]	[198]
Red mud Wheat crop	700 °C /2 h	MRBC	$SO_4•^-$ PFRs • OH	PDS activation by the active sites of MRBC, such as Fe (II) and PFRs	LFX [2]	[199]
Various sludges	300–900 °C 2 h	S-HPBC S-PBC S-HBC	$SO_4•^-$ PFRs •OH • O_2^-	Activation of PS by PFR-mediated electrons transferring activity Electrons transferring to Cr (VI) by PFRs	TC [2] Cr (VI) ⇒ Cr (III) [1]	[200]
Peanut shells	500 °C/4 h	BC-Ce	OFGs, CCPFRs	Electrons transferring to Cr (VI) by OFGs, CCPFRs; oxygen vacancies and graphitic structure in BC-Ce promoted by CeO_2	Cr (VI) ⇒ Cr (III) [1]	[201]
Rice husk	400 °C/1 h	BC400	OH• H_2O_2	(pH acid) Activation of O_2 by phenol OH and semiquinone types of PFRs	As (III) ⇒ As (V) [1]	[142]
			Semiquinone-type PFRs Quinoid C=O H_2O_2	(pH alkaline) Activation of O_2 by phenol OH and semiquinone types of PFRs		
Rice husk	550 °C	RH-BC	PRFs	Promotion of OCPFRs by BC-inducted Cr (VI) degradation	Cr (VI) ⇒ Cr (III) [1]	[143]
Stalk	450 °C/90 min	NBC	PFRs	N-doped BC-mediated evolution of PFRs for the transformation of Cr (VI)	Cr (VI) ⇒ Cr (III) [1]	[144]
Rice husk	500 °C/2h	MGBs	PFRs •OH • O_2^-	Efficient surface Fe (III)/Fe (II) cycling via electron transfer with the PFRs of magnetic greigite/BC composites (MGBs)	Cr (VI) ⇒ Cr (III) [1]	[145]

Table 9. *Cont.*

Biomass	Pyrolysis °C/Time	BC-Name	Active Radicals	Radical Mechanisms	Application [1] Degraded Compound [2]	Refs.
Sludge	220 °C/2h	BC	OCPFRs	UV-Vis photo-irradiation enhanced the production of PFRs Action of OCPFRs as electron donors to transform Cr (VI) into Cr (III)	Cr (VI) ⇒ Cr (III) [1]	[146]
Sludge	120 °C	SBC120	OCPFRs	OCPFR-mediated electrons transferring to Cr (VI) in neutral solutions	Cr (VI) ⇒ Cr (III) [1]	[147]
	270 °C	SBC270	CCPFRs	CCPFR-mediated electrons transferring to Cr (VI) in neutral solutions		
Rice husk	400 °C/1 h	rUBC, rDBC	Quinoid C=O PFRs	Quinoid C=O and PFR-mediated oxidation of As (III)	As (III) ⇒ As (V) [1]	[202]
Maize straw powder	500 °C/2 h	FhBC	PFRs • O_2^- •OH	Fe and PFR-mediated activation of O^2 and H_2O_2	As (III) ⇒ As (V) [1]	[203]
Sewage sludge	270 °C/2 h	SBC	SO_4•$^-$ PFRs •OH • O_2^-	Activation of PS by SBC via PFR-mediated electrons transferring	As (III) ⇒ As (V) [1]	[204]
Pinewood	600 °C/1 h	Fe/HC	• O_2^- •OH	Activation of O_2 and H_2O_2 by CCPFRs	Estrogens [2]	[205]
Rice straw	500 °C/1 h	BiPB	•OH PFRs	Generation of •OH by Bi/Bi_2O_3 and PFRs	Estrone [2,*]	[206]
Anaerobic digestion sludge	400 °C 600 °C 800 °C 1000 °C	ADSBC 400 ADSBC 600 ADSBC 800 ADSBC 1000	SO_4•$^-$ PFRs OH•	BC-mediated activation of PDS	Dyes [2] Estrogens [2] Sulfonamides [2] *E. coli* [2] Others [2]	[207]
Walnut shell	700 °C/1 h	BC700	PFRs	Oxidation by PFRs-mediated electron transfer	E1 [2] E2 [2] E3 [2]	[208]
Caragana korshinskii	650 °C/3 h	ACB-K-gC3N	PFRs h^+•OH• O_2^-	Electron photogeneration and PFR-mediated H_2O and O_2 activation	*S. aureus* [2] *E. coli* [2] RhB [2] TA [2] NOR [2] CAP [2]	[209]
Pinewood	600 °C	Ag^0-PBC	PFRs •OH • O_2^-	UV-light-promoted excitation of the electron-hole pairs and, subsequently, the production of ROS Enhanced ROS generation by PFRs	MB [2] *E. coli* [2]	[210]
Rice straw	400 °C, 700 °C 120 min	Nano-BC	PFRs •OH • O_2^-	Oxidation and damage by ROS	eDNA [2]	[211]

Table 9. *Cont.*

Biomass	Pyrolysis °C/Time	BC-Name	Active Radicals	Radical Mechanisms	Application [1] Degraded Compound [2]	Refs.
Rice straw	500 °C	RS-BC	Quinones Phenols PFRs	By electron acceptor capacity (EAC) By electron donor capacity (EDC)	⇆ Redox property [1] Electronic storage [1]	[212]

In the table, the voices with [1] are applications, while those with [2] are degraded compounds; BCs = biochar; MBCs = modified-biochar; BDOMs = biochar-derived dissolved organic matters; PMS = peroxy-mono-sulfate; BPA = bisphenol A; PS = persulfate; TDWW = textile dyeing wastewater; TC = tetracycline; (TDWW); SBC = sawdust biochar; AO-7 = acid orange 7; RhB = rhodamine B; DOM = dissolved organic matter in BC; CIP = ciprofloxacin; SMX = sulfamethoxazole; PDS = peroxydisulfate; CA = clofibric acid; WW = wastewater; ⇆ = reversible; PNP = p-nitrophenol (water pollutant); SMT = sulfamethazine; SDZ = sulfadiazine; OCPFRs = oxygen-centered environmental persistent free radicals; CCPFRs-O = carbon-centered environmental persistent free radicals with oxygen atoms; * photocatalytic; CCPFRs = carbon-centered environmental persistent free radicals; BiPB = bismuth-containing BC; PFX = pefloxacin; OTC = oxytetracycline; CTC = chlorotetracycline; OFX = ofloxacin; AZT = atrazine; TMP = trimethoprim; AOPs = advanced oxidation processes; ACT = acetaminophen; PFOA = perfluorooctanoic acid; 3 degradation efficiency in presence of ascorbic acid (AA); 2,4-DCP = 2,4-dichlorophenol; CAP = chloramphenicol; FF = florfenicol; TAP = thiamphenicol; CV = crystal violet dye; MB = methylene blue; MNZ = metronidazole; DOX = doxorubicin; xxx = refers to the temperature of pyrolysis process; NPA = N-phosphono methyl iminodiacetic acid (organophosphorus pesticide (OP); NOR = norfloxacin; E1 = estrone; E2 = 17-estradiol; E3 = estriol; PPCPs = pharmaceuticals and personal care products; TPhP = triphenyl phosphate; ENR = enrofloxacin; FLE = fleroxacin (FLE); $ HTC = hydro-thermal carbonization; # HTP = high temperature pyrolysis; PolyS = polystyrene; OFGs = oxygen-containing functional group; S-HPBC = S-doped hydrothermal + pyrocarbon BC; S-HBC = S-doped hydrochar, S-PBC = S-doped pyrocarbon.

Figure 11 shows the relative abundances of the types of PFR applications concerning the 72 case studies considered here.

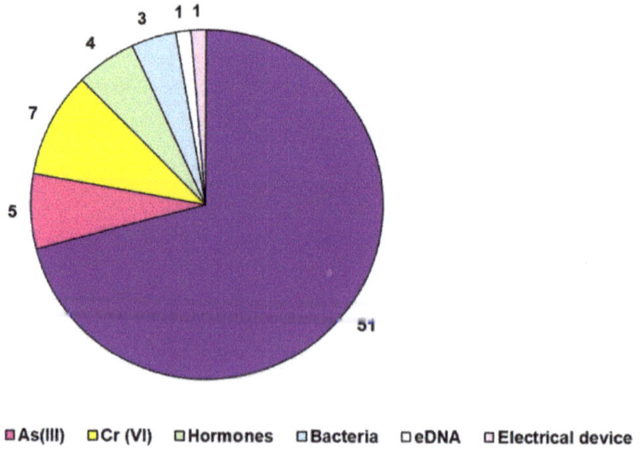

Figure 11. Number of publications for each type of EPR application found in the literature, considering a randomly selected population of 72 papers published from 2019 to 2023.

As for the mechanisms, many publications regarded the activation, which was sometimes photocatalytic, of PS, PMS, and PDS by BC. The employed BC was derived from different feedstock biomasses (bagasse powder, poplar and pine sawdust, cellulose, lignin, blue algae, waste straw, and other sources, as reported in Table 9), not doped, or doped with nitrogen atoms or different metals including Fe, Mn, Co, Ni, Zn. In these processes, the electron transfer promoted by metals and/or PFRs of a diverse nature, based on the pyrolysis conditions, generated ROS such as $SO_4\bullet^-$, $\bullet OH$, $\bullet O_2^-$, $\bullet O_2H$ and non-radical species ($^1O_2$), which carried out the oxidative degradation of different organic xenobiotics, including drugs, dyes, antibiotics, and hormones, as well as phenols or aromatic derivatives. Many other publications reported the use of BC to activate or photochemically activate O_2 or H_2O_2 (Fenton-like systems) via metal and/or PFRs-mediated single-electron transfer. The generated ROS (•OH, $•O_2^-$, $•O_2H$) and oxygen non-radical species ($^1O_2$) successfully oxidized several organic pollutants, degraded hormones, and eDNA and, in some cases, showed antibacterial effects against *E. coli* and *S. aureus*. Moreover, the capacity of BC to transfer electrons via transitional metals or PFRs was used to oxidize As (III) to As (V) or reduce Cr (VI) to Cr (III), thus resulting in helping to remove hazardous inorganic contaminants from industrial wastewater. A notable recent review reported on the efficiency of BC/layered double hydroxide composites as catalysts for the treatment of organic wastewater by advanced oxidation processes [213]. Several studies reported in this paper by Liu et al. demonstrated that degradation processes were based on radical reactions triggered by BC-associated PFRs [213].

3.2.2. BC-Associated PFRs Shadows: Cytotoxicity and Biotoxicity

Despite the plethora of possible beneficial applications of BC, PFRs, as well as other free radicals and the toxic substances that compose BC, such as heavy metals, PAHs, dioxins, and perfluorochemicals (PFCs), are released into the environment during the pyrolysis process, thus representing a potential risk to the environment and humans [214]. Additionally, as well as other contaminants, the possible carbon allotropes formed during pyrolysis are severe contaminants in air, water, and soil [215]. Black carbon, carbon black (CB), carbon nanotubes, graphene, quantum dots, and fullerenes can possess distinct toxicity that depends on many factors, including the type of allotrope, particle size, form, structural defects, coating molecules, and grade of functionalization [215]. Understanding the toxicity of carbon nanomaterials and nanoproducts that are possibly present in BC is essential for human and environmental health, safety, and public acceptance. In this regard, recent studies have focused their attention on the adverse effects of BC due to its particle size and the various interactions with the environment that could occur [216,217]. Upon its application, BC may produce harmful environmental effects due to aging by oxidative or biological processes, leading to changes in its properties [218,219]. Additionally, higher toxicity has been reported for BC with micro- or nano-dimensioned particles. It has been reported that the presence of micro-BC (MBC) or nano-BC (NBC) can promote the release of heavy metal ions into the medium when applied to soil [214]. Kim et al. (2018) observed that BC particles with a particle size of less than 0.45 μm could increase the release and mobility of As in soil [220]. Regarding the biotoxicity of MBC/NBC, it has been previously reported that particle-induced oxidative stress is a crucial mechanism of MBC/NBC cytotoxicity, which increases as the particle size decreases. Also, the PFR concentration on the surface of particles with an aerodynamic diameter of less than 1 μm is the highest [139,221]. While several reviews and studies exist on the production and modification of BC, the reaction mechanisms, and the beneficial active role of BC in environmental remediation, the adverse effects and potential risks of BC have only recently been evidenced. The comprehensive phenomena and mechanisms involved in BC toxicity still require elucidation, especially in environmental media different from soil, including water and the atmosphere. It is imperative to systematically study and discuss the possible adverse environmental effects of BC application concerning various media, including water and the atmosphere, by determining the corresponding occurrence, detection, assessment, and avoidance measures. Worryingly, the current knowledge concerning the possible adverse effects on the environment and biota deriving from the extensive application of PFRs originating in BC is even more limited [222]. Although they are emerging as contaminants of increasing concern, their formation, fate, toxicity, and health risks are poorly known [222]. Thermal treatment, a common remediation technique to clean industrial soils, induces the formation of PFRs; this could paradoxically increase soil toxicity, which is contrary to the original remediation objective. For example, there is still little knowledge on the formation and toxicity of PFRs in soils contaminated by polycyclic aromatic hydrocarbons (PAHs) [223]. BC-derived

PFRs, as well as those present in the environment and deriving from combustion and soil restoration, the burning of coal, wood, straw, cigarettes, oil, and other fuels, and from the restoration of organic contaminated soil, can enter the human body mainly through three pathways including the respiratory tract, from skin exposure, and via ingestion [149]. PFRs are not toxic to living beings and the environment, but they can stimulate the formation of other harmful substances and free radicals, including various types of ROS, when in the environment or in vivo [223]. As is well documented, ROS can interfere with the normal redox and metabolic processes, thus causing oxidative stress in biota [224]. Additionally, it has been reported that exposure to PFRs may induce cell degeneration or apoptosis and may affect the normal functions of the heart or lungs of humans [223]. So far, cytotoxicity and biotoxicity are the two categories of toxicity reported as attributable to PFRs (Table 10).

Table 10. Potential toxic hazards caused by PFR exposure.

Target	Danger	Material Source	Refs.
Cells	⇑ Lungs' T (Th1, Th2, Th17) cells	PM, DCB230, MCP230	[225,226]
	⇓ P450 activity	PM, MCP230	[227]
	Cardiomyocytes' apoptosis	DCB230	[228]
	⇓ Survival of gastric epithelial cells	BaP–Na montmorillonite	[229]
	Loss of normal morphology of pulmonary epithelial cells	DCB230	[230]
	Mitochondrial depolarization	DCB230	[226]
	Changes in VEGF	ZnO/MCB	[231]
Enzymes Proteins Genes	Altered expression activity of Cyp1a, Cyp2b, Cyp2e1, Cyp2d2, Cyp3a and other genes	DCB230, MCP230	[225]
	⇑ Expression levels of peroxiredoxin-6 Cofilin 1, annexin A8	MCP230, CGUFP, ZnO/MCB	[226]
	⇓ of GSH, GPx, SOD	ZnO/MCB	[232]
Organs and tissues	Altered normal renal hemodynamics and urodynamics	N.R.	[233]
	Liver damage	N.R.	[234]
	Impairment of left ventricular function	DCB230	[235]
	Airway hyperresponsiveness Lung inflammation	MCP230	[236]
Individuals	Abnormalities in zebrafish	DCB230	[237]
	⇓ Growth and reproduction of luminescent bacteria	PM	[238]
	Altered behavior of *Caenorhabditis elegans*	Biochar	[239]
	⇓ Energy consumption	MCP230	[240]
Disease	⇑ Severity of the flu	DCB230	[241]
	Asthma	MCP230	[226]
	Cardiovascular disease and dysfunction	DCB230	[228]
Other damage	Oxidative stress	DCB230, ZnO/MCB	[232]
	DNA damage	BaP	[46]
	Lipid peroxidation	MCP230	[242]

⇑ Improved, increased; ⇓ decreased, inhibited; PM = particulate matter; DCP230 = 1,2-dichlorobenzene at 230 °C; MCP230 = mono-chlorophenol at 230 °C; MCB = mono-chlorobenzene; CGUFP = combustion generated ultrafine particle; N.R. = not reported; GSH = glutathione; GPx = glutathione peroxidase; SOD = superoxide dismutase; VEGF = vascular endothelial growth factor.

Usually, toxicity tests are carried out in research laboratories using both environmental samples and lab-prepared PFRs, such as those generated by MCP230 (a mixture of CuO and chlorophenol at 230 °C), DCB230 (a mixture of CuO, 0.2 μm amorphous silica and 1,2-dichlorobenzene at 230 °C), CGUFP (combustion-generated ultrafine particle) or other mixtures of transitional metals and substituted aromatic compounds. For cytotoxicity experiments, cultured cells extracted from the bronchial epithelium and rats are used, while biotoxicity essays are carried out on plants, fishes, rabbits, and worms. Generally, it was observed that exposure to PFRs causes oxidative stress. More specifically, the cytotoxicity tests evidenced cell variation with decreased numbers and activity, a disparity in protein expression, and DNA damage. Biotoxicity experiments revealed abnormalities in development and behavior, disease, and organ and tissue damage. Although BC can serve as an environmentally sustainable soil amendment material due to its ability to enhance several chemical properties of soil, such as pH, electrical conductivity, CEC, and organic carbon content, thus contributing to the overall improvement of nutrient retention in the soil, BC amendments with high concentrations of PFRs negatively affect crop growth.

Additionally, it has been found that PFRs used in aquicultural solutions inhibited the germination rate of different crops by ROS induction [149]. The oxidative stress brought about by the production of ROS can also damage the plasma membrane of the root system and hinder plant root growth. Moreover, PFRs induced neurotoxicity in *Cryptobacterium hidradenoma*, transforming it into a neurotoxin for soil organisms and thereby posing a threat to their survival.

4. Future Challenges and Risk Prevention Strategies

This review has evidenced that BC and mainly the PFRs generated during the pyrolysis processes performed to produce it could be double-edged weapons. BC is reported to be an eco-friendly and low-cost black gold with many beneficial properties, including the capability to remove organic and inorganic pollutants from water by adsorption processes and/or through its PFRs. However, several studies have reported that PFRs can be very dangerous to the environment and humans through a ROS-dependent mechanism. In addition to being produced by various common xenobiotics, PFRs can easily be converted into secondary pollutants, causing further biotoxicity. The still too-little-studied transport and transformation of PFRs in the environment can also affect the behavior of other substances, leading to potential environmental hazards that are not yet fully understood. Therefore, further exploration of the ecological impact of PFRs and the development of prevention and control measures are necessary. In this regard, although some progress has been made in terms of environmental risk and biotoxicity studies concerning PFRs, research is still in the initial stages, and there is an urgent need for systematic and in-depth studies on the production and transport of PFRs. A more in-depth understanding of the influence of environmental conditions on their occurrence is needed to control the external factors for reducing the negative output of PFRs and promoting their degradative action on xenobiotics. More rational knowledge about their toxicity mechanisms is necessary to have more precise toxicological equivalence regulation. A better strategy to prevent the risks associated with PFRs could be to avoid exposure to them by reducing contact with combustion sources, such as vehicle exhaust and cigarette smoke. Proper air filtration systems removing PFRs from indoor air and wearing protective masks or respirators could lower the possibility of contact with PFRs in outdoor environments. Additionally, treatments for limiting the adverse health effects associated with PFR exposure, such as using antioxidants, which can neutralize ROS, could be another strategy to protect humans from the adverse impacts of PFR exposure.

5. Conclusions and Future Perspectives

The pivotal role of BC-related PFRs in BC catalytic efficiency in water and soil pollution remediation has been reviewed in this paper. The main mechanisms by which PFRs can originate and the main methods to detect them on BC have been discussed. The key roles of feedstock biomasses and pyrolysis conditions in the formation of PFRs has also been reported. Several recent case studies concerning the critical role of PFRs in the catalytic oxidative degradation by BC of organic pollutants and in the removal of Cr/As and other metal ions in aqueous phase have been reported and discussed. It has been evidenced that for organic pollutants remediation, PFRs act as activator agents of oxidant substrates, which are subsequently involved in the degradation process. Otherwise, for metal ion removal, the PFRs on BC act as electron transfers for the adsorption and concomitant reduction/oxidation of metals by BC. Finally, PFRs on BC could also help mediate the recycling of Fe^{3+}/Fe^{2+} in Fenton-like processes to enhance the efficiency of BC in pollutant removal. However, many aspects remain unclear concerning PFRs, including the influence of BC size on the evolution of PFRs, the exact relationship between the reactivity of BC and its size, and the overall roles and relative significance of PFRs, quinone moieties, and the carbon structure of BC in the activation of oxidizing agents and in the redox transformation of inorganic contaminants. Furthermore, while several studies about the laboratory applications of BC for wastewater remediation have been developed, future studies should concentrate on the up-scaled utilization of BC for the treatment of different real wastewaters, such as industrial wastewater and municipal wastewater. To this end, efforts are needed to design proper reactors and to develop methods for the large-scale production of the desired BC. Finally, to better contribute to the circular economy, the reutilization of spent BC should be given serious consideration.

Supplementary Materials: The following supporting information can be downloaded at: https://www.mdpi.com/article/10.3390/toxics12040245/s1, Table S1. Main sources and general application of BCs; Table S2. Fast and slow pyrolysis details; Table S3. Techniques typically used to characterize BCs in terms of their physicochemical, surface, and structural characterization; Table S4. Properties of BCs produced from various feedstocks at various production temperatures.

Author Contributions: The authors (S.A. and O.G.P.) contributed equally to this work. All authors have read and agreed to the published version of the manuscript.

Funding: This work was funded by the Fundação Carlos Chagas Filho Amparo à Pesquisa do Estado do Rio de Janeiro (FAPERJ project n.SEI-260003/001227/2020), the European Union—Next Generation EU–the Italian Ministry of University (MUR) (PRIN2022, project n. 2022JM3LZ3_(SUST-CARB)).

Institutional Review Board Statement: Not applicable.

Informed Consent Statement: Not applicable.

Data Availability Statement: Not applicable.

Conflicts of Interest: The authors declare no conflicts of interest.

References

1. Zhang, R.; Zhang, R.; Zimmerman, A.R.; Wang, H.; Gao, B. Applications, Impacts, and Management of Biochar Persistent Free Radicals: A Review. *Environ. Pollut.* **2023**, *327*, 121543. [CrossRef] [PubMed]
2. Wang, J.; Wang, S. Preparation, Modification, and Environmental Application of Biochar: A Review. *J. Clean. Prod.* **2019**, *227*, 1002–1022. [CrossRef]
3. Xiong, X.; Yu, I.K.M.; Cao, L.; Tsang, D.C.W.; Zhang, S.; Ok, Y.S. A Review of Biochar-Based Catalysts for Chemical Synthesis, Biofuel Production, and Pollution Control. *Bioresour. Technol.* **2017**, *246*, 254–270. [CrossRef] [PubMed]
4. Behera, B.; Selvam, S.M.; Dey, B.; Balasubramanian, P. Algal Biodiesel Production with Engineered Biochar as a Heterogeneous Solid Acid Catalyst. *Bioresour. Technol.* **2020**, *310*, 123392. [CrossRef] [PubMed]
5. Lee, J.; Kim, K.-H.; Kwon, E.E. Biochar as a Catalyst. *Renew. Sustain. Energy Rev.* **2017**, *77*, 70–79. [CrossRef]
6. Zhang, L.; Lim, E.Y.; Loh, K.-C.; Ok, Y.S.; Lee, J.T.E.; Shen, Y.; Wang, C.-H.; Dai, Y.; Tong, Y.W. Biochar Enhanced Thermophilic Anaerobic Digestion of Food Waste: Focusing on Biochar Particle Size, Microbial Community Analysis and Pilot-Scale Application. *Energy Convers. Manag.* **2020**, *209*, 112654. [CrossRef]
7. Supraja, K.V.; Kachroo, H.; Viswanathan, G.; Verma, V.K.; Behera, B.; Doddapaneni, T.R.K.C.; Kaushal, P.; Ahammad, S.Z.; Singh, V.; Awasthi, M.K.; et al. Biochar Production and Its Environmental Applications: Recent Developments and Machine Learning Insights. *Bioresour. Technol.* **2023**, *387*, 129634. [CrossRef] [PubMed]

8. Awasthi, M.K.; Wang, M.; Chen, H.; Wang, Q.; Zhao, J.; Ren, X.; Li, D.; Awasthi, S.K.; Shen, F.; Li, R.; et al. Heterogeneity of Biochar Amendment to Improve the Carbon and Nitrogen Sequestration through Reduce the Greenhouse Gases Emissions during Sewage Sludge Composting. *Bioresour. Technol.* **2017**, *224*, 428–438. [CrossRef] [PubMed]
9. Gupta, S.; Sireesha, S.; Sreedhar, I.; Patel, C.M.; Anitha, K.L. Latest Trends in Heavy Metal Removal from Wastewater by Biochar Based Sorbents. *J. Water Process Eng.* **2020**, *38*, 101561. [CrossRef]
10. Zhao, F.; Tang, L.; Jiang, H.; Mao, Y.; Song, W.; Chen, H. Prediction of Heavy Metals Adsorption by Hydrochars and Identification of Critical Factors Using Machine Learning Algorithms. *Bioresour. Technol.* **2023**, *383*, 129223. [CrossRef] [PubMed]
11. Min, L.; Zhongsheng, Z.; Zhe, L.; Haitao, W. Removal of Nitrogen and Phosphorus Pollutants from Water by FeCl3- Impregnated Biochar. *Ecol. Eng.* **2020**, *149*, 105792. [CrossRef]
12. Premarathna, K.S.D.; Rajapaksha, A.U.; Sarkar, B.; Kwon, E.E.; Bhatnagar, A.; Ok, Y.S.; Vithanage, M. Biochar-Based Engineered Composites for Sorptive Decontamination of Water: A Review. *Chem. Eng. J.* **2019**, *372*, 536–550. [CrossRef]
13. Chauhan, S.; Shafi, T.; Dubey, B.K.; Chowdhury, S. Biochar-Mediated Removal of Pharmaceutical Compounds from Aqueous Matrices via Adsorption. *Waste Dispos. Sustain. Energy* **2023**, *5*, 37–62. [CrossRef] [PubMed]
14. Manikandan, S.K.; Pallavi, P.; Shetty, K.; Bhattacharjee, D.; Giannakoudakis, D.A.; Katsoyiannis, I.A.; Nair, V. Effective Usage of Biochar and Microorganisms for the Removal of Heavy Metal Ions and Pesticides. *Molecules* **2023**, *28*, 719. [CrossRef] [PubMed]
15. Valizadeh, S.; Lee, S.S.; Choi, Y.J.; Baek, K.; Jeon, B.-H.; Andrew Lin, K.-Y.; Park, Y.-K. Biochar Application Strategies for Polycyclic Aromatic Hydrocarbons Removal from Soils. *Environ. Res.* **2022**, *213*, 113599. [CrossRef] [PubMed]
16. Saeed, M.; Ilyas, N.; Jayachandran, K.; Gaffar, S.; Arshad, M.; Sheeraz Ahmad, M.; Bibi, F.; Jeddi, K.; Hessini, K. Biostimulation Potential of Biochar for Remediating the Crude Oil Contaminated Soil and Plant Growth. *Saudi J. Biol. Sci.* **2021**, *28*, 2667–2676. [CrossRef] [PubMed]
17. He, P.; Yu, Q.; Zhang, H.; Shao, L.; Lü, F. Removal of Copper (II) by Biochar Mediated by Dissolved Organic Matter. *Sci. Rep.* **2017**, *7*, 7091. [CrossRef] [PubMed]
18. Beesley, L.; Moreno-Jiménez, E.; Gomez-Eyles, J.L.; Harris, E.; Robinson, B.; Sizmur, T. A Review of Biochars' Potential Role in the Remediation, Revegetation and Restoration of Contaminated Soils. *Environ. Pollut.* **2011**, *159*, 3269–3282. [CrossRef] [PubMed]
19. Jeffery, S.; Verheijen, F.G.A.; van der Velde, M.; Bastos, A.C. A Quantitative Review of the Effects of Biochar Application to Soils on Crop Productivity Using Meta-Analysis. *Agric. Ecosyst. Environ.* **2011**, *144*, 175–187. [CrossRef]
20. Kinney, T.J.; Masiello, C.A.; Dugan, B.; Hockaday, W.C.; Dean, M.R.; Zygourakis, K.; Barnes, R.T. Hydrologic Properties of Biochars Produced at Different Temperatures. *Biomass Bioenergy* **2012**, *41*, 34–43. [CrossRef]
21. Qin, Y.; Li, G.; Gao, Y.; Zhang, L.; Ok, Y.S.; An, T. Persistent Free Radicals in Carbon-Based Materials on Transformation of Refractory Organic Contaminants (ROCs) in Water: A Critical Review. *Water Res.* **2018**, *137*, 130–143. [CrossRef] [PubMed]
22. Vejerano, E.P.; Rao, G.; Khachatryan, L.; Cormier, S.A.; Lomnicki, S. Environmentally Persistent Free Radicals: Insights on a New Class of Pollutants. *Environ. Sci. Technol.* **2018**, *52*, 2468–2481. [CrossRef] [PubMed]
23. Volpe, R.; Bermudez Menendez, J.M.; Ramirez Reina, T.; Volpe, M.; Messineo, A.; Millan, M.; Titirici, M.-M. Free Radicals Formation on Thermally Decomposed Biomass. *Fuel* **2019**, *255*, 115802. [CrossRef]
24. Huang, Y.; Guo, X.; Ding, Z.; Chen, Y.; Hu, X. Environmentally Persistent Free Radicals in Biochar Derived from Laminaria Japonica Grown in Different Habitats. *J. Anal. Appl. Pyrolysis* **2020**, *151*, 104941. [CrossRef]
25. Ruan, X.; Sun, Y.; Du, W.; Tang, Y.; Liu, Q.; Zhang, Z.; Doherty, W.; Frost, R.L.; Qian, G.; Tsang, D.C.W. Formation, Characteristics, and Applications of Environmentally Persistent Free Radicals in Biochars: A Review. *Bioresour. Technol.* **2019**, *281*, 457–468. [CrossRef] [PubMed]
26. Wang, R.-Z.; Huang, D.-L.; Liu, Y.-G.; Zhang, C.; Lai, C.; Wang, X.; Zeng, G.-M.; Gong, X.-M.; Duan, A.; Zhang, Q.; et al. Recent Advances in Biochar-Based Catalysts: Properties, Applications and Mechanisms for Pollution Remediation. *Chem. Eng. J.* **2019**, *371*, 380–403. [CrossRef]
27. Odinga, E.S.; Waigi, M.G.; Gudda, F.O.; Wang, J.; Yang, B.; Hu, X.; Li, S.; Gao, Y. Occurrence, Formation, Environmental Fate and Risks of Environmentally Persistent Free Radicals in Biochars. *Environ. Int.* **2020**, *134*, 105172. [CrossRef] [PubMed]
28. Fang, G.; Liu, C.; Gao, J.; Dionysiou, D.D.; Zhou, D. Manipulation of Persistent Free Radicals in Biochar to Activate Persulfate for Contaminant Degradation. *Environ. Sci. Technol.* **2015**, *49*, 5645–5653. [CrossRef] [PubMed]
29. Li, F.; Duan, F.; Ji, W.; Gui, X. Biochar-activated persulfate for organic contaminants removal: Efficiency, mechanisms and influencing factors. *Ecotoxicol. Environ. Saf.* **2020**, *198*, 110653. [CrossRef]
30. Liu, F.; Zhou, H.; Pan, Z.; Liu, Y.; Yao, G.; Guo, Y.; Lai, B. Degradation of sulfamethoxazole by cobalt-nickel powder composite catalyst coupled with peroxymonosulfate: Performance, degradation pathways and mechanistic consideration. *J. Hazard. Mater.* **2020**, *400*, 123322. [CrossRef]
31. Liang, J.; Xu, X.; Qamar Zaman, W.; Hu, X.; Zhao, L.; Qiu, H.; Cao, X. Different Mechanisms between Biochar and Activated Carbon for the Persulfate Catalytic Degradation of Sulfamethoxazole: Roles of Radicals in Solution or Solid Phase. *Chem. Eng. J.* **2019**, *375*, 121908. [CrossRef]
32. Fang, G.; Gao, J.; Liu, C.; Dionysiou, D.D.; Wang, Y.; Zhou, D. Key Role of Persistent Free Radicals in Hydrogen Peroxide Activation by Biochar: Implications to Organic Contaminant Degradation. *Environ. Sci. Technol.* **2014**, *48*, 1902–1910. [CrossRef] [PubMed]
33. Fang, G.; Zhu, C.; Dionysiou, D.D.; Gao, J.; Zhou, D. Mechanism of Hydroxyl Radical Generation from Biochar Suspensions: Implications to Diethyl Phthalate Degradation. *Bioresour. Technol.* **2015**, *176*, 210–217. [CrossRef] [PubMed]

34. Fang, G.; Liu, C.; Wang, Y.; Dionysiou, D.D.; Zhou, D. Photogeneration of Reactive Oxygen Species from Biochar Suspension for Diethyl Phthalate Degradation. *Appl. Catal. B Environ.* **2017**, *214*, 34–45. [CrossRef]
35. Luo, K.; Yang, Q.; Pang, Y.; Wang, D.; Li, X.; Lei, M.; Huang, Q. Unveiling the Mechanism of Biochar-Activated Hydrogen Peroxide on the Degradation of Ciprofloxacin. *Chem. Eng. J.* **2019**, *374*, 520–530. [CrossRef]
36. Zhang, Y.; Xu, M.; He, R.; Zhao, J.; Kang, W.; Lv, J. Effect of Pyrolysis Temperature on the Activated Permonosulfate Degradation of Antibiotics in Nitrogen and Sulfur-Doping Biochar: Key Role of Environmentally Persistent Free Radicals. *Chemosphere* **2022**, *294*, 133737. [CrossRef] [PubMed]
37. Liao, S.; Pan, B.; Li, H.; Zhang, D.; Xing, B. Detecting Free Radicals in Biochars and Determining Their Ability to Inhibit the Germination and Growth of Corn, Wheat and Rice Seedlings. *Environ. Sci. Technol.* **2014**, *48*, 8581–8587. [CrossRef] [PubMed]
38. Ahmad, M.; Rajapaksha, A.U.; Lim, J.E.; Zhang, M.; Bolan, N.; Mohan, D.; Vithanage, M.; Lee, S.S.; Ok, Y.S. Biochar as a Sorbent for Contaminant Management in Soil and Water: A Review. *Chemosphere* **2014**, *99*, 19–33. [CrossRef] [PubMed]
39. Qin, J.; Chen, Q.; Sun, M.; Sun, P.; Shen, G. Pyrolysis Temperature-Induced Changes in the Catalytic Characteristics of Rice Husk-Derived Biochar during 1,3-Dichloropropene Degradation. *Chem. Eng. J.* **2017**, *330*, 804–812. [CrossRef]
40. Yang, J.; Pignatello, J.J.; Pan, B.; Xing, B. Degradation of P-Nitrophenol by Lignin and Cellulose Chars: H_2O_2-Mediated Reaction and Direct Reaction with the Char. *Environ. Sci. Technol.* **2017**, *51*, 8972–8980. [CrossRef]
41. Zhang, R.; Zimmerman, A.R.; Zhang, R.; Li, P.; Zheng, Y.; Gao, B. Persistent free radicals generated from a range of biochars and their physiological effects on wheat seedlings. *Sci. Total Environ.* **2024**, *908*, 168260. [CrossRef] [PubMed]
42. Alfei, S.; Pandoli, O.G. Bamboo-Based Biochar: A Still Too Little-Studied Black Gold and Its Current Applications. *J. Xenobiot.* **2024**, *14*, 416–451. [CrossRef] [PubMed]
43. Ginoble Pandoli, O.; Santos de Sá, D.; Nogueira Barbosa Junior, M.; Paciornik, S. Bamboo-Based Lignocellulose Biomass as Catalytic Support for Organic Synthesis and Water Treatments. In *Bamboo Science and Technology. Environmental Footprints and Eco-Design of Products and Processes*; Palombini, F.L., Nogueira, F.M., Eds.; Springer: Singapore, 2023; pp. 297–327. [CrossRef]
44. Liu, X.; Chen, Z.; Lu, S.; Shi, X.; Qu, F.; Cheng, D.; Wei, W.; Shon, H.K.; Ni, B.J. Persistent free radicals on biochar for its catalytic capability: A review. *Water Res.* **2024**, *250*, 120999. [CrossRef] [PubMed]
45. Luo, K.; Pang, Y.; Wang, D.; Li, X.; Wang, L.; Lei, M.; Huang, Q.; Yang, Q. A critical review on the application of biochar in environmental pollution remediation: Role of persistent free radicals (PFRs). *J. Environ. Sci.* **2021**, *108*, 201–216. [CrossRef] [PubMed]
46. Yi, J.F.; Lin, Z.Z.; Li, X.; Zhou, Y.Q.; Guo, Y. A short review on environmental distribution and toxicity of the environmentally persistent free radicals. *Chemosphere* **2023**, *340*, 139922. [CrossRef] [PubMed]
47. Jahirul, M.I.; Rasul, M.G.; Chowdhury, A.A.; Ashwath, N. Biofuels Production through Biomass Pyrolysis—A Technological Review. *Energies* **2012**, *5*, 4952–5001. [CrossRef]
48. Yaashikaa, P.R.; Senthil Kumar, P.; Varjani, S.J.; Saravanan, A. A Review on Photochemical, Biochemical and Electrochemical Transformation of CO_2 into Value-Added Products. *J. CO2 Util.* **2019**, *33*, 131–147. [CrossRef]
49. Saidur, R.; Abdelaziz, E.A.; Demirbas, A.; Hossain, M.S.; Mekhilef, S. A Review on Biomass as a Fuel for Boilers. *Renew. Sustain. Energy Rev.* **2011**, *15*, 2262–2289. [CrossRef]
50. Yaashikaa, P.R.; Kumar, P.S.; Varjani, S.; Saravanan, A. A Critical Review on the Biochar Production Techniques, Characterization, Stability and Applications for Circular Bioeconomy. *Biotechnol. Rep.* **2020**, *28*, e00570. [CrossRef] [PubMed]
51. Lehmann, J. Biochar for environmental management: An introduction. *Biochar for Environmental Management. Sci. Technol.* **2009**, *25*, 15801–15811.
52. Mohanty, S.K.; Valenca, R.; Berger, A.W.; Yu, I.K.M.; Xiong, X.; Saunders, T.M.; Tsang, D.C.W. Plenty of Room for Carbon on the Ground: Potential Applications of Biochar for Stormwater Treatment. *Sci. Total Environ.* **2018**, *625*, 1644–1658. [CrossRef] [PubMed]
53. Ayla Norris Smidth. A New Boost for Biochar as a Natural Climate Solution. 2023. Available online: https://blog.nature.org/science-brief/a-new-boost-for-biochar-as-a-natural-climate-solution/#:~:text=Biochar%20is%20a%20type%20of%20charcoal%20made%20from,other%20biomass%20would,%20leading%20to%20more%20carbon%20sequestration (accessed on 27 December 2023).
54. Yu, D.; Yu, Y.; Tang, J.; Li, X.; Ke, C.; Yao, Z. Application Fields of Kitchen Waste Biochar and Its Prospects as Catalytic Material: A Review. *Sci. Total Environ.* **2022**, *810*, 152171. [CrossRef] [PubMed]
55. Yu, S.; Zhang, W.; Dong, X.; Wang, F.; Yang, W.; Liu, C.; Chen, D. A Review on Recent Advances of Biochar from Agricultural and Forestry Wastes: Preparation, Modification and Applications in Wastewater Treatment. *J. Environ. Chem. Eng.* **2024**, *12*, 111638. [CrossRef]
56. Wijitkosum, S. Biochar Derived from Agricultural Wastes and Wood Residues for Sustainable Agricultural and Environmental Applications. *Int. Soil Water Conserv. Res.* **2022**, *10*, 335–341. [CrossRef]
57. Yao, Y.; Gao, B.; Inyang, M.; Zimmerman, A.R.; Cao, X.; Pullammanappallil, P.; Yang, L. Biochar Derived from Anaerobically Digested Sugar Beet Tailings: Characterization and Phosphate Removal Potential. *Bioresour. Technol.* **2011**, *102*, 6273–6278. [CrossRef] [PubMed]
58. Puettmann, M.; Sahoo, K.; Wilson, K.; Oneil, E. Life Cycle Assessment of Biochar Produced from Forest Residues Using Portable Systems. *J. Clean. Prod.* **2020**, *250*, 119564. [CrossRef]

59. Papageorgiou, A.; Azzi, E.S.; Enell, A.; Sundberg, C. Biochar Produced from Wood Waste for Soil Remediation in Sweden: Carbon Sequestration and Other Environmental Impacts. *Sci. Total Environ.* **2021**, *776*, 145953. [CrossRef] [PubMed]
60. Laird, D. Biochar Amendments Make the Harvesting of Crop Residue for Bioenergy Production Sustainable. *Nutr. Cycl. Agroecosystems* **2023**. [CrossRef]
61. Li, N.; He, M.; Lu, X.; Yan, B.; Duan, X.; Chen, G.; Wang, S.; Hou, L. Municipal Solid Waste Derived Biochars for Wastewater Treatment: Production, Properties and Applications. *Resour. Conserv. Recycl.* **2022**, *177*, 106003. [CrossRef]
62. Shi, W.; Wang, H.; Yan, J.; Shan, L.; Quan, G.; Pan, X.; Cui, L. Wheat Straw Derived Biochar with Hierarchically Porous Structure for Bisphenol A Removal: Preparation, Characterization, and Adsorption Properties. *Sep. Purif. Technol.* **2022**, *289*, 120796. [CrossRef]
63. Foong, S.Y.; Chan, Y.H.; Chin, B.L.F.; Lock, S.S.M.; Yee, C.Y.; Yiin, C.L.; Peng, W.; Lam, S.S. Production of Biochar from Rice Straw and Its Application for Wastewater Remediation—An Overview. *Bioresour. Technol.* **2022**, *360*, 127588. [CrossRef] [PubMed]
64. Gunamantha, M.; Widana, G.A.B. Characterization the Potential of Biochar from Cow and Pig Manure for Geoecology Applicationion. *IOP Conf. Ser. Earth Environ. Sci.* **2018**, *131*, 012055. [CrossRef]
65. Rathnayake, D.; Schmidt, H.-P.; Leifeld, J.; Mayer, J.; Epper, C.A.; Bucheli, T.D.; Hagemann, N. Biochar from Animal Manure: A Critical Assessment on Technical Feasibility, Economic Viability, and Ecological Impact. *GCB Bioenergy* **2023**, *15*, 1078–1104. [CrossRef]
66. Drané, M.; Zbair, M.; Hajjar-Garreau, S.; Josien, L.; Michelin, L.; Bennici, S.; Limousy, L. Unveiling the Potential of Corn Cob Biochar: Analysis of Microstructure and Composition with Emphasis on Interaction with NO_2. *Materials* **2024**, *17*, 159. [CrossRef] [PubMed]
67. Luo, L.; Wang, J.; Lv, J.; Liu, Z.; Sun, T.; Yang, Y.; Zhu, Y.-G. Carbon Sequestration Strategies in Soil Using Biochar: Advances, Challenges, and Opportunities. *Environ. Sci. Technol.* **2023**, *57*, 11357–11372. [CrossRef] [PubMed]
68. Brassard, P.; Godbout, S.; Lévesque, V.; Palacios, J.H.; Raghavan, V.; Ahmed, A.; Hogue, R.; Jeanne, T.; Verma, M. 4—Biochar for Soil Amendment. In *Char and Carbon Materials Derived from Biomass*; Jeguirim, M., Limousy, L., Eds.; Elsevier: Amsterdam, The Netherlands, 2019; pp. 109–146. ISBN 978-0-12-814893-8.
69. Godlewska, P.; Schmidt, H.P.; Ok, Y.S.; Oleszczuk, P. Biochar for Composting Improvement and Contaminants Reduction. A Review. *Bioresour. Technol.* **2017**, *246*, 193–202. [CrossRef] [PubMed]
70. Xiang, W.; Zhang, X.; Chen, J.; Zou, W.; He, F.; Hu, X.; Tsang, D.C.W.; Ok, Y.S.; Gao, B. Biochar Technology in Wastewater Treatment: A Critical Review. *Chemosphere* **2020**, *252*, 126539. [CrossRef] [PubMed]
71. Rawat, S.; Wang, C.-T.; Lay, C.-H.; Hotha, S.; Bhaskar, T. Sustainable Biochar for Advanced Electrochemical/Energy Storage Applications. *J. Energy Storage* **2023**, *63*, 107115. [CrossRef]
72. Qiu, B.; Shao, Q.; Shi, J.; Yang, C.; Chu, H. Application of Biochar for the Adsorption of Organic Pollutants from Wastewater: Modification Strategies, Mechanisms and Challenges. *Sep. Purif. Technol.* **2022**, *300*, 121925. [CrossRef]
73. Kalu, S.; Kulmala, L.; Zrim, J.; Peltokangas, K.; Tammeorg, P.; Rasa, K.; Kitzler, B.; Pihlatie, M.; Karhu, K. Potential of Biochar to Reduce Greenhouse Gas Emissions and Increase Nitrogen Use Efficiency in Boreal Arable Soils in the Long-Term. *Front. Environ. Sci.* **2022**, *10*, 914766. [CrossRef]
74. Jiang, T.; Wang, B.; Gao, B.; Cheng, N.; Feng, Q.; Chen, M.; Wang, S. Degradation of Organic Pollutants from Water by Biochar-Assisted Advanced Oxidation Processes: Mechanisms and Applications. *J. Hazard. Mater.* **2023**, *442*, 130075. [CrossRef] [PubMed]
75. Lyu, H.; Zhang, Q.; Shen, B. Application of Biochar and Its Composites in Catalysis. *Chemosphere* **2020**, *240*, 124842. [CrossRef] [PubMed]
76. Gayathri, R.; Gopinath, K.P.; Kumar, P.S. Adsorptive Separation of Toxic Metals from Aquatic Environment Using Agro Waste Biochar: Application in Electroplating Industrial Wastewater. *Chemosphere* **2021**, *262*, 128031. [CrossRef] [PubMed]
77. Hemavathy, R.V.; Kumar, P.S.; Kanmani, K.; Jahnavi, N. Adsorptive Separation of Cu(II) Ions from Aqueous Medium Using Thermally/Chemically Treated Cassia Fistula Based Biochar. *J. Clean. Prod.* **2020**, *249*, 119390. [CrossRef]
78. El-Naggar, A.; El-Naggar, A.H.; Shaheen, S.M.; Sarkar, B.; Chang, S.X.; Tsang, D.C.W.; Rinklebe, J.; Ok, Y.S. Biochar Composition-Dependent Impacts on Soil Nutrient Release, Carbon Mineralization, and Potential Environmental Risk: A Review. *J. Environ. Manag.* **2019**, *241*, 458–467. [CrossRef] [PubMed]
79. El-Naggar, A.; Lee, S.S.; Rinklebe, J.; Farooq, M.; Song, H.; Sarmah, A.K.; Zimmerman, A.R.; Ahmad, M.; Shaheen, S.M.; Ok, Y.S. Biochar Application to Low Fertility Soils: A Review of Current Status, and Future Prospects. *Geoderma* **2019**, *337*, 536–554. [CrossRef]
80. Hu, X.; Xu, J.; Wu, M.; Xing, J.; Bi, W.; Wang, K.; Ma, J.; Liu, X. Effects of Biomass Pre-Pyrolysis and Pyrolysis Temperature on Magnetic Biochar Properties. *J. Anal. Appl. Pyrolysis* **2017**, *127*, 196–202. [CrossRef]
81. Senthil Kumar, P.; Abhinaya, R.V.; Gayathri Lashmi, K.; Arthi, V.; Pavithra, R.; Sathyaselvabala, V.; Dinesh Kirupha, S.; Sivanesan, S. Adsorption of Methylene Blue Dye from Aqueous Solution by Agricultural Waste: Equilibrium, Thermodynamics, Kinetics, Mechanism and Process Design. *Colloid J.* **2011**, *73*, 651–661. [CrossRef]
82. Kumar, P.S.; Senthamarai, C.; Deepthi, A.S.L.S.; Bharani, R. Adsorption Isotherms, Kinetics and Mechanism of Pb(II) Ions Removal from Aqueous Solution Using Chemically Modified Agricultural Waste. *Can. J. Chem. Eng.* **2013**, *91*, 1950–1956. [CrossRef]
83. Varjani, S.; Kumar, G.; Rene, E.R. Developments in Biochar Application for Pesticide Remediation: Current Knowledge and Future Research Directions. *J. Environ. Manag.* **2019**, *232*, 505–513. [CrossRef]

84. Bridgwater, A.V. Review of Fast Pyrolysis of Biomass and Product Upgrading. *Biomass Bioenergy* **2012**, *38*, 68–94. [CrossRef]
85. Ng, W.C.; You, S.; Ling, R.; Gin, K.Y.-H.; Dai, Y.; Wang, C.-H. Co-Gasification of Woody Biomass and Chicken Manure: Syngas Production, Biochar Reutilization, and Cost-Benefit Analysis. *Energy* **2017**, *139*, 732–742. [CrossRef]
86. Cantrell, K.B.; Hunt, P.G.; Uchimiya, M.; Novak, J.M.; Ro, K.S. Impact of Pyrolysis Temperature and Manure Source on Physicochemical Characteristics of Biochar. *Bioresour. Technol.* **2012**, *107*, 419–428. [CrossRef] [PubMed]
87. Funke, A.; Ziegler, F. Hydrothermal Carbonization of Biomass: A Summary and Discussion of Chemical Mechanisms for Process Engineering. *Biofuels Bioprod. Biorefining* **2010**, *4*, 160–177. [CrossRef]
88. Klinghoffer, N.B.; Castaldi, M.J.; Nzihou, A. Influence of Char Composition and Inorganics on Catalytic Activity of Char from Biomass Gasification. *Fuel* **2015**, *157*, 37–47. [CrossRef]
89. Bergman, P.C.; Boersma, A.R.; Zwart, R.W.R.; Kiel, J.H. *Torrefaction for Biomass Co-Firing in Existing Coal-Fired Power Stations*; Report No. ECNC05013; Energy Research Centre of The Netherlands (ECN): Petten, The Netherlands, 2005; p. 71.
90. Nunoura, T.; Wade, S.R.; Bourke, J.P.; Antal, M.J. Studies of the Flash Carbonization Process. 1. Propagation of the Flaming Pyrolysis Reaction and Performance of a Catalytic Afterburner. *Ind. Eng. Chem. Res.* **2006**, *45*, 585–599. [CrossRef]
91. Basu, P. Chapter 5—Pyrolysis. In *Biomass Gasification, Pyrolysis and Torrefaction*, 3rd ed.; Basu, P., Ed.; Academic Press: Cambridge, MA, USA, 2018; pp. 155–187. ISBN 978-0-12-812992-0.
92. Devi, M.; Rawat, S.; Sharma, S. A Comprehensive Review of the Pyrolysis Process: From Carbon Nanomaterial Synthesis to Waste Treatment. *Oxf. Open Mater. Sci.* **2021**, *1*, itab014. [CrossRef]
93. Cha, J.S.; Park, S.H.; Jung, S.-C.; Ryu, C.; Jeon, J.-K.; Shin, M.-C.; Park, Y.-K. Production and Utilization of Biochar: A Review. *J. Ind. Eng. Chem.* **2016**, *40*, 1–15. [CrossRef]
94. Yaashikaa, P.R.; Senthil Kumar, P.; Varjani, S.J.; Saravanan, A. Advances in Production and Application of Biochar from Lignocellulosic Feedstocks for Remediation of Environmental Pollutants. *Bioresour. Technol.* **2019**, *292*, 122030. [CrossRef]
95. Neogi, S.; Sharma, V.; Khan, N.; Chaurasia, D.; Ahmad, A.; Chauhan, S.; Singh, A.; You, S.; Pandey, A.; Bhargava, P.C. Sustainable biochar: A facile strategy for soil and environmental restoration, energy generation, mitigation of global climate change and circular bioeconomy. *Chemosphere* **2022**, *293*, 133474. [CrossRef]
96. Swagathnath, G.; Rangabhashiyam, S.; Parthsarathi, K.; Murugan, S.; Balasubramanian, P. Modeling Biochar Yield and Syngas Production During the Pyrolysis of Agro-Residues. In Proceedings of the Green Buildings and Sustainable Engineering; Drück, H., Pillai, R.G., Tharian, M.G., Majeed, A.Z., Eds.; Springer: Singapore, 2019; pp. 325–336.
97. Tan, H.; Lee, C.T.; Ong, P.Y.; Wong, K.Y.; Bong CP, C.; Li, C.; Gao, Y. A Review on the Comparison Between Slow Pyrolysis and Fast Pyrolysis on the Quality of Lignocellulosic and Lignin-Based Biochar. *IOP Conf. Ser. Mater. Sci. Eng.* **2021**, *1051*, 012075. [CrossRef]
98. Ibitoye, S.E.; Mahamood, R.M.; Jen, T.-C.; Loha, C.; Akinlabi, E.T. An Overview of Biomass Solid Fuels: Biomass Sources, Processing Methods, and Morphological and Microstructural Properties. *J. Bioresour. Bioprod.* **2023**, *8*, 333–360. [CrossRef]
99. Kumar, A.; Bhattacharya, T. Biochar and its application. In Proceedings of the Conference: Biogeochemical Cycles and Climate Change, Dhanbad, India, 10–11 August 2018; pp. 1–17.
100. Tomczyk, A.; Sokołowska, Z.; Boguta, P. Biochar Physicochemical Properties: Pyrolysis Temperature and Feedstock Kind Effects. *Rev. Environ. Sci. Bio/Technol.* **2020**, *19*, 191–215. [CrossRef]
101. Ding, W.; Dong, X.; Ime, I.M.; Gao, B.; Ma, L.Q. Pyrolytic Temperatures Impact Lead Sorption Mechanisms by Bagasse Biochars. *Chemosphere* **2014**, *105*, 68–74. [CrossRef] [PubMed]
102. Banik, C.; Lawrinenko, M.; Bakshi, S.; Laird, D.A. Impact of Pyrolysis Temperature and Feedstock on Surface Charge and Functional Group Chemistry of Biochars. *J. Environ. Qual.* **2018**, *47*, 452–461. [CrossRef] [PubMed]
103. Zhao, S.-X.; Ta, N.; Wang, X.-D. Effect of Temperature on the Structural and Physicochemical Properties of Biochar with Apple Tree Branches as Feedstock Material. *Energies* **2017**, *10*, 1293. [CrossRef]
104. Suliman, W.; Harsh, J.B.; Abu-Lail, N.I.; Fortuna, A.-M.; Dallmeyer, I.; Garcia-Pérez, M. The Role of Biochar Porosity and Surface Functionality in Augmenting Hydrologic Properties of a Sandy Soil. *Sci. Total Environ.* **2017**, *574*, 139–147. [CrossRef] [PubMed]
105. Gontijo, L.O.; Junior, M.N.B.; de Sá, D.S.; Letichevsky, S.; Pedrozo-Peñafiel, M.J.; Aucélio, R.Q.; Bott, I.S.; Alves, H.D.L.; Fragneaud, B.; Maciel, I.O.; et al. 3D conductive monolithic carbons from pyrolyzed bamboo for microfluidic self-heating system. *Carbon* **2023**, *213*, 118214. [CrossRef]
106. Sudarsan, J.S.; Prasanna, K.; Shiam Babu, R.; Sai Krishna, V.M.V. Chapter 11—Biochar: A Sustainable Solution for Organic Waste Management a Way Forward towards Circular Economy. In *Recent Trends in Solid Waste Management*; Ravindran, B., Gupta, S.K., Bhat, S.A., Chauhan, P.S., Tyagi, N., Eds.; Elsevier: Amsterdam, The Netherlands, 2023; pp. 215–230. ISBN 978-0-443-15206-1.
107. Zhou, B.; Liu, Q.; Shi, L.; Liu, Z. Electron Spin Resonance Studies of Coals and Coal Conversion Processes: A Review. *Fuel Process. Technol.* **2019**, *188*, 212–227. [CrossRef]
108. Zhang, S.; Cui, J.; Wu, H.; Zheng, Q.; Song, D.; Wang, X.; Zhang, S. Organic Carbon, Total Nitrogen, and Microbial Community Distributions within Aggregates of Calcareous Soil Treated with Biochar. *Agric. Ecosyst. Environ.* **2021**, *314*, 107408. [CrossRef]
109. Ma, L.; Song, D.; Liu, M.; Li, Y.; Li, Y. Effects of Earthworm Activities on Soil Nutrients and Microbial Diversity under Different Tillage Measures. *Soil Tillage Res.* **2022**, *222*, 105441. [CrossRef]
110. Hu, W.; Gao, W.; Tang, Y.; Zhang, Q.; Tu, C.; Cheng, J. Remediation via Biochar and Potential Health Risk of Heavy Metal Contaminated Soils. *Environ. Earth Sci.* **2022**, *81*, 482. [CrossRef]

111. Kant Bhatia, S.; Palai, A.K.; Kumar, A.; Kant Bhatia, R.; Kumar Patel, A.; Kumar Thakur, V.; Yang, Y.-H. Trends in Renewable Energy Production Employing Biomass-Based Biochar. *Bioresour. Technol.* **2021**, *340*, 125644. [CrossRef] [PubMed]
112. Sakhiya, A.K.; Anand, A.; Kaushal, P. Production, Activation, and Applications of Biochar in Recent Times. *Biochar* **2020**, *2*, 253–285. [CrossRef]
113. Nguyen, H.N.; Pignatello, J.J. Laboratory Tests of Biochars as Absorbents for Use in Recovery or Containment of Marine Crude Oil Spills. *Environ. Eng. Sci.* **2013**, *30*, 374–380. [CrossRef]
114. Jothirani, R.; Kumar, P.S.; Saravanan, A.; Narayan, A.S.; Dutta, A. Ultrasonic Modified Corn Pith for the Sequestration of Dye from Aqueous Solution. *J. Ind. Eng. Chem.* **2016**, *39*, 162–175. [CrossRef]
115. Suganya, S.; Senthil Kumar, P.; Saravanan, A.; Sundar Rajan, P.; Ravikumar, C. Computation of Adsorption Parameters for the Removal of Dye from Wastewater by Microwave Assisted Sawdust: Theoretical and Experimental Analysis. *Environ. Toxicol. Pharmacol.* **2017**, *50*, 45–57. [CrossRef]
116. Saravanan, A.; Kumar, P.S.; Renita, A.A. Hybrid Synthesis of Novel Material through Acid Modification Followed Ultrasonication to Improve Adsorption Capacity for Zinc Removal. *J. Clean. Prod.* **2018**, *172*, 92–105. [CrossRef]
117. Luo, Z.; Yao, B.; Yang, X.; Wang, L.; Xu, Z.; Yan, X.; Tian, L.; Zhou, H.; Zhou, Y. Novel Insights into the Adsorption of Organic Contaminants by Biochar: A Review. *Chemosphere* **2022**, *287*, 132113. [CrossRef] [PubMed]
118. Lou, L.; Wu, B.; Wang, L.; Luo, L.; Xu, X.; Hou, J.; Xun, B.; Hu, B.; Chen, Y. Sorption and Ecotoxicity of Pentachlorophenol Polluted Sediment Amended with Rice-Straw Derived Biochar. *Bioresour. Technol.* **2011**, *102*, 4036–4041. [CrossRef] [PubMed]
119. Tan, X.; Liu, Y.; Zeng, G.; Wang, X.; Hu, X.; Gu, Y.; Yang, Z. Application of Biochar for the Removal of Pollutants from Aqueous Solutions. *Chemosphere* **2015**, *125*, 70–85. [CrossRef] [PubMed]
120. Yu, X.-Y.; Ying, G.-G.; Kookana, R.S. Reduced Plant Uptake of Pesticides with Biochar Additions to Soil. *Chemosphere* **2009**, *76*, 665–671. [CrossRef] [PubMed]
121. Liang, C.; Niu, H.-Y.; Guo, H.; Niu, C.-G.; Yang, Y.-Y.; Liu, H.-Y.; Tang, W.-W.; Feng, H.-P. Efficient Photocatalytic Nitrogen Fixation to Ammonia over Bismuth Monoxide Quantum Dots-Modified Defective Ultrathin Graphitic Carbon Nitride. *Chem. Eng. J.* **2021**, *406*, 126868. [CrossRef]
122. Kumbhar, D.; Palliyarayil, A.; Reghu, D.; Shrungar, D.; Umapathy, S.; Sil, S. Rapid Discrimination of Porous Bio-Carbon Derived from Nitrogen Rich Biomass Using Raman Spectroscopy and Artificial Intelligence Methods. *Carbon* **2021**, *178*, 792–802. [CrossRef]
123. Zhang, K.; Sun, P.; Faye, M.C.A.S.; Zhang, Y. Characterization of Biochar Derived from Rice Husks and Its Potential in Chlorobenzene Degradation. *Carbon* **2018**, *130*, 730–740. [CrossRef]
124. Chen, D.; Xu, J.; Ling, P.; Fang, Z.; Ren, Q.; Xu, K.; Jiang, L.; Wang, Y.; Su, S.; Hu, S.; et al. Formation and Evolution Mechanism of Persistent Free Radicals in Biochar during Biomass Pyrolysis: Insights from Biochar's Element Composition and Chemical Structure. *Fuel* **2024**, *357*, 129910. [CrossRef]
125. Wu, C.; Fu, L.; Li, H.; Liu, X.; Wan, C. Using Biochar to Strengthen the Removal of Antibiotic Resistance Genes: Performance and Mechanism. *Sci. Total Environ.* **2022**, *816*, 151554. [CrossRef] [PubMed]
126. Huang, C.; Qin, F.; Zhang, C.; Huang, D.; Tang, L.; Yan, M.; Wang, W.; Song, B.; Qin, D.; Zhou, Y.; et al. Effects of Heterogeneous Metals on the Generation of Persistent Free Radicals as Critical Redox Sites in Iron-Containing Biochar for Persulfate Activation. *ACS EST Water* **2023**, *3*, 298–310. [CrossRef]
127. Yu, J.; Zhu, Z.; Zhang, H.; Shen, X.; Qiu, Y.; Yin, D.; Wang, S. Persistent Free Radicals on N-Doped Hydrochar for Degradation of Endocrine Disrupting Compounds. *Chem. Eng. J.* **2020**, *398*, 125538. [CrossRef]
128. Ma, L.; Syed-Hassan, S.S.A.; Tong, Y.; Xiong, Z.; Chen, Y.; Xu, J.; Jiang, L.; Su, S.; Hu, S.; Wang, Y.; et al. Interactions of Cellulose- and Lignin-Derived Radicals during Pyrolysis: An in-Situ Electron Paramagnetic Resonance (EPR) Study. *Fuel Process. Technol.* **2023**, *239*, 107536. [CrossRef]
129. Retcofsky, H.L.; Hough, M.R.; Clarkson, R.B. Nature of the Free Radicals in Coals, Pyrolyzed Coals, and Liquefaction Products. *Am. Chem. Soc. Div. Fuel Chem. Prepr.* **1979**, *24*, 177.
130. Liu, J.; Jiang, X.; Shen, J.; Zhang, H. Chemical Properties of Superfine Pulverized Coal Particles. Part 1. Electron Paramagnetic Resonance Analysis of Free Radical Characteristics. *Adv. Powder Technol.* **2014**, *25*, 916–925. [CrossRef]
131. Zhang, Y.; Sun, X.; Bian, W.; Peng, J.; Wan, H.; Zhao, J. The Key Role of Persistent Free Radicals on the Surface of Hydrochar and Pyrocarbon in the Removal of Heavy Metal-Organic Combined Pollutants. *Bioresour. Technol.* **2020**, *318*, 124046. [CrossRef] [PubMed]
132. Qin, J.; Cheng, Y.; Sun, M.; Yan, L.; Shen, G. Catalytic Degradation of the Soil Fumigant 1,3-Dichloropropene in Aqueous Biochar Slurry. *Sci. Total Environ.* **2016**, *569–570*, 1–8. [CrossRef] [PubMed]
133. Huang, D.; Luo, H.; Zhang, C.; Zeng, G.; Lai, C.; Cheng, M.; Wang, R.; Deng, R.; Xue, W.; Gong, X.; et al. Nonnegligible Role of Biomass Types and Its Compositions on the Formation of Persistent Free Radicals in Biochar: Insight into the Influences on Fenton-like Process. *Chem. Eng. J.* **2019**, *361*, 353–363. [CrossRef]
134. Wang, Y.; Gu, X.; Huang, Y.; Ding, Z.; Chen, Y.; Hu, X. Insight into Biomass Feedstock on Formation of Biochar-Bound Environmentally Persistent Free Radicals under Different Pyrolysis Temperatures. *RSC Adv.* **2022**, *12*, 19318–19326. [CrossRef] [PubMed]
135. Tao, W.; Duan, W.; Liu, C.; Zhu, D.; Si, X.; Zhu, R.; Oleszczuk, P.; Pan, B. Formation of Persistent Free Radicals in Biochar Derived from Rice Straw Based on a Detailed Analysis of Pyrolysis Kinetics. *Sci. Total Environ.* **2020**, *715*, 136575. [CrossRef] [PubMed]

136. Xiang, C.; Liu, Q.; Shi, L.; Liu, Z. A Study on the New Type of Radicals in Corncob Derived Biochars. *Fuel* **2020**, *277*, 118163. [CrossRef]
137. Chintala, R.; Subramanian, S.; Fortuna, A.-M.; Schumacher, T.E. Examining Biochar Impacts on Soil Abiotic and Biotic Processes and Exploring the Potential for Pyrosequencing Analysis. Chapter 6. In *Biochar Application Essential Soil Microbial Ecology*; Komang Ralebitso-Senior, T., Orr, C.H., Eds.; Elsevier: Amsterdam, The Netherlands, 2016; pp. 133–162. ISBN 9780128034330. [CrossRef]
138. Takeshita, A.; Uemura, Y.; Onoe, K. Quantification of Persistent Free Radicals (PFRs) Formed in Thermally Decomposed Cellulose and Its Correlation with Residual Carbon Amount. *J. Anal. Appl. Pyrolysis* **2020**, *150*, 104883. [CrossRef]
139. Pan, B.; Li, H.; Lang, D.; Xing, B. Environmentally Persistent Free Radicals: Occurrence, Formation Mechanisms and Implications. *Environ. Pollut.* **2019**, *248*, 320–331. [CrossRef]
140. Lomnicki, S.M.; Truong, H.; Vejerano, E.P.; Dellinger, B. Copper Oxide-Based Model of Persistent Free Radical Formation on Combustion-Derived Particulate Matter. *Environ. Sci. Technol.* **2008**, *42*, 4982–4988. [CrossRef] [PubMed]
141. Gasim, M.F.; Choong, Z.-Y.; Koo, P.-L.; Low, S.-C.; Abdurahman, M.-H.; Ho, Y.-C.; Mohamad, M.; Suryawan, I.W.; Lim, J.-W.; Oh, W.-D. Application of Biochar as Functional Material for Remediation of Organic Pollutants in Water: An Overview. *Catalysts* **2022**, *12*, 210. [CrossRef]
142. Zhong, D.; Jiang, Y.; Zhao, Z.; Wang, L.; Chen, J.; Ren, S.; Liu, Z.; Zhang, Y.; Tsang, D.C.W.; Crittenden, J.C. pH Dependence of Arsenic Oxidation by Rice-Husk-Derived Biochar: Roles of Redox-Active Moieties. *Environ. Sci. Technol.* **2019**, *53*, 9034–9044. [CrossRef] [PubMed]
143. Zhang, K.; Sun, P.; Zhang, Y. Decontamination of Cr(VI) Facilitated Formation of Persistent Free Radicals on Rice Husk Derived Biochar. *Front. Environ. Sci. Eng.* **2019**, *13*, 22. [CrossRef]
144. Zhu, S.; Huang, X.; Yang, X.; Peng, P.; Li, Z.; Jin, C. Enhanced Transformation of Cr(VI) by Heterocyclic-N within Nitrogen-Doped Biochar: Impact of Surface Modulatory Persistent Free Radicals (PFRs). *Environ. Sci. Technol.* **2020**, *54*, 8123–8132. [CrossRef] [PubMed]
145. Wang, X.; Xu, J.; Liu, J.; Liu, J.; Xia, F.; Wang, C.; Dahlgren, R.A.; Liu, W. Mechanism of Cr(VI) Removal by Magnetic Greigite/Biochar Composites. *Sci. Total Environ.* **2020**, *700*, 134414. [CrossRef] [PubMed]
146. Tang, Z.; Kong, Y.; Zhao, S.; Jia, H.; Vione, D.; Kang, Y.; Gao, P. Enhancement of Cr(VI) Decontamination by Irradiated Sludge Biochar in Neutral Conditions: Evidence of a Possible Role of Persistent Free Radicals. *Sep. Purif. Technol.* **2021**, *277*, 119414. [CrossRef]
147. Zhu, Y.; Wei, J.; Li, J. Decontamination of Cr(VI) from Water Using Sewage Sludge-Derived Biochar: Role of Environmentally Persistent Free Radicals. *Chin. J. Chem. Eng.* **2023**, *56*, 97–103. [CrossRef]
148. Baltrėnaitė-Gedienė, E.; Lomnicki, S.; Guo, C. Impact of Biochar, Fertilizers and Cultivation Type on Environmentally Persistent Free Radicals in Agricultural Soil. *Environ. Technol. Innov.* **2022**, *28*, 102755. [CrossRef]
149. Tai, Y.; Sun, J.; Tian, H.; Liu, F.; Han, B.; Fu, W.; Liu, Z.; Yang, X.; Liu, Q. Efficient Degradation of Organic Pollutants by S-$NaTaO_3$/Biochar under Visible Light and the Photocatalytic Performance of a Permonosulfate-Based Dual-Effect Catalytic System. *J. Environ. Sci.* **2023**, *125*, 388–400. [CrossRef]
150. Kelley, M.A.; Hebert, V.Y.; Thibeaux, T.M.; Orchard, M.A.; Hasan, F.; Cormier, S.A.; Thevenot, P.T.; Lomnicki, S.M.; Varner, K.J.; Dellinger, B.; et al. Model Combustion-Generated Particulate Matter Containing Persistent Free Radicals Redox Cycle to Produce Reactive Oxygen Species. *Chem. Res. Toxicol.* **2013**, *26*, 1862–1871. [CrossRef] [PubMed]
151. Squadrito, G.L.; Cueto, R.; Dellinger, B.; Pryor, W.A. Quinoid Redox Cycling as a Mechanism for Sustained Free Radical Generation by Inhaled Airborne Particulate Matter. *Free Radic. Biol. Med.* **2001**, *31*, 1132–1138. [CrossRef] [PubMed]
152. Alburquerque, J.A.; Salazar, P.; Barrón, V.; Torrent, J.; del Campillo, M.d.C.; Gallardo, A.; Villar, R. Enhanced Wheat Yield by Biochar Addition under Different Mineral Fertilization Levels. *Agron. Sustain. Dev.* **2013**, *33*, 475–484. [CrossRef]
153. Jiang, S.-F.; Ling, L.-L.; Chen, W.-J.; Liu, W.-J.; Li, D.-C.; Jiang, H. High Efficient Removal of Bisphenol A in a Peroxymonosulfate/Iron Functionalized Biochar System: Mechanistic Elucidation and Quantification of the Contributors. *Chem. Eng. J.* **2019**, *359*, 572–583. [CrossRef]
154. Luo, H.; Lin, Q.; Zhang, X.; Huang, Z.; Liu, S.; Jiang, J.; Xiao, R.; Liao, X. New Insights into the Formation and Transformation of Active Species in nZVI/BC Activated Persulfate in Alkaline Solutions. *Chem. Eng. J.* **2019**, *359*, 1215–1223. [CrossRef]
155. Pi, Z.; Li, X.; Wang, D.; Xu, Q.; Tao, Z.; Huang, X.; Yao, F.; Wu, Y.; He, L.; Yang, Q. Persulfate Activation by Oxidation Biochar Supported Magnetite Particles for Tetracycline Removal: Performance and Degradation Pathway. *J. Clean. Prod.* **2019**, *235*, 1103–1115. [CrossRef]
156. Zhang, X.; Huang, R.; Show, P.L.; Mahlknecht, J.; Wang, C. Degradation of Tetracycline by Nitrogen-Doped Biochar as a Peroxydisulfate Activator: Nitrogen Doping Pattern and Non-Radical Mechanism. *Sustain. Horiz.* **2024**, *10*, 100091. [CrossRef]
157. He, J.; Xiao, Y.; Tang, J.; Chen, H.; Sun, H. Persulfate Activation with Sawdust Biochar in Aqueous Solution by Enhanced Electron Donor-Transfer Effect. *Sci. Total Environ.* **2019**, *690*, 768–777. [CrossRef] [PubMed]
158. Danping, W.U.; Fangfang, L.I.; Jing, Z.; Peng, W.; Min, W.U. Photocatalysis degradation of rhodamine B by dissolved organic matter of biochars. *Chin. J. Environ. Eng.* **2019**, *13*, 2562–2569. [CrossRef]
159. Wang, H.; Guo, W.; Yin, R.; Du, J.; Wu, Q.; Luo, H.; Liu, B.; Sseguya, F.; Ren, N. Biochar-Induced Fe(III) Reduction for Persulfate Activation in Sulfamethoxazole Degradation: Insight into the Electron Transfer, Radical Oxidation and Degradation Pathways. *Chem. Eng. J.* **2019**, *362*, 561–569. [CrossRef]

160. Zhu, K.; Wang, X.; Geng, M.; Chen, D.; Lin, H.; Zhang, H. Catalytic Oxidation of Clofibric Acid by Peroxydisulfate Activated with Wood-Based Biochar: Effect of Biochar Pyrolysis Temperature, Performance and Mechanism. *Chem. Eng. J.* **2019**, *374*, 1253–1263. [CrossRef]
161. Li, L.; Lai, C.; Huang, F.; Cheng, M.; Zeng, G.; Huang, D.; Li, B.; Liu, S.; Zhang, M.; Qin, L.; et al. Degradation of Naphthalene with Magnetic Bio-Char Activate Hydrogen Peroxide: Synergism of Bio-Char and Fe–Mn Binary Oxides. *Water Res.* **2019**, *160*, 238–248. [CrossRef] [PubMed]
162. Luo, K.; Pang, Y.; Yang, Q.; Wang, D.; Li, X.; Wang, L.; Lei, M.; Liu, J. Enhanced Ciprofloxacin Removal by Sludge-Derived Biochar: Effect of Humic Acid. *Chemosphere* **2019**, *231*, 495–501. [CrossRef] [PubMed]
163. Deng, R.; Luo, H.; Huang, D.; Zhang, C. Biochar-Mediated Fenton-like Reaction for the Degradation of Sulfamethazine: Role of Environmentally Persistent Free Radicals. *Chemosphere* **2020**, *255*, 126975. [CrossRef] [PubMed]
164. Liu, G.; Zhang, Y.; Yu, H.; Jin, R.; Zhou, J. Acceleration of Goethite-Catalyzed Fenton-like Oxidation of Ofloxacin by Biochar. *J. Hazard. Mater.* **2020**, *397*, 122783. [CrossRef] [PubMed]
165. Liu, B.; Guo, W.; Wang, H.; Si, Q.; Zhao, Q.; Luo, H.; Ren, N. Activation of Peroxymonosulfate by Cobalt-Impregnated Biochar for Atrazine Degradation: The Pivotal Roles of Persistent Free Radicals and Ecotoxicity Assessment. *J. Hazard. Mater.* **2020**, *398*, 122768. [CrossRef] [PubMed]
166. Grilla, E.; Vakros, J.; Konstantinou, I.; Manariotis, I.D.; Mantzavinos, D. Activation of Persulfate by Biochar from Spent Malt Rootlets for the Degradation of Trimethoprim in the Presence of Inorganic Ions. *J. Chem. Technol. Biotechnol.* **2020**, *95*, 2348–2358. [CrossRef]
167. Feng, Z.; Zhou, B.; Yuan, R.; Li, H.; He, P.; Wang, F.; Chen, Z.; Chen, H. Biochar Derived from Different Crop Straws as Persulfate Activator for the Degradation of Sulfadiazine: Influence of Biomass Types and Systemic Cause Analysis. *Chem. Eng. J.* **2022**, *440*, 135669. [CrossRef]
168. Kim, D.-G.; Ko, S.-O. Effects of Thermal Modification of a Biochar on Persulfate Activation and Mechanisms of Catalytic Degradation of a Pharmaceutical. *Chem. Eng. J.* **2020**, *399*, 125377. [CrossRef]
169. He, W.; Zhu, Y.; Zeng, G.; Zhang, Y.; Wang, Y.; Zhang, M.; Long, H.; Tang, W. Efficient Removal of Perfluorooctanoic Acid by Persulfate Advanced Oxidative Degradation: Inherent Roles of Iron-Porphyrin and Persistent Free Radicals. *Chem. Eng. J.* **2020**, *392*, 123640. [CrossRef]
170. Cao, W.; Zeng, C.; Guo, X.; Liu, Q.; Zhang, X.; Mameda, N. Enhanced Electrochemical Degradation of 2,4-Dichlorophenol with the Assist of Hydrochar. *Chemosphere* **2020**, *260*, 127643. [CrossRef] [PubMed]
171. Xu, H.; Zhang, Y.; Li, J.; Hao, Q.; Li, X.; Liu, F. Heterogeneous Activation of Peroxymonosulfate by a Biochar-Supported Co_3O_4 Composite for Efficient Degradation of Chloramphenicols. *Environ. Pollut.* **2020**, *257*, 113610. [CrossRef] [PubMed]
172. Li, X.; Jia, Y.; Zhou, M.; Su, X.; Sun, J. High-Efficiency Degradation of Organic Pollutants with Fe, N Co-Doped Biochar Catalysts via Persulfate Activation. *J. Hazard. Mater.* **2020**, *397*, 122764. [CrossRef] [PubMed]
173. Xie, Y.; Hu, W.; Wang, X.; Tong, W.; Li, P.; Zhou, H.; Wang, Y.; Zhang, Y. Molten Salt Induced Nitrogen-Doped Biochar Nanosheets as Highly Efficient Peroxymonosulfate Catalyst for Organic Pollutant Degradation. *Environ. Pollut.* **2020**, *260*, 114053. [CrossRef] [PubMed]
174. Yan, J.; Yang, L.; Qian, L.; Han, L.; Chen, M. Nano-Magnetite Supported by Biochar Pyrolyzed at Different Temperatures as Hydrogen Peroxide Activator: Synthesis Mechanism and the Effects on Ethylbenzene Removal. *Environ. Pollut.* **2020**, *261*, 114020. [CrossRef] [PubMed]
175. Mian, M.M.; Liu, G.; Zhou, H. Preparation of N-Doped Biochar from Sewage Sludge and Melamine for Peroxymonosulfate Activation: N-Functionality and Catalytic Mechanisms. *Sci. Total Environ.* **2020**, *744*, 140862. [CrossRef] [PubMed]
176. da Silva Veiga, P.A.; Schultz, J.; da Silva Matos, T.T.; Fornari, M.R.; Costa, T.G.; Meurer, L.; Mangrich, A.S. Production of High-Performance Biochar Using a Simple and Low-Cost Method: Optimization of Pyrolysis Parameters and Evaluation for Water Treatment. *J. Anal. Appl. Pyrolysis* **2020**, *148*, 104823. [CrossRef]
177. Sun, P.; Zhang, K.-K.; Zhang, Y.; Zhang, Y. Sunflower-Straw-Derived Biochar-Enhanced Fe(III)/$S_2O_8^{2-}$ System for Degradation of Benzoic Acid. *Environ. Sci.* **2020**, *41*, 2301–2309.
178. Zeng, L.; Chen, Q.; Tan, Y.; Lan, P.; Zhou, D.; Wu, M.; Liang, N.; Pan, B.; Xing, B. Dual Roles of Biochar Redox Property in Mediating 2,4-Dichlorophenol Degradation in the Presence of Fe^{3+} and Persulfate. *Chemosphere* **2021**, *279*, 130456. [CrossRef] [PubMed]
179. Yang, F.; Zhu, Q.; Gao, Y.; Jian, H.; Wang, C.; Sun, H. Effects of Biochar-Dissolved Organic Matter on the Photodegradation of Sulfamethoxazole and Chloramphenicol in Biochar Solutions as Revealed by Oxygen Reduction Performances and Free Radicals. *Sci. Total Environ.* **2021**, *781*, 146807. [CrossRef]
180. Min, L.; Zhang, P.; Fan, M.; Xu, X.; Wang, C.; Tang, J.; Sun, H. Efficient Degradation of P-Nitrophenol by Fe@pomelo Peel-Derived Biochar Composites and Its Mechanism of Simultaneous Reduction and Oxidation Process. *Chemosphere* **2021**, *267*, 129213. [CrossRef] [PubMed]
181. Mer, K.; Sajjadi, B.; Egiebor, N.O.; Chen, W.Y.; Mattern, D.L.; Tao, W. Enhanced Degradation of Organic Contaminants Using Catalytic Activity of Carbonaceous Structures: A Strategy for the Reuse of Exhausted Sorbents. *J. Environ. Sci.* **2021**, *99*, 267–273. [CrossRef] [PubMed]

182. Cao, Y.; Jiang, S.; Kang, X.; Zhang, H.; Zhang, Q.; Wang, L. Enhancing Degradation of Atrazine by Fe-Phenol Modified Biochar/Ferrate(VI) under Alkaline Conditions: Analysis of the Mechanism and Intermediate Products. *Chemosphere* **2021**, *285*, 131399. [CrossRef] [PubMed]
183. Zhang, R.; Zheng, X.; Zhang, D.; Niu, X.; Ma, J.; Lin, Z.; Fu, M.; Zhou, S. Insight into the Roles of Endogenous Minerals in the Activation of Persulfate by Graphitized Biochar for Tetracycline Removal. *Sci. Total Environ.* **2021**, *768*, 144281. [CrossRef] [PubMed]
184. Yi, Y.; Luo, J.; Fang, Z. Magnetic Biochar Derived from Eichhornia Crassipes for Highly Efficient Fenton-like Degradation of Antibiotics: Mechanism and Contributions. *J. Environ. Chem. Eng.* **2021**, *9*, 106258. [CrossRef]
185. Zhang, Y.; Xu, M.; Liang, S.; Feng, Z.; Zhao, J. Mechanism of Persulfate Activation by Biochar for the Catalytic Degradation of Antibiotics: Synergistic Effects of Environmentally Persistent Free Radicals and the Defective Structure of Biochar. *Sci. Total Environ.* **2021**, *794*, 148707. [CrossRef] [PubMed]
186. Wang, X.; Zhang, P.; Wang, C.; Jia, H.; Shang, X.; Tang, J.; Sun, H. Metal-Rich Hyperaccumulator-Derived Biochar as an Efficient Persulfate Activator: Role of Intrinsic Metals (Fe, Mn and Zn) in Regulating Characteristics, Performance and Reaction Mechanisms. *J. Hazard. Mater.* **2022**, *424*, 127225. [CrossRef] [PubMed]
187. Luo, K.; Yang, C.; Li, X.; Pang, Y.; Yang, Q. Mn-Doped Biochar Derived from Sewage Sludge for Ciprofloxacin Degradation. *J. Environ. Eng.* **2022**, *148*, 04022048. [CrossRef]
188. Li, H.; Liu, Y.; Jiang, F.; Bai, X.; Li, H.; Lang, D.; Wang, L.; Pan, B. Persulfate Adsorption and Activation by Carbon Structure Defects Provided New Insights into Ofloxacin Degradation by Biochar. *Sci. Total Environ.* **2022**, *806*, 150968. [CrossRef] [PubMed]
189. Dai, Z.; Zhao, L.; Peng, S.; Yue, Z.; Zhan, X.; Wang, J. Removal of Oxytetracycline Promoted by Manganese-Doped Biochar Based on Density Functional Theory Calculations: Comprehensive Evaluation of the Effect of Transition Metal Doping. *Sci. Total Environ.* **2022**, *806*, 150268. [CrossRef]
190. Rangarajan, G.; Farnood, R. Role of Persistent Free Radicals and Lewis Acid Sites in Visible-Light-Driven Wet Peroxide Activation by Solid Acid Biochar Catalysts—A Mechanistic Study. *J. Hazard. Mater.* **2022**, *438*, 129514. [CrossRef] [PubMed]
191. An, X.; Chen, Y.; Ao, M.; Jin, Y.; Zhan, L.; Yu, B.; Wu, Z.; Jiang, P. Sequential Photocatalytic Degradation of Organophosphorus Pesticides and Recovery of Orthophosphate by Biochar/α-Fe_2O_3/MgO Composite: A New Enhanced Strategy for Reducing the Impacts of Organophosphorus from Wastewater. *Chem. Eng. J.* **2022**, *435*, 135087. [CrossRef]
192. Jiang, X.; Xiao, Y.; Xiao, J.; Zhang, W.; Qiu, R. The Effect of Persistent Free Radicals in Sludge Derived Biochar on the Removal of P-Chlorophenol. Available online: https://ssrn.com/abstract=3992617 (accessed on 4 January 2024).
193. Pan, Y.; Peng, Z.; Liu, Z.; Shao, B.; Liang, Q.; He, Q.; Wu, T.; Zhang, X.; Zhao, C.; Liu, Y.; et al. Activation of Peroxydisulfate by Bimetal Modified Peanut Hull-Derived Porous Biochar for the Degradation of Tetracycline in Aqueous Solution. *J. Environ. Chem. Eng.* **2022**, *10*, 107366. [CrossRef]
194. Yin, Q.; Yan, H.; Liang, Y.; Jiang, Z.; Wang, H.; Nian, Y. Activation of Persulfate by Blue Algae Biochar Supported FeO_X Particles for Tetracycline Degradation: Performance and Mechanism. *Sep. Purif. Technol.* **2023**, *319*, 124005. [CrossRef]
195. Badiger, S.M.; Nidheesh, P.V. Applications of Biochar in Sulfate Radical-Based Advanced Oxidation Processes for the Removal of Pharmaceuticals and Personal Care Products. *Water Sci. Technol.* **2023**, *87*, 1329–1348. [CrossRef] [PubMed]
196. Yu, Y.; Zhong, Z.; Guo, H.; Yu, Y.; Zheng, T.; Li, H.; Chang, Z. Biochar–Goethite Composites Inhibited/Enhanced Degradation of Triphenyl Phosphate by Activating Persulfate: Insights on the Mechanism. *Sci. Total Environ.* **2023**, *858*, 159940. [CrossRef] [PubMed]
197. Pei, S.; Zhao, Y.; Li, W.; Qu, C.; Ren, Y.; Yang, Y.; Liu, J.; Wu, C. Critical Impact of Pyrolysis Temperatures on Biochars for Peroxymonosulfate Activation: Structural Characteristics, Degradation Performance and Mechanism. *Chem. Eng. J.* **2023**, *477*, 147271. [CrossRef]
198. Zhang, Y.; He, R.; Zhao, J.; Zhang, X.; Bildyukevich, A.V. Effect of Aged Biochar after Microbial Fermentation on Antibiotics Removal: Key Roles of Microplastics and Environmentally Persistent Free Radicals. *Bioresour. Technol.* **2023**, *374*, 128779. [CrossRef] [PubMed]
199. Yang, Z.; An, Q.; Deng, S.; Xu, B.; Li, Z.; Deng, S.; Zhao, B.; Ye, Z. Efficient Activation of Peroxydisulfate by Modified Red Mud Biochar Derived from Waste Corn Straw for Levofloxacin Degradation: Efficiencies and Mechanisms. *J. Environ. Chem. Eng.* **2023**, *11*, 111609. [CrossRef]
200. Zhang, Y.; He, R.; Zhao, J. Removal Mechanism of Tetracycline-Cr(VI) Combined Pollutants by Different S-Doped Sludge Biochars: Role of Environmentally Persistent Free Radicals. *Chemosphere* **2023**, *317*, 137856. [CrossRef]
201. Luo, K.; Lin, N.; Li, X.; Pang, Y.; Wang, L.; Lei, M.; Yang, Q. Efficient Hexavalent Chromium Removal by Nano-Cerium Oxide Functionalized Biochar: Insight into the Role of Reduction. *J. Environ. Chem. Eng.* **2023**, *11*, 110004. [CrossRef]
202. Zhong, D.; Zhao, Z.; Jiang, Y.; Yang, X.; Wang, L.; Chen, J.; Guan, C.-Y.; Zhang, Y.; Tsang, D.C.W.; Crittenden, J.C. Contrasting Abiotic As(III) Immobilization by Undissolved and Dissolved Fractions of Biochar in Ca^{2+}-Rich Groundwater under Anoxic Conditions. *Water Res.* **2020**, *183*, 116106. [CrossRef] [PubMed]
203. Huang, Y.; Gao, M.; Deng, Y.; Khan, Z.H.; Liu, X.; Song, Z.; Qiu, W. Efficient Oxidation and Adsorption of As(III) and As(V) in Water Using a Fenton-like Reagent, (Ferrihydrite)-Loaded Biochar. *Sci. Total Environ.* **2020**, *715*, 136957. [CrossRef] [PubMed]
204. Zhu, Y.; Wei, J.; Li, J. Biochar-Activated Persulfate Oxidation of Arsenic(III): Nonnegligible Roles of Environmentally Persistent Free Radicals. *J. Environ. Chem. Eng.* **2023**, *11*, 111033. [CrossRef]

205. Yu, J.; Zhu, Z.; Zhang, H.; Chen, T.; Qiu, Y.; Xu, Z.; Yin, D. Efficient Removal of Several Estrogens in Water by Fe-Hydrochar Composite and Related Interactive Effect Mechanism of H_2O_2 and Iron with Persistent Free Radicals from Hydrochar of Pinewood. *Sci. Total Environ.* **2019**, *658*, 1013–1022. [CrossRef] [PubMed]
206. Zhu, N.; Li, C.; Bu, L.; Tang, C.; Wang, S.; Duan, P.; Yao, L.; Tang, J.; Dionysiou, D.D.; Wu, Y. Bismuth Impregnated Biochar for Efficient Estrone Degradation: The Synergistic Effect between Biochar and Bi/Bi2O3 for a High Photocatalytic Performance. *J. Hazard. Mater.* **2020**, *384*, 121258. [CrossRef] [PubMed]
207. Chen, Y.; Duan, X.; Zhang, C.; Wang, S.; Ren, N.; Ho, S.-H. Graphitic Biochar Catalysts from Anaerobic Digestion Sludge for Nonradical Degradation of Micropollutants and Disinfection. *Chem. Eng. J.* **2020**, *384*, 123244. [CrossRef]
208. Xu, H.; Han, Y.; Wang, G.; Deng, P.; Feng, L. Walnut Shell Biochar Based Sorptive Remediation of Estrogens Polluted Simulated Wastewater: Characterization, Adsorption Mechanism and Degradation by Persistent Free Radicals. *Environ. Technol. Innov.* **2022**, *28*, 102870. [CrossRef]
209. Wang, T.; Zheng, J.; Cai, J.; Liu, Q.; Zhang, X. Visible-Light-Driven Photocatalytic Degradation of Dye and Antibiotics by Activated Biochar Composited with K^+ Doped g-C3N4: Effects, Mechanisms, Actual Wastewater Treatment and Disinfection. *Sci. Total Environ.* **2022**, *839*, 155955. [CrossRef]
210. Shi, J.; Wang, J.; Liang, L.; Xu, Z.; Chen, Y.; Chen, S.; Xu, M.; Wang, X.; Wang, S. Carbothermal Synthesis of Biochar-Supported Metallic Silver for Enhanced Photocatalytic Removal of Methylene Blue and Antimicrobial Efficacy. *J. Hazard. Mater.* **2021**, *401*, 123382. [CrossRef] [PubMed]
211. Lian, F.; Yu, W.; Zhou, Q.; Gu, S.; Wang, Z.; Xing, B. Size Matters: Nano-Biochar Triggers Decomposition and Transformation Inhibition of Antibiotic Resistance Genes in Aqueous Environments. *Environ. Sci. Technol.* **2020**, *54*, 8821–8829. [CrossRef] [PubMed]
212. Zeng, L.; Wu, M.; Wu, G.-J. Electron exchange capacity of rice biochar at different preparation temperatures. *China Environ. Sci.* **2019**, *39*, 4329–4336.
213. Liu, G.; Zhang, X.; Liu, H.; He, Z.; Show, P.L.; Vasseghian, Y.; Wang, C. Biochar/Layered Double Hydroxides Composites as Catalysts for Treatment of Organic Wastewater by Advanced Oxidation Processes: A Review. *Environ. Res.* **2023**, *234*, 116534. [CrossRef] [PubMed]
214. Xiang, L.; Liu, S.; Ye, S.; Yang, H.; Song, B.; Qin, F.; Shen, M.; Tan, C.; Zeng, G.; Tan, X. Potential Hazards of Biochar: The Negative Environmental Impacts of Biochar Applications. *J. Hazard. Mater.* **2021**, *420*, 126611. [CrossRef] [PubMed]
215. Kharisov, B.I.; Kharissova, O.V. Carbon Allotropes in the Environment and Their Toxicity. In *Carbon Allotropes: Metal-Complex Chemistry, Properties and Applications*; Springer: Cham, Switzerland, 2019. [CrossRef]
216. El-Naggar, A.; Lee, M.-H.; Hur, J.; Lee, Y.H.; Igalavithana, A.D.; Shaheen, S.M.; Ryu, C.; Rinklebe, J.; Tsang, D.C.W.; Ok, Y.S. Biochar-Induced Metal Immobilization and Soil Biogeochemical Process: An Integrated Mechanistic Approach. *Sci. Total Environ.* **2020**, *698*, 134112. [CrossRef] [PubMed]
217. Cui, H.; Li, D.; Liu, X.; Fan, Y.; Zhang, X.; Zhang, S.; Zhou, J.; Fang, G.; Zhou, J. Dry-Wet and Freeze-Thaw Aging Activate Endogenous Copper and Cadmium in Biochar. *J. Clean. Prod.* **2021**, *288*, 125605. [CrossRef]
218. Rombolà, A.G.; Fabbri, D.; Baronti, S.; Vaccari, F.P.; Genesio, L.; Miglietta, F. Changes in the Pattern of Polycyclic Aromatic Hydrocarbons in Soil Treated with Biochar from a Multiyear Field Experiment. *Chemosphere* **2019**, *219*, 662–670. [CrossRef] [PubMed]
219. Joseph, S.; Camps-Arbestain, M.; Lin, Y.; Munroe, P.; Chia, C.H.; Hook, J.M.; Zwieten, L.V.; Kimber, S.; Cowie, A.L.; Singh, B.P.; et al. An Investigation into the Reactions of Biochar in Soil. *Soil Res.* **2010**, *48*, 501–515. [CrossRef]
220. Kim, H.-B.; Kim, S.-H.; Jeon, E.-K.; Kim, D.-H.; Tsang, D.C.W.; Alessi, D.S.; Kwon, E.E.; Baek, K. Effect of Dissolved Organic Carbon from Sludge, Rice Straw and Spent Coffee Ground Biochar on the Mobility of Arsenic in Soil. *Sci. Total Environ.* **2018**, *636*, 1241–1248. [CrossRef]
221. Jia, C.; Luo, J.; Fan, J.; Clark, J.H.; Zhang, S.; Zhu, X. Urgently Reveal Longly Hidden Toxicant in a Familiar Fabrication Process of Biomass-Derived Environment Carbon Material. *J. Environ. Sci.* **2021**, *100*, 250–256. [CrossRef]
222. Liu, J.; Gao, N.; Wen, X.; Jia, H.; Lichtfouse, E. Plant and Algal Toxicity of Persistent Free Radicals and Reactive Oxygen Species Generated by Heating Anthracene-Contaminated Soils from 100 to 600 °C. *Environ. Chem. Lett.* **2021**, *19*, 2695–2703. [CrossRef]
223. Xu, Y.; Lu, X.; Su, G.; Chen, X.; Meng, J.; Li, Q.; Wang, C.; Shi, B. Scientific and Regulatory Challenges of Environmentally Persistent Free Radicals: From Formation Theory to Risk Prevention Strategies. *J. Hazard. Mater.* **2023**, *456*, 131674. [CrossRef] [PubMed]
224. Jaligama, S.; Patel, V.S.; Wang, P.; Sallam, A.; Harding, J.; Kelley, M.; Mancuso, S.R.; Dugas, T.R.; Cormier, S.A. Radical Containing Combustion Derived Particulate Matter Enhance Pulmonary Th17 Inflammation via the Aryl Hydrocarbon Receptor. *Part. Fibre Toxicol.* **2018**, *15*, 20. [CrossRef] [PubMed]
225. Harmon, A.C.; Hebert, V.Y.; Cormier, S.A.; Subramanian, B.; Reed, J.R.; Backes, W.L.; Dugas, T.R. Particulate Matter Containing Environmentally Persistent Free Radicals Induces AhR-Dependent Cytokine and Reactive Oxygen Species Production in Human Bronchial Epithelial Cells. *PLoS ONE* **2018**, *13*, e0205412. [CrossRef] [PubMed]
226. Reed, J.R.; dela Cruz, A.L.N.; Lomnicki, S.M.; Backes, W.L. Environmentally Persistent Free Radical-Containing Particulate Matter Competitively Inhibits Metabolism by Cytochrome P450 1A2. *Toxicol. Appl. Pharmacol.* **2015**, *289*, 223–230. [CrossRef] [PubMed]
227. Chuang, G.C.; Xia, H.; Mahne, S.E.; Varner, K.J. Environmentally Persistent Free Radicals Cause Apoptosis in HL-1 Cardiomyocytes. *Cardiovasc. Toxicol.* **2017**, *17*, 140–149. [CrossRef] [PubMed]

228. Zhao, S.; Miao, D.; Zhu, K.; Kelin, T.; Wang, C.; Sharma, V.; Jia, H. Interaction of Benzo[a]Pyrene with Cu(II)-Montmorillonite: Generation and Toxicity of Environmentally Persistent Free Radicals and Reactive Oxygen Species. *Environ. Int.* **2019**, *129*, 154–163. [CrossRef] [PubMed]
229. Thevenot, P.; Saravia, J.; Jin, N.; Giaimo, J.; Chustz, R.; Mahne, S.; Kelley, M.; Hebert, V.; Dellinger, B.; Dugas, T.; et al. Radical-Containing PM0.2 Initiates Epithelial-to-Mesenchymal Transitions in Airway Epithelial Cells. *Am. J. Respir. Cell Mol. Biol.* **2012**, *48*, 188–197. [CrossRef] [PubMed]
230. Zhang, X.; Gu, W.; Ma, Z.; Liu, Y.; Ru, H.; Zhou, J.; Zang, Y.; Xu, Z.; Qian, G. Short-Term Exposure to ZnO/MCB Persistent Free Radical Particles Causes Mouse Lung Lesions via Inflammatory Reactions and Apoptosis Pathways. *Environ. Pollut.* **2020**, *261*, 114039. [CrossRef]
231. Balakrishna, S.; Saravia, J.; Thevenot, P.; Ahlert, T.; Lominiki, S.; Dellinger, B.; Cormier, S.A. Environmentally Persistent Free Radicals Induce Airway Hyperresponsiveness in Neonatal Rat Lungs. *Part. Fibre Toxicol.* **2011**, *8*, 11. [CrossRef]
232. Huang, H.-S.; Ma, M.-C.; Chen, J.; Chen, C.-F. Changes in Renal Hemodynamics and Urodynamics in Rats with Chronic Hyperoxaluria and after Acute Oxalate Infusion: Role of Free Radicals. *Neurourol. Urodyn.* **2003**, *22*, 176–182. [CrossRef] [PubMed]
233. Reinke, L.A.; Moore, D.R.; Nanji, A.A. Pronounced Hepatic Free Radical Formation Precedes Pathological Liver Injury in Ethanol-Fed Rats. *Alcohol. Clin. Exp. Res.* **2000**, *24*, 332–335. [CrossRef] [PubMed]
234. Burn, B.R.; Varner, K.J. Environmentally Persistent Free Radicals Compromise Left Ventricular Function during Ischemia/Reperfusion Injury. *Am. J. Physiol.-Heart Circ. Physiol.* **2015**, *308*, H998–H1006. [CrossRef] [PubMed]
235. Wang, P.; You, D.; Saravia, J.; Shen, H.; Cormier, S.A. Maternal Exposure to Combustion Generated PM Inhibits Pulmonary Th1 Maturation and Concomitantly Enhances Postnatal Asthma Development in Offspring. *Part. Fibre Toxicol.* **2013**, *10*, 29. [CrossRef] [PubMed]
236. Guan, X.; Truong, L.; Lomnicki, S.; Tanguay, R.; Cormier, S. Developmental Hazard of Environmentally Persistent Free Radicals and Protective Effect of TEMPOL in Zebrafish Model. *Toxics* **2021**, *9*, 12. [CrossRef] [PubMed]
237. Zhang, Y.; Guo, X.; Si, X.; Yang, R.; Zhou, J.; Quan, X. Environmentally Persistent Free Radical Generation on Contaminated Soil and Their Potential Biotoxicity to Luminous Bacteria. *Sci. Total Environ.* **2019**, *687*, 348–354. [CrossRef] [PubMed]
238. Zhang, X.; Chen, Y.; Hu, D.; Zhao, L.; Wang, L.; Wu, M. Neurotoxic effect of environmental persistent free radicals in rice biochar to Caenorhabditis elegans. *China Environ. Sci.* **2019**, *39*, 2644–2651.
239. Stephenson, E.J.; Ragauskas, A.; Jaligama, S.; Redd, J.R.; Parvathareddy, J.; Peloquin, M.J.; Saravia, J.; Han, J.C.; Cormier, S.A.; Bridges, D. Exposure to Environmentally Persistent Free Radicals during Gestation Lowers Energy Expenditure and Impairs Skeletal Muscle Mitochondrial Function in Adult Mice. *Am. J. Physiol. -Endocrinol. Metab.* **2016**, *310*, E1003–E1015. [CrossRef] [PubMed]
240. Lee, G.I.; Saravia, J.; You, D.; Shrestha, B.; Jaligama, S.; Hebert, V.Y.; Dugas, T.R.; Cormier, S.A. Exposure to Combustion Generated Environmentally Persistent Free Radicals Enhances Severity of Influenza Virus Infection. *Part Fibre Toxicol* **2014**, *11*, 57. [CrossRef]
241. Briedé, J.J.; Godschalk, R.W.L.; Emans, M.T.G.; de Kok, T.M.C.M.; van Agen, E.; van Maanen, J.M.S.; van Schooten, F.-J.; Kleinjans, J.C.S. In Vitro and In Vivo Studies on Oxygen Free Radical and DNA Adduct Formation in Rat Lung and Liver during Benzo[a]Pyrene Metabolism. *Free Radic. Res.* **2004**, *38*, 995–1002. [CrossRef]
242. Balakrishna, S.; Lomnicki, S.; McAvey, K.M.; Cole, R.B.; Dellinger, B.; Cormier, S.A. Environmentally Persistent Free Radicals Amplify Ultrafine Particle Mediated Cellular Oxidative Stress and Cytotoxicity. *Part. Fibre Toxicol.* **2009**, *6*, 11. [CrossRef] [PubMed]

MDPI AG
Grosspeteranlage 5
4052 Basel
Switzerland
Tel.: +41 61 683 77 34

Toxics Editorial Office
E-mail: toxics@mdpi.com
www.mdpi.com/journal/toxics

www.ingramcontent.com/pod-product-compliance
Lightning Source LLC
Chambersburg PA
CBHW061515210926
51850CB00008B/1925
9783725857456